S0-CAH-290

Synthesis of Natural Products: Problems of Stereoselectivity

Volume I

Editors

Pavel Kočovský
Institute of Organic Chemistry and Biochemistry
Czechoslovak Academy of Sciences
Prague, Czechoslovakia

František Tureček
The J. Heyrovsky Institute of Physical Chemistry
and Electrochemistry
Czechoslovak Academy of Sciences
Prague, Czechoslovakia

Josef Hájíček
Research Institute for Pharmacy and Biochemistry
Prague, Czechoslovakia

CRC Press, Inc.
Boca Raton, Florida

Library of Congress Cataloging-in-Publication Data

Kočovský, Pavel.
Synthesis of natural products.

Includes bibliographies and indexes.
1. Natural products. 2. Chemistry, Organic--Synthesis. I. Tureček, František. II. Hájíček, Josef. III. Title.
QD415.K62 1986 547.7′0459 85-29943
ISBN 0-8493-6406-X (v. 1)
ISBN 0-8493-6407-8 (v. 2)

This book represents information obtained from authentic and highly regarded sources. Reprinted material is quoted with permission, and sources are indicated. A wide variety of references are listed. Every reasonable effort has been made to give reliable data and information, but the author and the publisher cannot assume responsibility for the validity of all materials or for the consequences of their use.

Direct all inquiries to CRC Press, Inc., 2000 Corporate Blvd., N.W., Boca Raton, Florida, 33431.

International Standard Book Number 0-8493-6406-X (v. 1)
International Standard Book Number 0-8493-6407-8 (v. 2)

Library of Congress Card Number 85-29943
Printed in the United States

PREFACE

The artificial synthesis of natural compounds is one of the principal domains of organic chemistry. Since the times of Wöhler's preparation of urea the organic synthesis has made great progress, hand in hand with the growing knowledge of structure and reactivity of organic compounds. Simultaneously, organic synthesis has developed into an important tool for determining structure of new compounds isolated from the natural material, and even nowadays this aspect remains important to complement modern spectral methods.

Following the pioneering work of Emil Fischer who turned chemists' attention to detailed investigation of the spatial arrangement of organic molecules, and in the wake of Robinson's mechanistically important synthesis of tropinone, organic synthesis had experienced her salad days in the fifties, owing to the personality of R. B. Woodward and his students and co-workers. So as the Robinson's synthesis of tropinone and Ingold's mechanistic studies in the thirties represented the landmarks of the electron theory, the later work of Barton on steroids and terpenes set fundamentals to conformational analysis and sparked the interest of other chemists in stereochemistry in general. The knowledge gathered from studying model synthetic compounds had tremendous importance for accomplishing now-classical, stereoselective syntheses of alkaloids, steroids, and terpenes. The chemistry of steroids has especially been fruitful in providing new procedures and reagents, and one is often surprised to find out how many common reagents have originated from this special area.

The chemistry of prostaglandins that made the chemists face the problem of efficiency in synthesis has contributed with a flood of new synthetic procedures and culminated with the introduction of Corey's retrosynthetic analysis that rationized the oldtime "art & craftsmanship" approach.

The parallel development of methodologies in the chemistry of peptides and nucleotides has brought and tested a pleiade of protecting and activating groups, together with the novel concept of solid-phase synthesis.

The mutually stimulating interaction of the synthetic and physical organic chemistry blossomed into the formulation of the principles of orbital symmetry control, in a way a "by-product" crowning the long synthetic way to vitamine B_{12}. Recently the synthetic efforts have successfully stretched out to the realm of polyether and macrolide antibiotics, a result of the further improvement in the selectivity and efficiency of synthetic procedures.

The contemporary trends look into rationalization of synthetic procedures by introducing new strategies and avoiding the lengthy protecting-deprotecting steps. This is achieved by employing "tailored" building blocks which are connected and modified with the help of selective reagents and transformations.

In the late sixties, G. Stork pioneered the selective generation of reactive species that served as intermediates of new multi-step sequences performed as "one-pot" reactions. The syntheses of new generations are characterized by convergent strategies that use two or more building blocks assembled via highly selective tactical steps. The chemo-, regio-, and stereoselectivity plays the key role in syntheses of polyfunctional natural compounds. At the very beginning stands Stork's preparation of cantharidine (1953), considered as the first stereoselective synthesis. With the progress in the selectivity of constructing stereogenous elements, some modern methods reach the ultimate level set by enzymatic transformations, reaction steps with 90% d.e. or e.e. being more a rule than an exception.

Total syntheses of natural compounds have been treated in a number of monographs, e.g. ApSimon's *The Total Synthesis of Natural Compounds,* Volumes 1 to 6, which dealt with selected classes, such as antibiotics, alkaloids, terpenes, steroids, etc. There also exist numerous monographs on the synthesis of prostaglandins, peptides and carbohydrates. All these books share a common feature, they classify the material by structural type. With the exception of Warren's undergraduate textbook *Organic Synthesis: The Disconnection Ap-*

proach, there is no comprehensive and systematic approach treating the synthesis of natural compounds from a methodological point of view. The present book is an attempt to fill this gap. We have based the text on the classification and analysis of synthetic approaches according to typical structural subunits occurring in natural products. The syntheses are then classified by the mode of formation of strategic bonds. With cyclic systems this leads to discerning mono- and ditopic annulations, rearrangements, and fragmentations. The ring-forming reactions are further classified as to the site of the closure, taking into consideration different build-up strategies. Modifications of cyclic systems are treated separately as ring expansions and contractions, introduction of substituents and conversions of carbocycles into heterocycles. With acyclic systems we distinguish *de novo* syntheses of the chain, its modifications, and fragmentation methods converting cyclic systems to acyclic ones. We stress the importance of the "chiral pool" in convergent syntheses. When the situation makes it necessary, we present and discuss reaction mechanisms, especially with acyclic systems. For the sake of clarity, the chiral centers are discerned by distinct symbols, e.g. conventional asterisks (*) for residing centers and squares (■) for those newly formed. In convergent syntheses that utilize more chiral synthons, the residing centers are denoted by triangles (▲) or (▼). Stereoselective syntheses of olefins and compounds with axial chirality have also been included. Some elegant, curious, or serendipitous examples have been selected in a divertimento in order to underline the esthetic aspects of the synthesis of natural products.

Of necessity, the topic could not be treated comprehensively otherwise we would have had to compile an encyclopedy on organic chemistry. Instead we have attempted to select the most representative examples that, on the other hand, introduced some subjectivity into our approach. Here we apologize to all authors whose synthetic work does not appear in the book. We do hope that we have partially remedied this flaw by inserting numerous references to specialized review articles.

The book is based partially on the lectures delivered by one of us (P. K.) for advanced courses on organic chemistry at the Institute of Organic Chemistry and Biochemistry of the Czechoslovak Academy of Sciences and at Charles University in Prague. We believe that the largely extended text could serve as a supplementary source for advanced courses on organic synthesis.

We are fully aware that writing a book on a topic that undergoes perpetual and buoyant development is a difficult and unrewarding task, because the text becomes obsolete almost as rapidly as it is written. The references included in the book were truncated by the first half of 1984. In order to catch up with the latest trends, we have inserted some 1985 references as notes added in proofs.

This book would not have been written without help and assistance of many of our colleagues. We thank Drs. V. Pouzar, P. Kuzmič, I. Stibor, I. Valterová, P. Majer, M. Smrčina, K. Bláha (all Prague), and J. S. Hallock (Ithaca) for their comments on different parts of the manuscript. Our special thanks are due to Dr. V. Černý (Prague) who carefully read the whole manuscript, and to Dr. C. N. Hodge (Ithaca) and N. Hill (Lausanne) who helped us to keep the English within acceptable limits. We highly appreciate the technical assistance of J. Doležalová, J. Pachová, O. Turečková, Z. Leblová, J. Hájičková, I. Stieborová, and M. Bárová. The responsibility for the correctness lies, of course, on the authors themselves.

Prague, Ithaca, Zurich, Lausanne, 1985
Pavel Kočovský
František Tureček
Josef Hájíček

THE EDITORS

Pavel Kočovský is presently a Research Associate at the Institute of Organic Chemistry and Biochemistry of the Czechoslovak Academy of Sciences, Prague, Czechoslovakia. He received a degree in chemistry in 1974 from the Prague Technical University and a Ph.D. in 1977 from the Czechoslovak Academy of Sciences. In the same year he joined the Institute of Organic Chemistry and Biochemistry. He spent an academic year 1983/1984 at Cornell University, Ithaca, New York. He serves as a part-time lecturer at Charles University, Prague.

His research interest includes stereochemistry, reaction mechanisms, neighboring group participation, steroid chemistry and, most recently, application of organometallics to organic synthesis. He has published over 60 papers and co-authored two textbooks on organic synthesis and natural products.

Dr. Kočovský is a member of the Czechoslovak Chemical Society and American Chemical Society.

František Tureček is currently a Research Associate at the J. Heyrovský Institute of Physical Chemistry and Electrochemistry, Prague, Czechoslovakia. He received a Ph.D. from Charles University, Prague in 1977 in organic chemistry and joined the Heyrovský Institute in the same year. He spent a postdoctoral year at Cornell University, Ithaca, New York, in 1981 and was a Visiting Fellow at Ecole Polytechnique Fédérale, Lausanne, Switzerland, in 1984.

His research interests are mainly physical organic chemistry, reaction mechanisms, stereochemistry, mass spectrometry and structure elucidation by spectral methods. He has published over 90 papers and co-authored a book on mass spectrometry.

Dr. Tureček is a member of the Czechoslovak Chemical Society and American Chemical Society.

Josef Hájíček has been working as a Research Associate at the Research Institute of Pharmacy and Biochemistry, Prague, Czechoslovakia. He received a degree in organic chemistry in 1972 from the Prague Technical University and joined the Research Institute in the same year. He received a Ph.D. in 1980 again from the Prague Technical University. He spent a postdoctoral year at the University of Zúrich, Switzerland in 1983/1984.

His research interests include mainly stereochemistry and chemistry of natural products, especially alkaloids. He published over 15 papers and holds about 30 patents.

Dr. Hájíček is a member of the Czechoslovak Chemical Society.

Dedicated to Professor Karel Wiesner

TABLE OF CONTENTS

Volume I

Introduction
Syntheses of Cyclic Systems
ortho-Condensed Systems
Spirocyclic Compounds
Bridged Systems
Medium and Large Rings
Stereoselective Substitution in Cyclic Systems
Chiral Synthons
Index

Volume II

Heterocyclic Compounds
Acyclic Systems
Compounds with Axial Chirality
Divertimento
Index

TABLE OF CONTENTS

Chapter 1
Introduction 1
I. Introduction 1

II. Steric Control 2
A. Kinetic and Thermodynamic Control 2
B. Conformational Kinetic and Thermodynamic Control 2

III. Formation of New Elements of Chirality 3
A. Center → Center Chirality Induction 4
B. Center → Axis Chirality Induction 5
C. Axis → Center Chirality Induction 5
D. Plane → Center Chirality Induction 6

IV. The Problem of Stereoselectivity 6
A. General Remarks 6
B. The Control Element 9
References 12

Chapter 2
Syntheses of Cyclic Systems 15
I. Introduction 15

II. Ring Formation 15

III. Systems With One Ring 16

IV. General Aspects of Ring-Closure Reactions 17

V. Polycyclic Systems 18

VI. Strategy for Stereoselective Substitution in Cyclic Systems 19

VII. Notes Added in Proof 20

References 20

Chapter 3
***ortho*-Condensed Systems** 23
I. Cyclizations at the Annulation Site 23
A. Monotopic Annulations 23
1. Three-Carbon Polar Annulation (Cyclopentane Annulation) 24
2. Four-Carbon Polar Annulation (Robinson Annulation) 27
3. Five-Carbon Polar Annulation 33
4. Carbocation Addition to the Double Bond 33
5. Ene Reaction 36
6. Thermal Cyclization of α-Alkynones 38

B. Ditopic Annulations ... 38
1. [2 + 1] Cycloadditions ... 39
2. [2 + 2] Cycloadditions ... 39
3. [3 + 2] Cycloadditions ... 43
4. [4 + 2] Cycloadditions ... 44
a. Heterosubstituted Dienes ... 46
b. Asymmetric Diels-Alder Reaction ... 48
c. Intramolecular Diels-Alder Reaction ... 48

II. Cyclization Outside the Annulation Sites ... 52

III. Modifications of Cyclic Systems ... 54
A. Ring Contractions ... 54
1. Favorskii Rearrangement ... 54
2. Wolf Rearrangement ... 55
3. Demyanov Rearrangement of Vicinal Aminoalcohols ... 55
4. Rearrangement of α,β-Epoxy Ketones ... 55
5. Tandem Oxidative Cleavage of the Double Bond Aldolization ... 55
B. Ring Expansions ... 55
1. Reactions of Ketones with Diazocompounds ... 56
2. Tiffeneu-Demyanov Ring Expansion ... 56
3. Tandem Oxidative Cleavage of the Double Bond Aldolization ... 57
4. Rearrangements of Vinylcyclopropane Derivatives ... 58
C. Combined Ring Contraction and Expansion ... 59
1. Cationic Rearrangements in [x.y.0] Systems ... 59
2. Cationic Rearrangements in [x.y.1] Systems ... 60
D. Fragmentation of a Cross-Piece Bond in the Ring ... 60
E. Cope Rearrangement ... 61

IV. *cis, trans*-Annulation of *ortho*-Condensed Systems ... 63
A. *cis*- and *trans*-Annulated Hydrindanes ... 63
B. *cis*- and *trans*-Annulated Decalins ... 63
C. *cis*- and *trans*-Annulated Hydroazulenes ... 65
D. Introduction of the Angular Methyl Group ... 66

V. Notes Added in Proof ... 69

References ... 70

Chapter 4
Spirocyclic Compounds ... 89
I. General Considerations ... 89

II. Alkylation Methods ... 89
A. α,α′-Dialkylation ... 89
B. α-Alkylation ... 89
C. *Ipso* Alkylations and Oxidative Coupling of Phenols ... 92

III. Inverse Alkylations and Related Methods ... 94

IV. Cycloadditions 96
A. [2 + 1] Cycloadditions 96
B. [2 + 2] Cycloadditions 96
C. [4 + 2] Cycloadditions 97
D. Conia, Ene, and Related Cyclizations 99

V. Cyclizations Outside the Spiroatom 99

VI. Rearrangements 100

VII. Notes Added in Proof 102

References 102

Chapter 5
Bridged Systems 107
I. Bicyclic Bridged Systems 107
A. Cyclization at Bridgehead Positions 107
1. Monotopic Cyclizations 107
2. Ditopic Cyclizations 110
3. Rearrangements in the Synthesis of Bridged Systems 112
a. Rearrangements of *ortho*-Condensed Systems to Bridged Systems 113
b. Rearrangements in Bridged Systems 113
4. Fragmentation Methods in the Synthesis of Bridged Systems 113
B. Cyclization Outside the Annulation Sites 114

II. Tricyclic Bridged Systems 114
A. Syntheses Including Formation of the Central Ring 117
1. Monotopic Cyclization 117
2. Ditopic Cyclization 122
B. Syntheses by Bridging a Central Ring 125
C. Rearrangements in the Synthesis of Tricyclic Bridged Systems 127

III. Polycyclic Bridged Systems 128

IV. Notes Added in Proof 130

References 130

Chapter 6
Medium and Large Rings 135
I. Cyclization Reactions 135

II. Fragmentation Reactions 138

III. "Zip" Reactions 140

IV. Notes Added in Proof 142

References 142

Chapter 7
Stereoselective Substitution in Cyclic Systems 145
I. *De novo* Construction of a Stereospecifically Substituted Ring 145
A. Cyclopropane Derivatives 145
B. Cyclobutane Derivatives...145
C. Cyclopentane Derivatives..146
D. Cyclohexane Derivatives 146

II. Stereoselective Attachment of Substituents.....................................148
A. Stereoselective Introduction of Substituents by Addition to Double Bonds 148
B. Stereoselective Introduction of Substituents in Vicinal Positions 150
C. [3 + 3] Sigmatropic Rearrangements 153
D. Neighboring Groups as Control Elements 156
E. Auxiliary Rings as Precursors of Substituents 157
1. Introduction of Substituents via Cleavage of *ortho*-Condensed Systems...........158
2. Introduction of Substituents via Cleavage of Bridged Systems...161
F. Transposition of Substituents 164
1. 1,2-Transpositions 164
2. 1,3-Transpositions 165
G. Inversion of Configuration 166
H. Stereoselective Substitution in Medium and Large Rings.......................167

III. Notes Added in Proof..170

References..172

Chapter 8
Chiral Synthons 183
I. Introduction...183

II. Terpenes 184
A. Limonene (11)..184
B. Carvone (20) 185
C. Pulegone (24) 186
D. β-Citronellol (35) 187
E. β-Pinene (38)..187
F. β-Carene (40) 188
G. Camphor (51) 188

III. Amino Acids...189
A. Norvaline (53)...189
B. Phenylalanine (61)...190
C. Glutamic Acid (65)...190
D. Leucine (89)...193
E. Cysteine (96)..193

IV. Hydroxy Acids...194
A. Malic Acid (98) 194
B. Lactic Acid (102) 194
C. Tartaric Acid (117)..196
D. Other Hydroxy Acids and Related Compounds.......................................197

V. Carbohydrates....197
A. Glyceraldehyde (127)....199
B. Glucose (131)....199
C. Mannose (144)....201
D. Other Carbohydrates....202

VI. Miscellaneous Chiral Synthons....202

VII. Syntheses Involving Several Chiral Synthons....205
A. The Synthesis of the Sexual Pheromone of Pine Sawflies....205
B. The Synthesis of Brefeldin A....206
C. The Synthesis of the Molecular Fragment of Griseoviridine (205)....206
D. The Synthesis of Cytochalasin B (210)....207
E. The Synthesis of Lasalocid A (217)....208
F. The Synthesis of Monensin (228)....209
G. The Synthesis of Vitamin B_{12}....209

VIII. Notes Added in Proof....209

References....213

Index....227

Chapter 1

I. INTRODUCTION

The stereochemistry of individual synthetic steps undoubtedly plays the key role in designing a total or partial synthesis of a complex natural product (for definition of stereoselectivity see References 1 and 2). The stereoselective synthesis is a procedure in which one stereoisomer of certain structure is formed preferentially over other stereoisomers.[2] Such a product need not be optically active. In general, the synthesis can start with a racemate or compounds lacking stereogenous elements. In such a case stereoselective synthesis results in a racemic target compound which, however, contains the stereogenous groups in a defined relative configuration. If we start our synthesis with an optically active compound (chiral synthon[3-5]), the product will also have definite absolute configuration.

As a simple example of a stereoselective synthesis let us consider the preparation of (−)-ephedrine (1) and (+)-pseudoephedrine (2) shown in Figure 1. (−)-Ephedrine is a pharmacodynamically active component of the Chinese drug Ma-Huang that is used for treating asthma. The *threo*-isomer, (+)-pseudoephedrine (2) which also occurs in nature, is less valuable from the pharmacological point of view. The effective synthesis of 1 should therefore avoid coformation of 2 and should lead directly to the desired enantiomer (−)-1 or, at least, produce a racemic *erythro*-isomer (±)-1 which would be resolved to antipodes.

A possible synthetic approach to 1 comprises reducing the racemic amino ketone 3. The reduction with sodium amalgam affords a mixture of (±)-1 and (±)-2, the latter slightly prevailing.[6,7] Clearly, this method is not stereoselective and appears unsuitable for preparation of the desired isomer (±)-1. In contrast, catalytic hydrogenation of 3 gives almost exclusively (±)-1, which can be easily isolated from the reaction mixture in a yield exceeding 90%.[8] The second method meets the requirements to prepare the *erythro*-form but the product still has to be resolved in order to obtain pure enantiomers.

It is often more economical to prepare the optically active product at the early stage of the synthesis. This can be achieved in two ways: either by an enantioselective synthesis that introduces a first chiral element into the substrate molecule,[9-11] or by using a chiral synthon available synthetically or from natural sources.[3-5] Then in the further synthetic steps one works with the desired enantiomer, so that half of the necessary reagents are saved, and problems of how to separate, utilize, or discard the undesirable enanatiomer of the target compound are avoided. The importance of syntheses starting with chiral synthons will apparently be growing in the future, especially in pharmaceutical chemistry.[12]

Beside the preparative aspects, the synthesis starting from a compound of known absolute configuration is also of importance for determining the absolute configuration of the target product[4,5,13] or for mechanistic studies of biochemical transformations,[14] etc. In some cases the quantity of the natural product isolated is so small that it is impossible to measure the optical activity. For instance, classical methods could not be applied to elucidate the absolute configuration of insect products which were obtained in minute amounts and which, moreover, were rather volatile. In such cases the only practical way is to synthesize the required compound from chiral synthons of known absolute configuration.[13] The synthetic and the natural product then can be correlated by using a suitable chiral shift reagent in the NMR spectra[15] even without knowing the optical rotation of the natural product. Conversely, this technique also makes it possible to determine the enantiomeric purity of both the synthetic and the natural product. Determination of the relative configuration of stereogenous elements in a natural product may be even less demanding as to the purity and amount of material, for it can be achieved by comparing the natural product with a synthetic standard by gas chromatography-mass spectrometry.[16]

The major part of this book will be devoted to problems of stereoselectivity in the synthesis

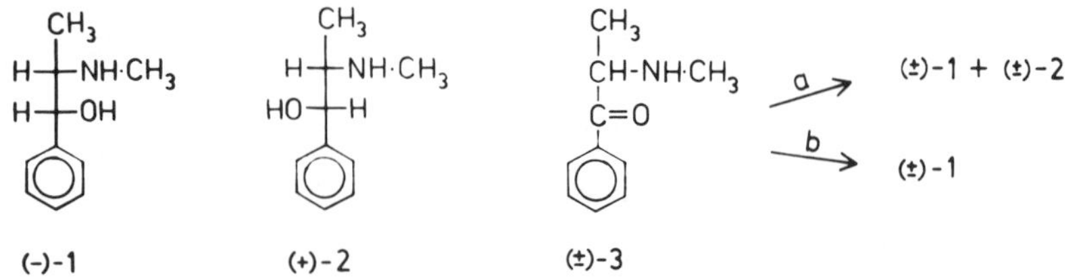

FIGURE 1. (a) Na – Hg; (b) H_2, Pt, HCl.

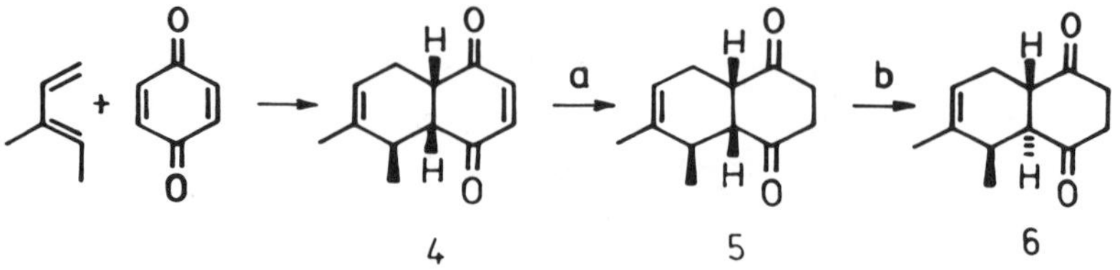

FIGURE 2. (a) Zn, AcOH; (b) CH_3ONa.

of natural products possessing a cyclic frame. The synthesis of compounds with a chiral aliphatic chain, itself an exciting and rapidly growing area of organic chemistry, will be given less space, for this topic has recently been covered in several excellent reviews. Syntheses of peptides and oligosaccharides have been omitted completely; in the former case the problem of stereoselectivity is to suppress undesirable racemization,[17] while in the latter case the problem is mostly reduced to stereoselective formation of the glycosidic bond. The stereochemistry of saccharides has also been treated in detail elsewhere.[18-20]

Some of the methods included in this book have not yet been utilized in the synthesis of natural products, but they may evolve into a valuable synthetic tool in the future. Several examples lacking stereogenous elements or not yielding products with sufficient stereoselectivity will be presented for illustrative purposes or for their close relationship to other synthetic procedures. Stereoselective enzymatic reactions will be presented only briefly, because this topic has been frequently covered in more specialized monographs.[14,21-24]

II. STERIC CONTROL

A. Kinetic and Thermodynamic Control

The stereoselective synthesis is based on steric control of particular reaction steps. This control can be of a kinetic or a thermodynamic nature,[2,25] as illustrated by the synthesis of the decalin system in Figure 2.[26,27] The Diels-Alder reaction, used to build up the skeleton, affords *cis*-annulated decalin derivative 4 (kinetic control), which is selectively reduced at the conjugated double bond to afford 5. If the desired annulation of the rings were *trans*, the *cis*-derivative 5 may be readily transformed to the more stable *trans*-isomer 6 by base-catalyzed equilibration (thermodynamic control).

B. Conformational Kinetic and Thermodynamic Control

In the frame of the general principle presented in the preceding paragraph, we will consider a conformational control that again can be of kinetic or thermodynamic nature. The Woodward synthesis of prostaglandins[28] may serve as an example of conformational kinetic control (Figure 3). Diazotation of the amino alcohol 7, having the cyclohexane ring in boat conformation with axial amine and hydroxyl groups, would lead to the undesirable epoxide 8. Flipping to the chair conformation (7 → 9) results in a diequatorial arrangement of both

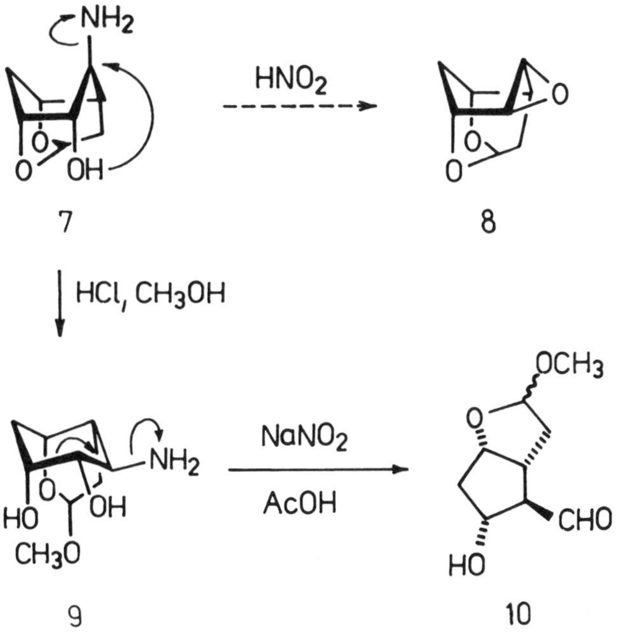

FIGURE 3.

groups, and the diazotation of 9 affords the aldehyde 10 which already contains the cyclopentane ring of the target prostaglandin and a correct stereochemistry of substituents. Similarly, a change of conformation can affect the course of other reactions, for instance the cleavage of the oxirane ring[29] (see also References 30 and 31).

The conformational thermodynamic control makes use of the different stability of substituents (e.g. equatorial and axial ones in cyclohexanes). This effect can be utilized for transforming the configuration of a substituent to a more stable one by equilibration. It may happen, however, that the kinetic and/or thermodynamic factors force the system in question into a conformation in which the substituent assumes an undesirable, thermodynamically stable configuration. This unfavorable effect can be overcome provided the skeleton is flexible so that one stable conformation (e.g. the chair form of the cyclohexane ring) can be transformed to another stable form. To illustrate this, let us consider a hypothetical case[2] shown in Figure 4. In 11 the substituent X is in the undesirable *trans*-configuration with respect to the carboxyl group. A simple equilibration would not be successful because the equatorial position of X in 11 is more stable than the axial one in the desired *cis*-isomer 12. By closing the lactone ring (11 $\rightarrow$ 13), the cyclohexane moiety is forced to assume an inverted chair conformation in which X becomes axial. Equilibration now will lead to the more stable equatorial isomer (13 $\rightarrow$ 14) in which X and the carboxyl are mutually *cis* oriented. Finally, hydrolysis of the lactone ring in 14 affords the hydroxy acid 12 in which X is axial and has the correct *cis* configuration with respect to the carboxyl group.[2] This tactic has been employed by Woodward et al.[29] in the synthesis of reserpine (see also Figure 13).

III. FORMATION OF NEW ELEMENTS OF CHIRALITY

In the synthesis of chiral compounds having either central (C), axial (A), or planar (P) chirality we usually employ methods that induce the chirality by means of agents or factors already involving the proper element of chirality. This is referred to as chirality induction

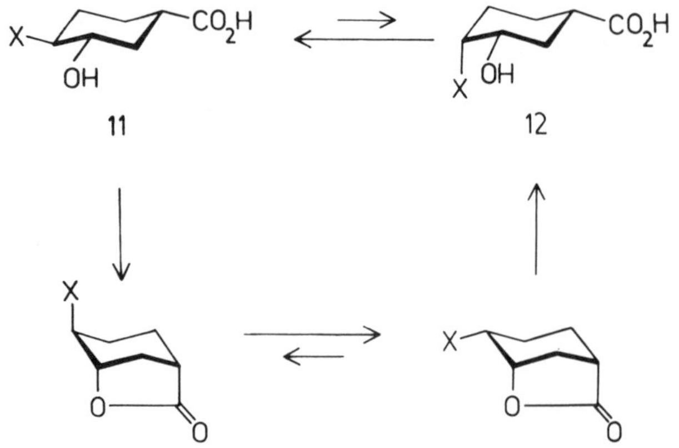

FIGURE 4.

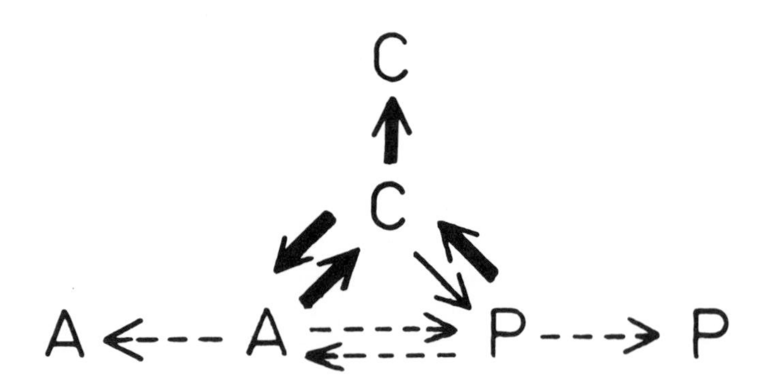

FIGURE 5. A . . . axial chirality, C . . . central chirality, P . . . planar chirality.

or asymmetric synthesis. In the following we shall deal only with "chemical" factors; "physical" factors such as polarized light[2] will not be discussed.

The formation of a new chirality element in the synthesized compound can be achieved through intramolecular factors; this means that one or more chirality elements in the precursor molecule determine the configuration of the newly formed chirality element. The intermolecular factors mostly comprise chiral reagents or, less frequently, chiral solvents. The parent element(s) can be preserved during the formation of the new one, or may be simultaneously destroyed (self-immolative process).[32,33] In this Chapter we will discuss only processes which are of importance for stereoselective synthesis of natural products, especially those that lead to formation of a new chiral center. The others can be found in Reference 33. Figure 5 shows the modes of chirality induction that have been described to date. The bold arrows denote the modes that we will discuss here; the full arrows mean the other known processes.[33] The broken arrows denote processes that, to our knowledge, have not yet been described.

A. Center → Center Chirality Induction

Intramolecular chirality induction of the C → C type with retention of the parent chirality center is exemplified by the classic work of Cram[34] (Figure 6). The reaction of aldehyde 15 with a Grignard reagent proceeds diastereoselectively, affording mostly the *threo*-isomer 16 (the new chirality center is denoted by ■). Note that the starting aldehyde may be a racemate; then of course the product would be a racemic *threo*-isomer.

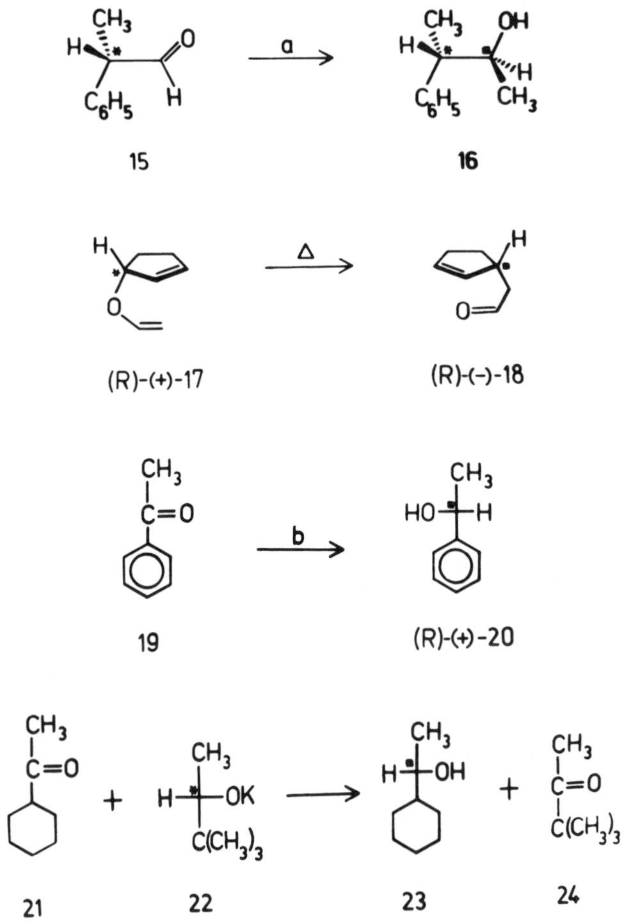

FIGURE 6. (C → C); (a) CH_3MgI; (b) $LiAlH_3(OR^*)$, Et_2O.

The Claisen rearrangement in the vinylether 17[32,33] is also stereoselective, but the parent chirality center vanishes during the intramolecular chirality transcription (self-immolative process).[32]

The intermolecular induction of chirality proceeding with retention of the reagent chirality is illustrated with the enantioselective reduction of acetophenone (19) with a chiral hydride.[35] The hydride reagent was prepared by partial solvolysis of lithium aluminum hydride with a chiral alcohol (quinine in this case).

Finally, the Meerwein-Ponndorf reduction of cyclohexylmethyl ketone (21) with chiral alkoxide 22 (Figure 6) represents an example of the intermolecular chirality induction in which the original chirality is destroyed.[36]

B. Center → Axis Chirality Induction

Although rather rarely, compounds with axial chirality occur in nature and therefore we will present here an example of synthesis of an optically active allene (Figure 7). The key rearrangement proceeds with concomitant destruction of the parent chirality center.[37] Other examples of the axial chirality induction (via intra- or intermolecular processes) can be found in Reference 33.

C. Axis → Center Chirality Induction

During the past years several efficient methods have appeared that have involved inter-

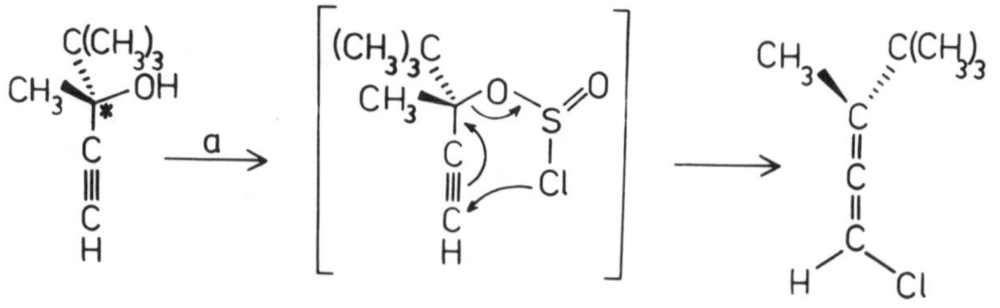

FIGURE 7. (C → A); (a) $SOCl_2$.

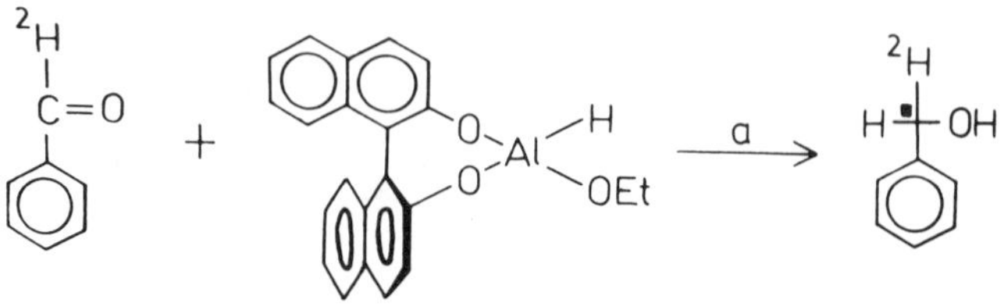

FIGURE 8. (A → C); (a), THF, −100°C 3 hr, then −78°C 16 hr.

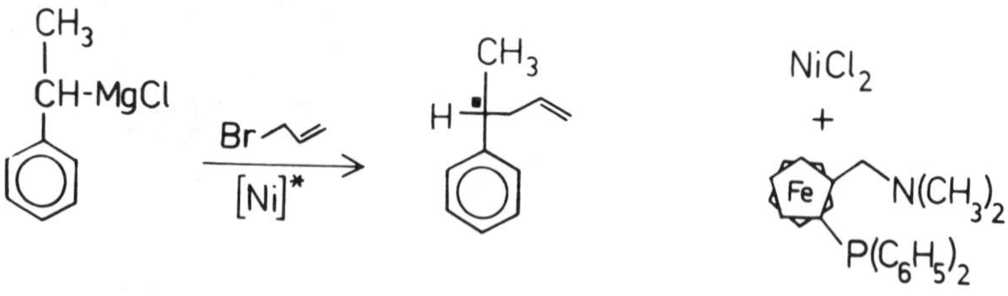

FIGURE 9. (P → C).

molecular chirality induction of the A → C type. The methods using axially chiral reagents usually achieve very high stereoselectivity which is rarely matched by reagents having central chirality. The A → C chirality induction is illustrated with reduction of carbonyl compounds with an axially chiral hydride (Figure 8).[38]

D. Plane → Center Chirality Induction

The last example that we present in this section shows an induction of central chirality in the catalyzed cross-coupling of a Grignard reagent with allyl bromide (Figure 9). The catalyst which possesses planar chirality is a sandwich complex prepared from nickel chloride and a chiral ferrocene.[39] The original chirality of the catalyst is preserved during the reaction.

IV. THE PROBLEM OF STEREOSELECTIVITY

A. General Remarks

In stereoselective syntheses we often encounter a situation in which the desired stereoisomer arises in low yield and it is therefore necessary to increase its yield relative to the other isomer. The problem is not always so simple as in Figure 2 where the desired *trans*-annulated isomer could be obtained conveniently by acid-base equilibration. In general, we can use several approaches to solve problems of stereoselectivity[25] as shown in Figure 10. Let us aim our stereoselective synthesis at a compound denoted as 25 with two chirality elements. For the sake of simplicity, we have chosen the target compound to be an alcohol (X=OH).

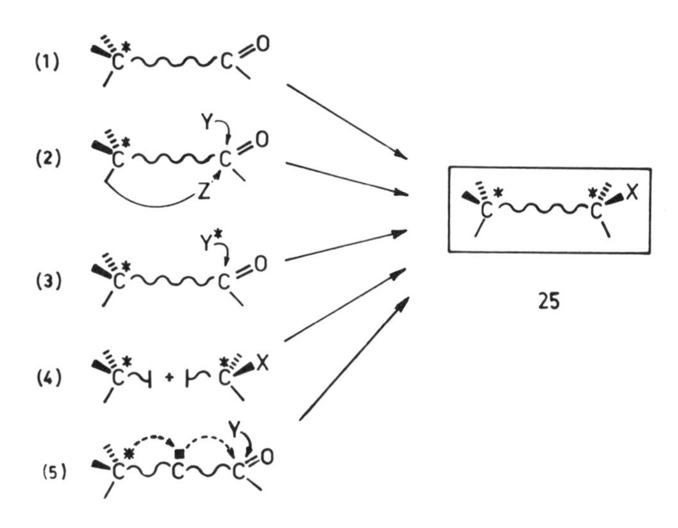

FIGURE 10.

1. Evidently, the simplest synthetic approach would be to reduce a carbonyl group. If the arising chirality center is located close to the parent chirality element (ideally in the vicinal position) or if there are some favorable conformational effects in the precursor molecule, then we can expect that the reduction would afford one diastereoisomer in excess. However, if the interaction between the centers is weak, e.g. if the carbonyl group is too remote from the parent chirality center, we would obtain a mixture of both diastereoisomers at worst in a 1:1 ratio. The separation of diastereoisomers with remote chirality elements may be difficult and often it is more advantageous to choose a different synthetic approach.
2. The remote chirality center may be connected with an auxiliary group which is able to reach to the reaction center. Such a group, referred to as control element, transduces[40] the chirality of the remote center and under favorable circumstances, may affect the formation of one diastereoisomer in excess.
3. A new chiral center may also be introduced by intervention of a chiral reagent, e.g. a chiral hydride which determines the desired absolute configuration on the center being formed, regardless of the configuration of the parent element.
4. Another approach relies on building up the target molecule from two chiral synthons of a proper absolute configuration.
5. Introduction of an auxiliary chirality element which then functions as in Procedure 1 represents another solution of the problem.

Procedures 1 to 4 can be documented by numerous examples from the chemistry of prostaglandins. The pivotal problem is to create the natural ''15α''-configuration of the hydroxyl group in the side-chain, which is necessary for biological activity (Figure 11).

In the first procedure (1),[41] reduction with zinc borohydride induces a negligible chirality and leads to a nearly equimolar mixture of both epimers. In an improved procedure, (2), the 11α-hydroxyl group is transformed to an ester or urethane that serves as a control element. At low temperature this specially designed element adopts a conformation that hinders the free rotation of the side chain and simultaneously directs the approach of the bulky reducing reagent from the more accessible side of the substrate molecule. These combined effects result in preferential formation of the natural ''15α''-alcohol.[42] An alternative approach (3) is based on reduction of the keto group with a chiral hydride which yields the desired ''15α''-epimer in an excellent optical yield.[43] The fourth example (4) shows how two chiral synthons are connected by the Wittig reaction.[44]

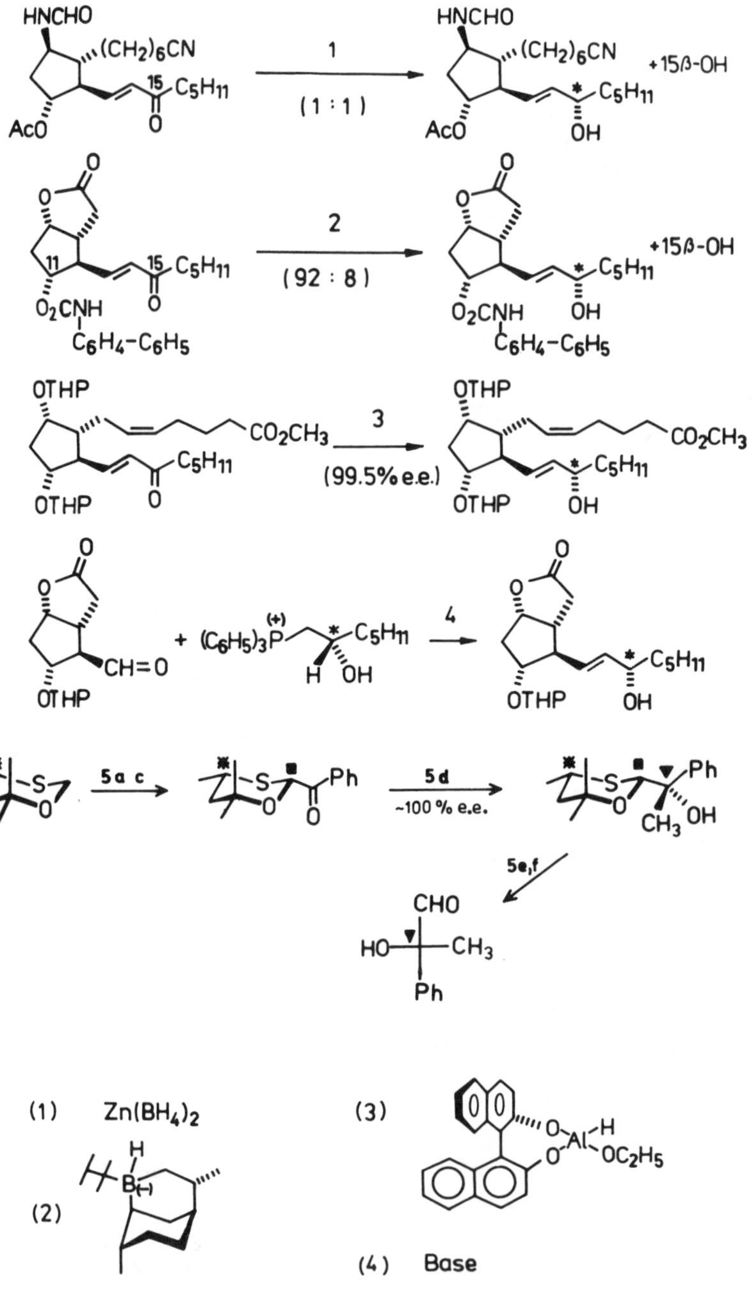

FIGURE 11. (a) BuLi; (b) C_6H_5CHO; (c) Me_2SO, CF_3CO_2H; (d) CH_3MgI; (e) CH_3I, $CaCO_3$, CH_3CN; (f) CrO_3, Me_2CO, H_2SO_4.

The last approach, the formation of an auxiliary chirality center (5), is illustrated with the stereoselective reaction of the carbonyl group with a Grignard reagent.[45] In this case the formation of the auxiliary chirality center is assisted by the conformational thermodynamic control. In the second step the metal atom of the Grignard reagent is coordinated to the ring heteroatoms, which directs the reagent approach from one side.

Note that in Procedures 1 and 2 we do not need any optically active starting compound, for the synthesis is aimed at creating a proper *relative* configuration. On the contrary in

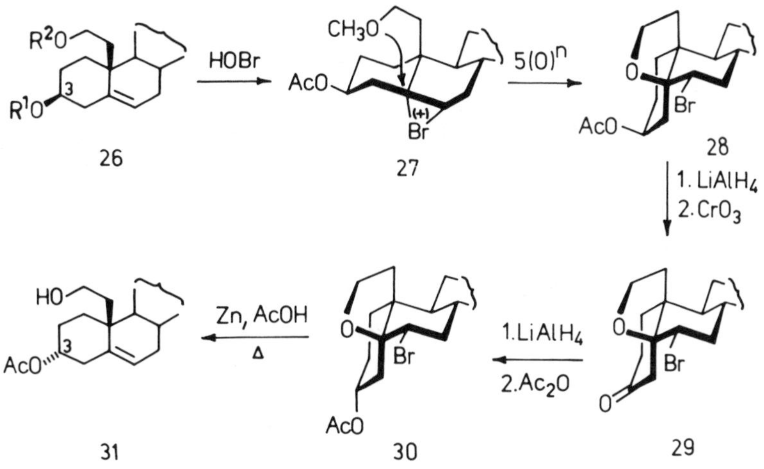

FIGURE 12.

Procedures 3 and 4, a chiral starting compound is essential, because the chiral elements are introduced independent of each other. In the last procedure, we have to start with an optically active compound to assure the correct absolute configuration of the third chiral center. If the synthesis were aimed at creating a particular relative configuration of the substituents, the starting compound could have been racemic.

B. The Control Element

Let us return to the role of the control element that has been introduced to our molecule in order to affect the steric course of some synthetic steps. The control element can work in three ways:

1. It may enforce a certain reaction conformation in which the molecule reacts without immediate ''chemical'' participation of the control element.
2. It may act in a ''physical way'', i.e. through nonbonding interactions, by hindering the reagent approach from one direction, but without forming new chemical bonds.
3. It may interact covalently with the reaction center or with the reagent. This is referred to as neighboring group participation.[46]

An example of conformational kinetic control, (1), in which the control element does not enter the reaction, but only imposes a proper conformation, is illustrated in Figure 12.[47] The purpose of this synthesis was to transform the steroid diol 26 (R^1 = R^2 = H) into the 3α-epimer 31 and simultaneously to differentiate by selective protection the reactivity of either hydroxyl group, which was needed for further reaction steps. In this case a simple S_N2 inversion at C-3 could not be used, for the substitution of a leaving group (e.g. a mesylate) by a nucleophile is accompanied by participation by the homoallylic double bond and depending on the reaction conditions, the reaction results in retention of configuration or a rearrangement.[44] Another approach, i.e. the oxidation of the 3β-hydroxyl to an oxo-group, followed by stereoselective reduction to epimeric 3α-alcohol, was complicated by the presence of the 5,6-double bond which is known to isomerize easily to form the α,β-enone system.[48] It was therefore necessary to protect the double bond first and then perform the configurational inversion at C-3. This was achieved by the following procedure (Figure 12).[47] Addition of hypobromous acid to 26 (R^1 = Ac, R^2 = CH_3), which proceeds with participation by the 19a-methoxyl group, afforded the bromoepoxide 28. Thereby the A ring

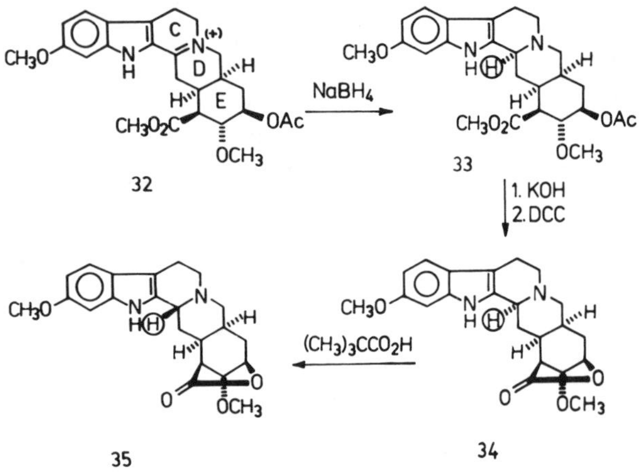

FIGURE 13.

is flipped to the other chair conformation due to the *cis* annulation of AB rings, and simultaneously the original equatorial 3β-acetate assumes an axial position (without changing its configuration!). The acetate 28 was subsequently hydrolyzed and the hydroxyl group oxidized to give the ketone 29 which further afforded the equatorial alcohol upon hydride reduction. The 3β-alcohol which arises as a minor by-product could be recycled by the same oxidation-reduction procedure. Acetylation of the 3α-hydroxyl group, followed by reductive cleavage of the bromoether group, afforded the desired product 31. The bromoepoxide moiety thus had three functions: (1) it changed the conformation of the A ring in such a way that the reduction of the C-3 ketone led predominantly to the 3α-alcohol; (2) it protected the 5,6-double bond, maintaining its position and avoiding the undesired isomerization; and (3) it protected temporarily the 19a-hydroxyl group and made it possible to differentiate the reactivity of the C-3 and C-19a hydroxyls by different protection. Other examples of the utilization of control elements can be found in References 28 to 31 and 49.

An example of the conformational thermodynamic control in which the control element is remote from the reaction center and does not enter the reaction is shown in Figure 13.[29] Reduction of the immonium salt 32 affords the product 33 with *trans*-annulated CD rings and an axial hydrogen atom at C-3. Unfortunately, this configuration is opposite to that in natural reserpine which was the synthetic target in this case. Note that the other asymmetric centers already have the correct, natural configuration. The desired configuration at C-3 was established with the help of thermodynamic conformational control, according to the principles outlined in Figure 4. The thermodynamically stable compound 33 was transformed to lactone 34, which resulted in flipping the C, D, and E rings to another conformation. Simultaneously, this changed the conformation of all substituents on the D and E rings and, most importantly, the C-2–C-3 bond became axial with respect to the D-ring. On acid-catalyzed equilibration (34 → 35) the latter bond assumes the more stable equatorial position and the product 35 now has all asymmetric centers in the correct relative configuration. Further synthetic steps then led to reserpine without stereochemical problems.

An illustrative example of kinetic steric control, (2), in which the control element takes part in the reaction through a ''physical interaction'', is depicted in Figure 11. The reduction of the C-15 ketone group in the prostaglandin side chain is affected by the biphenyl group anchored by a urethane link to the chirality center at C-11. At low temperatures the molecule is frozen in a conformation in which the biphenyl system comes close to the ketone group, possibly due to the π–π and van der Waals contact between the arene and the s-*cis*-enone systems (Figure 14).[42] As a result, the bulky hydride reagent used for the reduction prefers

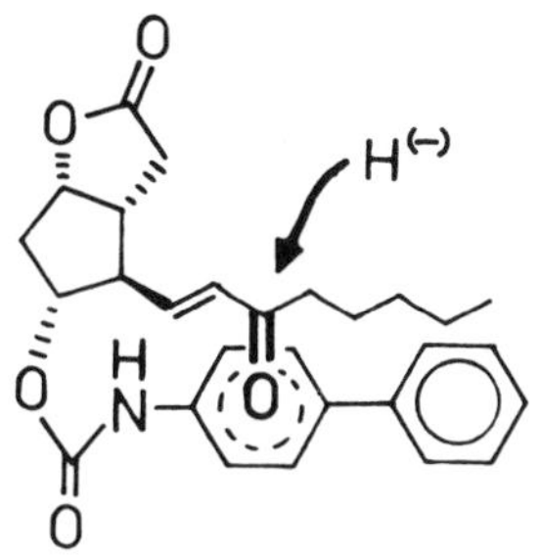

FIGURE 14.

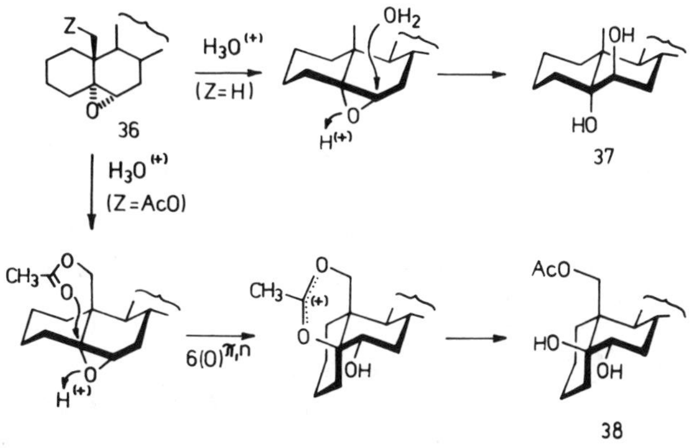

FIGURE 15.

to approach the carbonyl group from the opposite side (Figure 14) which leads to the formation of one stereoisomer in excess.

Finally, an example is presented of a control element which forms a transient covalent bond with the reaction center, and in this way affects the reaction course, (3). The steroid epoxide 36 (Z = H)[48] is cleaved in an acidic medium to give the diaxial diol 37, according to the Furst-Plattner rule (Figure 15). Introducing an acetoxy group into C-19 (36, Z = AcO) changes the reaction course, despite the fact that the conformation of the B ring remains unchanged. The protonated oxirane ring is cleaved by the carbonyl oxygen of the acetoxy group[50] (an intramolecular nucleophile) to give rise to equatorial diol 38. Further examples of the impact which control elements serving as participating groups have on the course of electrophilic addition are depicted in Figure 16. The steroid olefin 39 having the 5,6-double bond is attacked by an electrophile mainly from the α-side of the skeleton. The nucleophile then approaches from the β-side which, according to the Fürst-Plattner rule, but in contrast to the Markovnikov rule, gives a diaxial product,[48,49] the electrophile being oriented 5α, and the nucleophile 6β. The presence of a methoxymethylene group at C-19 (40) does not alter the approach of the electrophile but substantially changes the course of the second reaction step. Instead of an external nucleophile, the reaction center is attacked by the C-19a methoxyl group, thereby changing the overall regioselectivity of the reaction.[51,52] In the resulting cyclic ether the electrophile is bonded equatorially at 6α, whereas the nucleophile is linked to C-5 in the β-configuration. An acyloxy group located at C-19 (e.g. the 19-acetate 41)[53-55] affects the regioselectivity of the electrophilic addition in a similar way. In addition, the participating groups are capable of changing the reaction stereochemistry as well. For instance, hindering the α-side of the skeleton by 3α- or 7α-acetoxy groups (42,43) reverses the electrophile approach and the products are formed by

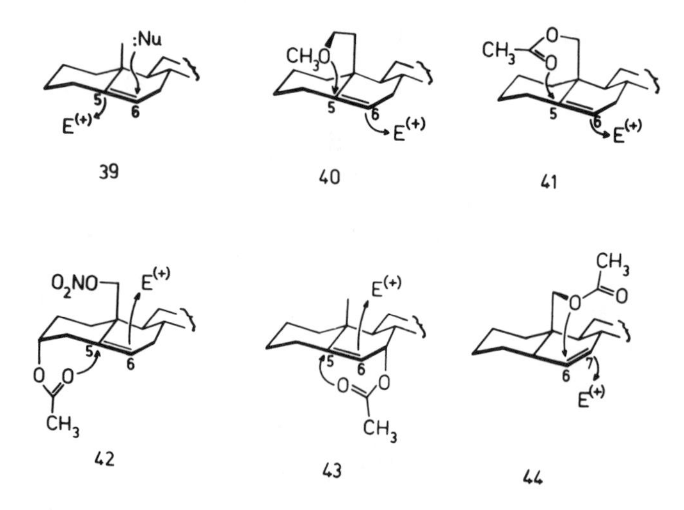

FIGURE 16.

attack from the β-side.[56,57] It should be borne in mind, however, that the ester groups can work as ambident nucleophiles, participating either with the carbonyl or the ether oxygen, and the actual course of the reaction is therefore very sensitive to the structure of the starting olefin. To illustrate this point, let us consider electrophilic addition to the unsaturated steroids 41 and 44. While in the $\Delta^{5,6}$-derivative 41 the 19-acetoxy group participates with the carbonyl oxygen, the $\Delta^{6,7}$-isomer 44 reacts with participation by the ether oxygen, giving rise to the corresponding cyclic ether.[58] For similar effects of neighboring group participation on the reaction course see Reference 46.

REFERENCES

1. **Zimmerman, H. E., Singer, L., and Thyagarajan, B. S.,** Overlap control carbanion stereoselectivity in alkaline epoxidation, *J. Am. Chem. Soc.*, 81, 108, 1959, footnote 16.
2. **Eliel, E. L.,** *Stereochemistry of Carbon Compounds,* McGraw-Hill, New York, 1962.
3. **Seebach, D. and Kalinowski, H. O.,** Enantiomerenreine Naturstoffe un Pharmaka aus billigen Vorlaüfern (Chiral pool), *Nachr. Chem. Techn.*, 24, 415, 1976.
4. **Kočovský, P.,** Chiral synthons and intermediates in organic synthesis I, *Chem. Listy,* 76, 1148, 1982 (in Czech).
5. **Kočovský, P. and Černý, M.,** Chiral synthons and intermediates in organic synthesis II, *Chem. Listy,* 77, 373, 1983 (in Czech).
6. **Eberhard, A.,** Über das Ephedrin und verwandte Verbindungen, *Arch. Pharm.*, 253, 62, 1915.
7. **Eberhard, A.,** Über die Synthese des inaktiven Ephedrins bezw. Pseudoephedrins, *Arch. Pharm.*, 258, 97, 1920.
8. **Hyde, J. F., Browning, E., and Adams, R.,** Synthetic homologs of *d,l*-ephedrine, *J. Am. Chem. Soc.*, 50, 2287, 1928.
9. **Morrison, J. D. and Mosher, H. S.,** *Asymmetric Organic Reactions,* Prentice Hall, New York, 1971.
10. **Izumi, Y. and Tai, A.,** *Stereo-Differentiating Reactions,* Academic Press, New York, 1977.
11. **ApSimon, J. W. and Seguin, R. P.,** Recent advances in asymmetric synthesis, *Tetrahedron,* 35, 2797, 1979.
12. **Seebach, D. and Hungerbühler, E.,** Modern Synthetic Methods, Vol. 2, Schefold, R., Ed., Salle & Sauerlander, Frankfurt, 1980.
13. **Mori, K.,** The synthesis of insect pheromones, in *The Total Synthesis of Natural Products,* ApSimon, J., Ed., Wiley-Interscience, New York, 1981.
14. **Rétey, J. and Robinson, J. A.,** *Stereospecificity in Organic Chemistry and Enzymology,* Verlag Chemie, Weinheim, 1982.

15. **Allinger, N. L. and Eliel, E. L.,** *Topics in Stereochemistry,* Vol. 2., Wiley-Interscience, New York, 1967.
16. **Vrkoč, J. and Ubik, K.,** 1-Nitro-trans-1-pentadecene as the defensive compound of termites, *Tetrahed. Lett.,* 1463, 1974.
17. **Ugi, I.,** The four component synthesis, in *The Peptides,* Vol. 2, Gross, E. and Meinhofer, J., Eds., Academic Press, New York, 1980, 365.
18. **Staněk, J., Černý, M., Kocourek, J., and Pacák, J.,** *The Monosaccharides,* Academia, Prague, 1963.
19. **Stoddart, J. F.,** *Stereochemistry of Carbohydrates,* Wiley-Interscience, London, 1971.
20. **Pigman, W. and Horton, D.,** *The Carbohydrates,* 2nd ed., Academic Press, New York, 1970.
21. **Kieslich, K.,** *Microbial Transformations of Non-steroid Cyclic Compounds,* Georg Thieme Verlag, Stuttgart, 1976.
22. **Čapek, A., Hanč, O., and Tadra, M.,** *Microbial Transformations of Steroids,* Academia, Prague, 1966.
23. **Akhrem, A. A. and Titov, Yu. A.,** *Steroids and Microorganisms,* Nauka, Moscow, 1970, (in Russian).
24. **Jones, E. R. H.,** The microbial hydroxylation of steroids and related compounds, *Pure Appl. Chem.,* 33, 39, 1973.
25. **Kočovský, P.,** Problems of stereoselectivity in the synthesis of natural products, *Chem. Listy,* 77, 800, 1983 (in Czech).
26. **Nazarov, I. N., Verkholetova, G. P., and Torgov, I. G.,** The synthesis of steroids and related compounds 42. Condensation of 1-vinyl-9-methyl-Δ^1-6-octalone with quinone, *Izv. Akad. Nauk SSSR, Ser. Khim.,* 283, 1959, (in Russian).
27. **Fringuelli, F., Pizzo, F., Taticchi, A., Halls, T. D. J., and Wenkert, E.,** Diels-Alder reactions of cycloalkenones. 1. Preparation and structure of the adducts, *J. Org. Chem.,* 47, 5056, 1982.
28. **Woodward, R. B., Gosteli, J., Ernest, I., Triary, R. J., Nestler, G., Raman, H., Sitrin, R., Suter, C., and Whitesell, J. K.,** A novel synthesis of prostaglandin $F_{2\alpha}$, *J. Am. Chem. Soc.,* 95, 6853, 1973.
29. **Woodward, R. B., Bader, F. E., Bickel, H., Frey, A. J., and Kierstead, R. W.,** The total synthesis of reserpine, *Tetrahedron,* 2, 1, 1958.
30. **Guthrie, R. D. and Murphy, D.,** Nitrogen-containing carbohydrate derivatives. Some azido- and epiminosugars, *J. Chem. Soc.,* 5288, 1963.
31. **Černý, M., Elbert, T., and Pacak, J.,** Preparation of 1,6-anhydro-2,3-dideoxy-2,3-epimino-β-D-mannopyranose and its conversion to 2-amino-1,6-anhydro-2-deoxy-β-D-mannopyranose, *Collect Czech. Chem. Commun.,* 39, 1752, 1974.
32. **Mislow, K.,** *Introduction to Stereochemistry,* W. A. Benjamin, New York, 1965.
33. **Sokolov, V. I.,** *Introduction to Theoretical Stereochemistry,* Nauka, Moscow, 1979 (in Russian).
34. **Cram, D. J. and Ehlhafez, F. A.,** Studies in stereochemistry. The rule of "steric" control of asymmetric induction in the synthesis of acyclic systems, *J. Am. Chem. Soc.,* 74, 5828, 1952.
35. **Červinka, O., and Bělovský, O.,** Some factors influencing the course of asymmetric reduction by optically active alkoxy lithium aluminum hydrides, *Collect. Czech. Chem. Commun.,* 30, 3897, 1967.
36. **Červinka, O., Suchan, V., and Masař, B.,** Absolute configuration of diaryl carbinols, *Collect. Czech. Chem. Commun.,* 30, 1693, 1965.
37. **Evans, R. J. D. and Landor, S. R.,** The absolute configuration of 3,4,4-trimethylpent-1-yn-3-ol and 1-chloro-3,4,4-trimethylpenta-1,2-dienes, *J. Chem. Soc.,* 2553, 1965.
38. **Noyori, R., Tomino, I., and Tanimoto, I.,** Virtually complete enantioface differentiation in carbonyl group reduction by a complex aluminium hydride reagent, *J. Am. Chem. Soc.,* 101, 3129, 1979.
39. **Hayashi, T., Tajika, M., Kumada, M., and Tamao, K.,** High stereoselectivity in asymmetric Grignard cross-coupling catalyzed by nickel complexes of chiral (aminoalkylferrocenyl) phosphines, *J. Am. Chem. Soc.,* 98, 3718, 1976.
40. **Corey, E. J.,** General methods for the construction of complex molecules, *Pure Appl. Chem.,* 14, 19, 1967.
41. **Corey, E. J., Andersen, N. H., Carlson, R. M., Paust, J., Vedejs, E., Vlattas, I., and Winter, R. E. K.,** Total synthesis of prostaglandins. Synthesis of the pure d,l-E_1, -$F_{1\alpha}$, -$F_{1\beta}$, -A_1 and -B_1 hormones, *J. Am. Chem. Soc.,* 90, 3245, 1968.
42. **Corey, E. J., Backer, K. B., and Varma, R. K.,** Efficient generation of the 15S configuration in prostaglandin synthesis. Attractive interactions in stereochemical control of carbonyl reduction, *J. Am. Chem. Soc.,* 94, 8616, 1972.
43. **Noyori, R., Tomino, I., and Nishizawa, M.,** A highly efficient synthesis of prostaglandin intermediates posessing the 15S configuration, *J. Am. Chem. Soc.,* 101, 5843, 1979.
44. **Corey, E. J.,** Prostaglandins, *Ann. N.Y. Acad. Sci.,* 180, 24, 1971.
45. **Eliel, E. L., Koskinies, J. K., and Lohri, B.,** A virtually completely asymmetric synthesis, *J. Am. Chem. Soc.,* 100, 1614, 1978.
46. **Capon, B. and McManus, S. P.,** *Neighboring Group Participation,* Plenum Publishing, London, 1976.
47. **Kočovský, P.,** Synthesis of some unsaturated 19-homocholestane derivatives, *Collect. Czech. Chem. Commun.,* 48, 3597, 1983.

48. **Kirk, D. N. and Hartshorn, M. P.,** *Steroid Reaction Mechanisms*, Elsevier, Amsterdam, 1968.
49. **Černý, M. and Staněk, J., Jr.,** 1,6-Anhydro derivatives of aldohexoses, *Adv. Carbohydr. Chem.*, 34, 23, 1977.
50. **Kočovský, P., Tureček, F., and Černý, V.,** Mechanism of acid cleavage of some steroid epoxides. Competition between neighboring group participation and external nucleophile attack, *Collect. Czech. Chem. Commun.*, 47, 124, 1982.
51. **Kočovský, P.,** Competition of 5(O)^n and 6(O)^n participation by 19a-substituent in hypobromous acid addition to 2,3- and 5,6-unsaturated 19-homocholestane derivatives, *Collect. Czech. Chem. Commun.*, 48, 3606, 1983.
52. **Kočovský, P. and Tureček, F.,** Mechanism and structural effects in bromolactonization, *Tetrahedron*, 39, 3621, 1983.
53. **Kočovský, P., Černý, V., and Synáčková, M.,** Participation of 19-ester groups in hypobromous acid addition to 2,3- and 5,6-unsaturated steroids, *Collect. Czech. Chem. Commun.*, 44, 1483, 1979.
54. **Kočovský, P., Tureček, F., and Černý, V.,** Mechanism of hypobromous acid addition to unsaturated steroids. Competition between neighboring group participation and external nucleophile attack, *Collect. Czech. Chem. Commun.*, 47, 117, 1982.
55. **Kočovský, P.,** Synthesis of 14-deoxy-14α-strophantidin, *Tetrahed. Lett.*, 21, 555, 1980.
56. **Kočovský, P.,** Acetoxy group as control element in electrophilic addition: Participation by acetoxy group and its competition with other participating groups in hypobromous acid addition to some 5-cholestene derivatives, *Collect. Czech. Chem. Commun.*, 48, 3629, 1983.
57. **Kočovský, P., Starý, I., Zajíček, J., and Vašíčková, S.,** unpublished results.
58. **Kočovský, P., Kohout, L., and Černy, V.,** Participation of 19-substituents in electrophilic additions to 6,7-unsaturated 5α-cholestane and B-homocholestane derivatives; a case of competition between the participation of an ambident neighboring group and external nucleophile attack, *Collect. Czech. Chem. Commun.*, 45, 559, 1980.

Chapter 2

SYNTHESES OF CYCLIC SYSTEMS

I. INTRODUCTION

Mother Nature provides us with a bountiful variety of compounds that contain at least one ring as a typical structural feature. There are natural substances containing carbocyclic and heterocyclic rings of different sizes, starting with a three-membered ring and ending with macrocycles. This Chapter will be focused on the general aspects of syntheses leading to cyclic systems. The following chapters will then deal with the particular strategies, depending on the structure, number of rings, and substitution pattern of the compounds to be synthesized.

II. RING FORMATION

In principle, a new cyclic system can be created in two ways: Either by a *de novo* synthesis from an acyclic precursor, or by modifying an already existing ring system, e.g. by contraction or expansion. The first route is depicted in Figure 1. Here we may distinguish a one-synthon ring closure (a) and a two- (or more) synthon ring closure (b). The latter can be further specified as (m + n), according to the number of heavy atoms composing the skeletons of the components.

The ring closure a simply consists of connecting two ends of a chain. We will call it a "monotopic" cyclization, because the new bond is formed in a single position of the future ring, though of course, between two atoms.[1]

The route b (Figure 1) affords more possibilities, since the two fragments can be joined in two different modes. A stepwise mode (b_1) consists of linking first the molecular fragments (synthons) by one bond, disregarding for the moment whether this is a single or a multiple bond, and then closing the ring in a second step. Let us note that the whole procedure may be performed as a one-pot synthesis. Cyclizations a and b_1 can be accomplished by standard "monotopic" methods, such as nucleophilic substitution, addition to a double bond, aldol, acyloin, Dieckmann or Thorpe condensations, ene reaction, McMurry coupling, etc. which will be elaborated in the following chapters. Let us note here that heterocyclic rings can be closed either by forming the carbon-heteroatom bond or by establishing a carbon-carbon bond, provided the chain already contains a heteroatom (Figure 2).

A second mode of ring closure (b_2 in Figure 1) consists of linking two synthons in one step by a concerted reaction, e.g. Diels-Alder or [2 + 2] cycloaddition. This mode will be called a "ditopic" ring closure, since it involves one-step bond formation at two different loci on the ring.[1] Both synthetic strategies (b_1 and b_2) are in principle interchangeable and it often depends on the structure of the target compound, availability of the starting material, complexity of the whole procedure or, last but not least, the personality of the author, which approach will be eventually chosen to solve the synthetic problem.

A different approach to the synthesis of cyclic systems rests on modifications of already existing cycles. A number of cyclic compounds are available as a starting material in large quantities, and these may represent a convenient starting point for a chemist to perform a synthesis. According to the modification carried out in the cyclic system we will distinguish three reaction types:

1. Ring contractions (Figure 3) that can be achieved via e.g. Favorskii, Wolf, Demyanov, or Cope rearrangement;
2. ring expansions based on the reaction of ketones with diazocompounds, heteroatom insertions (Beckmann and Baeyer-Villiger reaction), etc. (Figure 3). In polycyclic

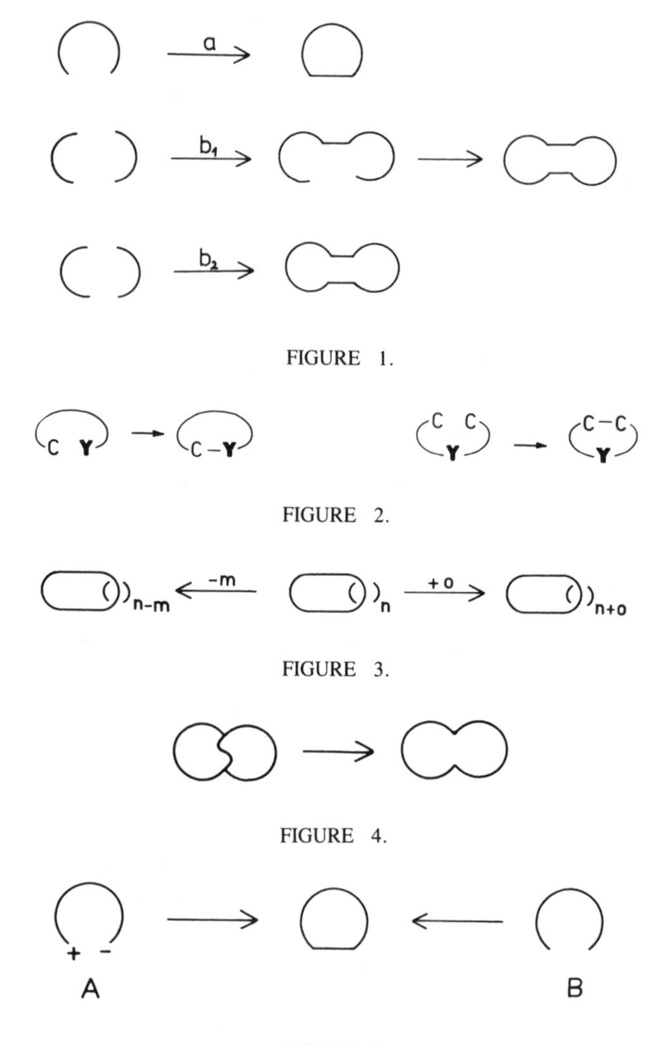

FIGURE 1.

FIGURE 2.

FIGURE 3.

FIGURE 4.

FIGURE 5.

systems the two methods can be combined to achieve a simultaneous expansion of one ring and a contraction of another ring. These methods represent an elegant means of modifying cyclic systems, and they can also be used for an interconversion of ortho-condensed, bridged, and spirocyclic systems, as will be demonstrated later; and

3. fragmentation of a cross-piece bond in a bicyclic system (Figure 4), which has been used frequently in syntheses of medium-ring compounds. Nevertheless, the method is also usable for the construction of six-membered rings, and syntheses of this type have been described, based on fragmentation of the bicyclo[2.2.0] hexane system.

III. SYSTEMS WITH ONE RING

Let us consider the nature of synthons used for a ring closure. If we have to close a ring by connecting the terminae of an aliphatic chain (Figure 5), two reaction classes come into consideration: a heterolytic cyclization in a dipolar synthon A or a homolytic bond formation in synthon B. The same applies to a ring construction involving two synthons (Figure 6). To make the heterolytic link we may combine two dipolar synthons, each carrying one positive and one negative charge (A_1), or we can use a combination of a doubly positive

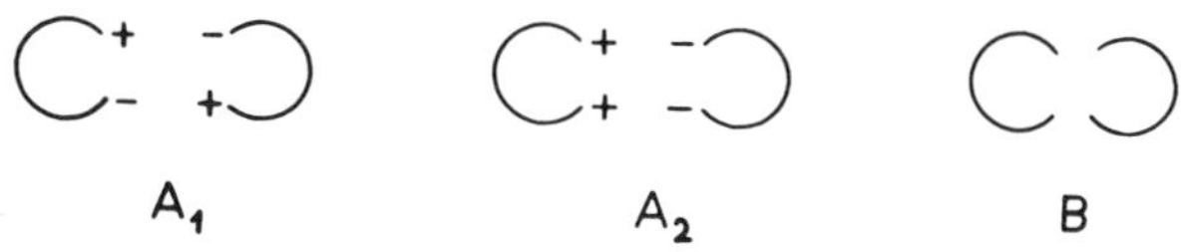

FIGURE 6.

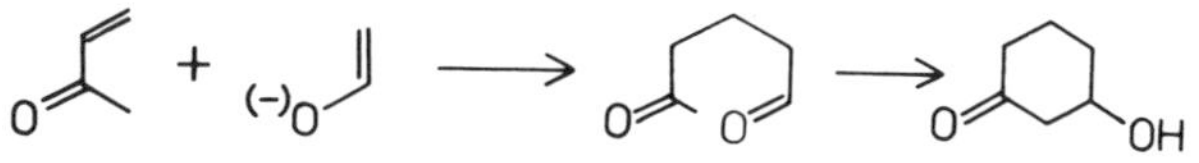

FIGURE 7.

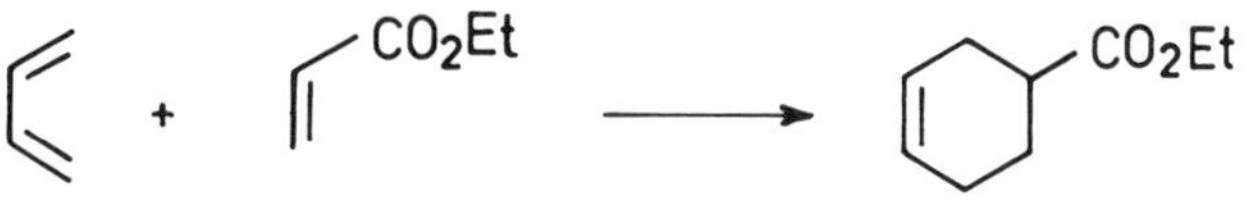

FIGURE 8.

FIGURE 9.

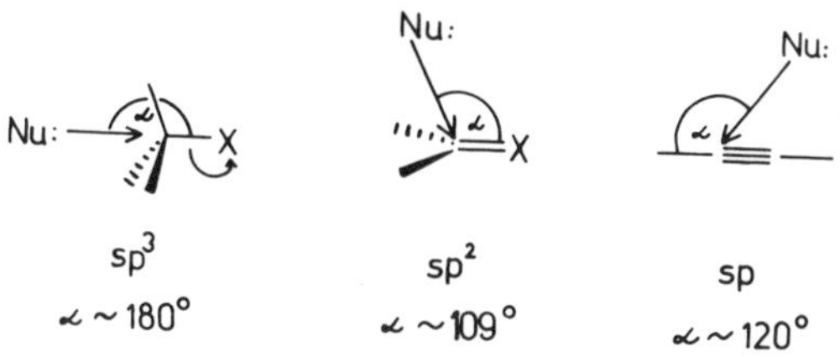

FIGURE 10.

and a doubly negative synthon (A_2). In the homolytic (B) mode, the most frequently used methods are [4 + 2] or [2 + 2] cycloadditions. Examples of these strategies (A and B) are given in Figures 7 and 8. Both the syntheses aim at a cyclohexane derivative. While the first procedure joins two synthons by the Michael addition of an enolate to an α,β-unsaturated ketone, and then forms the ring by aldol condensation, the second procedure creates the ring in a single step by the Diels-Alder reaction.

IV. GENERAL ASPECTS OF RING-CLOSURE REACTIONS

Ring-closure reactions play a key role in syntheses of both carbo- and heterocyclic systems. As mentioned above, beside the ditopic methods there are monotopic ring-forming reactions based on nucleophilic substitution and addition to multiple bonds (Figure 9). In order to proceed smoothly and give preparatively useful results, these reactions require that certain steric conditions be fulfilled in the transition state, e.g. colinearity of the attacking and leaving group in S_N2 substitution accompanied by the Walden inversion (Figure 10), etc. In an intramolecular reaction, it is not always possible to achieve an optimum geometrical arrangement, due to restrictions imposed by the chain connecting the reaction centers. The

Table 1
THE BALDWIN RULES[2,3]

Hybridization	Breaking bond	Ring size 3	4	5	6	7
Tet (sp^3)	*exo*	F[a]	F	F	F	F
	endo	D[b]	D	D	D	D
Trig (sp^2)	*exo*	F	F	F	F	F
	endo	D	D	D	F	F
Dig (sp)	*exo*	D	D	F	F	F
	endo	F	F	F	F	F

[a] F = favored.
[b] D = disfavored.

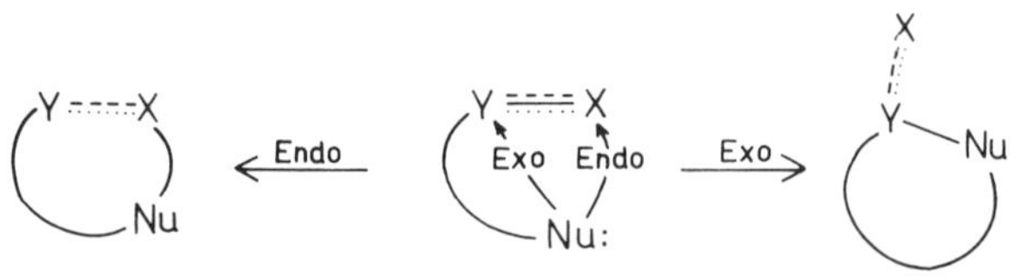

FIGURE 11.

steric and stereoelectronic effects on the course of the cyclization reactions have been summarized by Baldwin into a set of rules[2-5] that classify the reactions as to "favored" and "disfavored" ones, depending on the size of the ring to be closed (Table 1). The prefix *endo* or *exo* in Table 1 denotes whether the bond to be broken is *endo*- or *exocyclic* with respect to the smallest ring to be formed (Figure 11). The size of the ring is denoted by the corresponding number, generally n. The hybridization on the center that is attacked by the nucleophile is denoted by Tet (for sp^3), Trig (for sp^2), or Dig (for sp). It should be noted that the reactions classified as "disfavored" are not excluded *a priori*, and we shall see later that sometimes such a reaction can proceed, even though a "success" here may also indicate a different mechanism. In any case, the rules predict that difficulties might be encountered in some cyclization steps and therefore they should be taken into account to locate possible bottlenecks in the synthetic strategy. Let us note that the rules may also be applied to cationic and radical processes (for a discussion see References 2 to 6).

Ditopic processes are even more rigorously classified by the orbital-symmetry rules as "forbidden" and "allowed" (Table 2).[7] Thus, $[_{\pi}2_s + _{\pi}2_s]$ reactions are forbidden in the ground state (while allowed in the excited state), whereas $[_{\pi}4 + _{\pi}2_s]$ reactions (e.g. the Diels-Alder reaction) are allowed in the ground state (for a detailed discussion see References 7 and 8). It is pertinent to note, however, that these rules may be violated by intervention of transition metal catalysts.[9-11]

V. POLYCYCLIC SYSTEMS

Systems containing more than one ring can be, in principle, constructed in two manners (Figure 12). We can join another ring to an already existing system (annulation, a), or bisect a ring by a cross-piece bond to enlarge the number of cycles by one (b). Let us note that the former method is used more frequently, although the latter may also become advantageous under certain conditions.

Figure 13 shows the general modes of construction of orthocondensed, spirocyclic, and bridged bicyclic systems (or subsystems) by means of annulation. To make the bonds denoted by dotted lines in Figure 13, one has at his disposal an arsenal of mono- and ditopic methods which will be discussed in detail later.

Table 2
WOODWARD-HOFFMANN RULES FOR CYCLOADDITIONS[7]

m + n[a]	Allowed[b]	Forbidden[b]
4q	$m_s + n_a$	$m_s + n_s$
	$m_a + n_s$	$m_a + n_a$
4q + 2	$m_s + n_s$	$m_s + n_a$
	$m_a + n_a$	$m_a + n_s$

[a] m and n = number of electrons participating in the reaction.
[b] The rules refer to the ground electronic state, and are reversed in the first excited state.

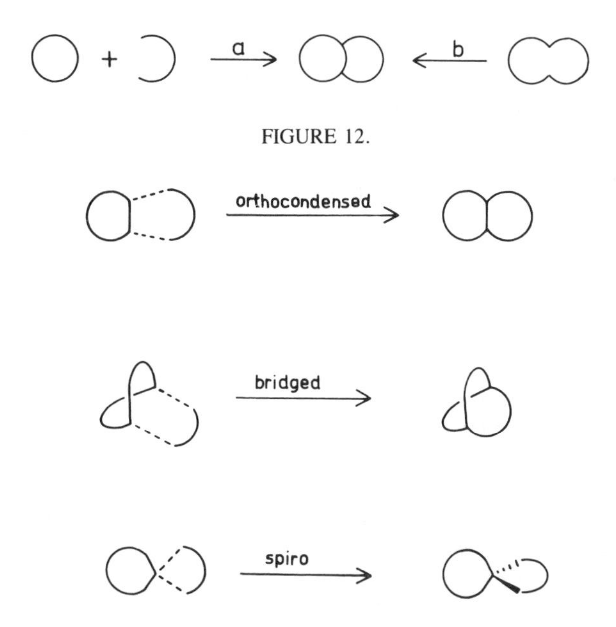

FIGURE 12.

FIGURE 13.

VI. STRATEGY FOR STEREOSELECTIVE SUBSTITUTION IN CYCLIC SYSTEMS

The majority of mono- and polycyclic natural products carry one or more substituents attached to the skeleton in a defined relative or absolute configuration. There are essentially three synthetic routes to such substituted systems:

1. The system is built up by cyclization of a precursor which already contains all the stereogenous elements (Figure 14). This method is commonly used in syntheses of macrocyclic compounds.
2. The substituents are introduced into a preformed ring (Figure 15) via addition or substitution; this is common with five- or six-membered rings.
3. The substituents are formed by cleaving auxiliary rings in the precursor molecule (Figure 16).

The individual procedures (Figures 14 to 16) can be of course, combined.

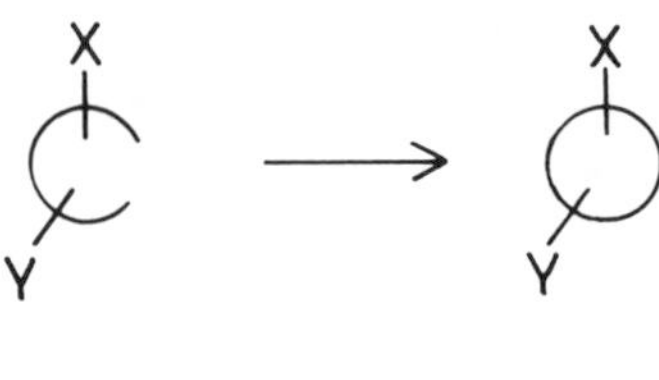

FIGURE 14.

FIGURE 15.

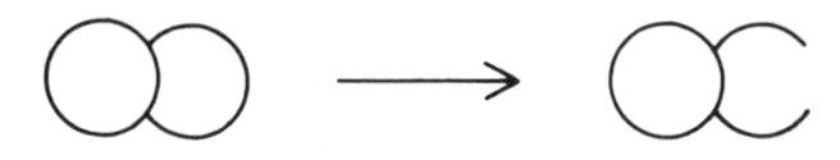

FIGURE 16.

In this Chapter we have attempted to give an outline of the strategic principles which are behind the tactics (i.e. methods and procedures) used for the construction of cyclic systems. These will be dealt with in detail in the following chapters. The general approaches to acyclic compounds are simpler and will be discussed in Chapter 9.

VII. NOTES ADDED IN PROOF

Insertion of a heteroatom into the spacer connecting the olefinic double bond with the attacking species may sometimes alter the regioselectivity of the ring closure. Thus the usual 6-endo-trig course of radical cyclization was reversed to 5-exo-trig by insertion of nitrogen atom.[12] Vinylogous Dieckmann condensation has been shown to obey Baldwin's rules.[13] 5-Endo-trig ring closures have been observed with several unsaturated sulfones.[14]

REFERENCES

1. **Kočovský, P.,** Problem of stereoselectivity in the synthesis of natural products, *Chem. Listy,* 77, 800, 1983 (in Czech).
2. **Baldwin, J. E.,** Rules for ring closure, *J. Chem. Soc. Chem. Commun.,* 734, 1976.
3. **Baldwin, J. E.,** Approach vector analysis: A stereochemical approach to reactivity, *J. Chem. Soc. Chem. Commun.,* 738, 1976.
4. **Baldwin, J. E. and Kruse, L. I.,** Rules, for ring closure. Stereoelectronic control in the endocyclic alkylation of ketone enolates, *J. Chem. Soc. Chem. Commun.,* 233, 1977.
5. **Baldwin, J. E.,** Rules for ring closure in *Further Perspectives in Organic Chemistry,* CIBA Foundation Symposium 53 (New Series), 2nd printing, Excerpta Medica, Amsterdam, 1979.
6. **Deslongchamps, P.,** *Stereoelectronic Effects in Organic Chemistry,* Pergamon Press, Oxford, 1983.
7. **Woodward, R. B. and Hoffmann, R.,** *Die Erhaltung der Orbitalsymmetrie,* Verlag Chemie, Weinheim, 1970.
8. **Fleming, I.,** *Frontier Orbitals and Organic Chemical Reactions,* J. Wiley & Sons, Chichester, 1976.
9. **Pearson, R. G.,** Orbital symmetry rules for unimolecular reactions, *J. Am. Chem. Soc.,* 94, 8287, 1972.

10. **Mach, K., Antropiusová, H., Petrusová, L., Sedmera, P., Hanuš, V., and Tureček, F.,** [6 + 2] Cycloadditions catalyzed by titanium complexes, *Tetrahedron,* 40, 3295, 1984.
11. **Mango, F. D.,** Transition metal catalysis of pericyclic reactions, *Coord. Chem. Rev.,* 15, 109, 1975.
12. **Padwa, A., Nimmesgern, H., and Wong, G. S. K.,** Radical cyclization as an approach toward the synthesis of pyrrolidines, *Tetrahed. Lett.,* 26, 957, 1985.
13. **Kodpinid, M. and Thebtaranonth, Y.,** Vinylogous Dieckmann condensation: an application of Baldwin's rules, *Tetrahed. Lett.,* 25, 2509, 1984.
14. **Auvray, P., Knochel, P., and Normant, J. F.,** 5-Endo-trigonal ring closures of unsaturated sulfones, *Tetrahed. Lett.,* 26, 4455, 1985.

Chapter 3

ortho-CONDENSED SYSTEMS

Many polycyclic natural products contain an *ortho*-condensed system; of all bicyclic subsystems composing the molecular framework, the decalin (or [4.4.0]), hydrindane (or [4.3.0]) and perhydroazulene (or [5.3.0]) skeletons appear most frequently. It is therefore not surprising that the synthesis of these structural blocks has received considerable attention by organic chemists. On the other hand, in the last years a variety of new synthetic methodologies has been developed that led to the specific construction of other *ortho*-condensed systems that have been found in nature, namely, [3.3.0], [3.1.0], [4.2.0], etc. The synthetic approach to these systems may employ cyclization at or outside the annulation sites, for which we can use methods of monotopic or ditopic annulation. In addition to *de novo* methods of construction of cyclic systems, one can start from a compound which already contains the basic polycyclic skeleton and then modify the existing rings by enlargement or contraction.

In this Chapter we shall demonstrate the principles and most important methods for building up the various ring systems. It should be emphasized that these methods can often be applied not only to *ortho*-condensed skeletons, but also to monocyclic compounds and bridged systems, as will be shown in the following chapters.

I. CYCLIZATIONS AT THE ANNULATION SITE

Depending on the key step in the synthetic strategy, the cyclization reactions of this type may be divided to monotopic and ditopic annulations, as already mentioned in Chapter 2. Another classification will distinguish one carbon annulations (leading to [n.1.0] systems) and similarly two-, three-, etc. carbon annulations creating generally [n.m.0] systems. One- and two-carbon annulation are usually ditopic processes such as carbene additions and photochemical [2 + 2] cycloadditions. By contrast, the three-carbon annulation is mostly achieved by monotopic procedures, though a ditopic reaction has also been applied. The synthetic arsenal for the four-carbon annulation is mostly represented by the Robinson annulation (monotopic) and the Diels-Alder reaction (ditopic). *ortho*-Condensed systems can also be built up by connecting two carbon centers in a larger ring, or by using the Cope rearrangement.

In this Chapter we will classify the individual methods according to the size of the ring to be closed. Some methods are applicable to rings of different size; those will be dealt with according to their mechanism. Other strategies are based on a stepwise construction of the ring; for instance, a four-carbon annulation can be realized by successive insertion of a one-carbon and a three-carbon synthon, or it is possible to construct a [n.3.0] system first, and then enlarge the five-membered ring. Methods of this type will also be covered in this Chapter.

A. Monotopic Annulations

Let us consider a two-step process in which the first cycle is connected to a side chain, which is then cyclized to form a new ring (Figures 1 and 13 in Chapter 2). Formally, this involves connecting two pairs of electrophilic and nucleophilic centers (Figure 6 in Chapter 2). As already discussed in Chapter 2, the corresponding synthons may each contain one electrophilic and one nucleophilic center, or one synthon may carry both nucleophilic centers whereas the other would bring the electrophilic ones. In such a two-step process, the first step would involve alkylation at the future annulation site. For this purpose we can utilize S_N alkylation (usually at the α-position to the carbonyl group),[1] Grignard reaction,[1] Michael

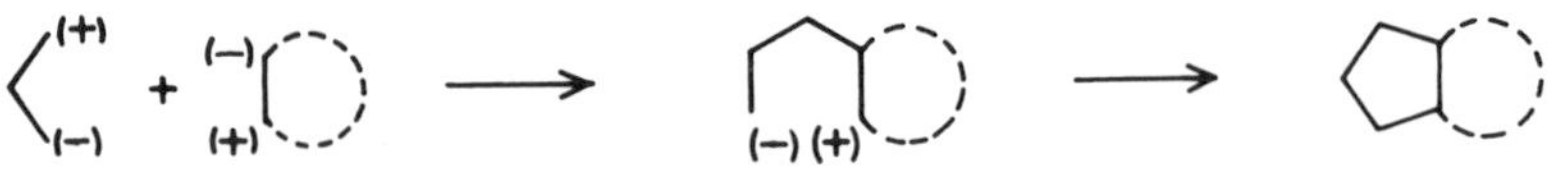

FIGURE 1.

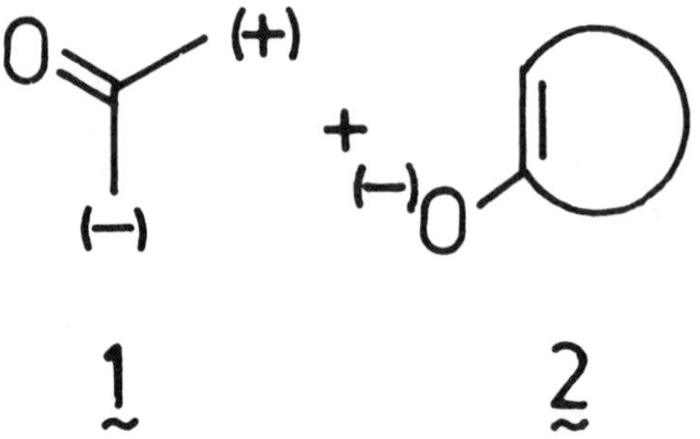

FIGURE 2.

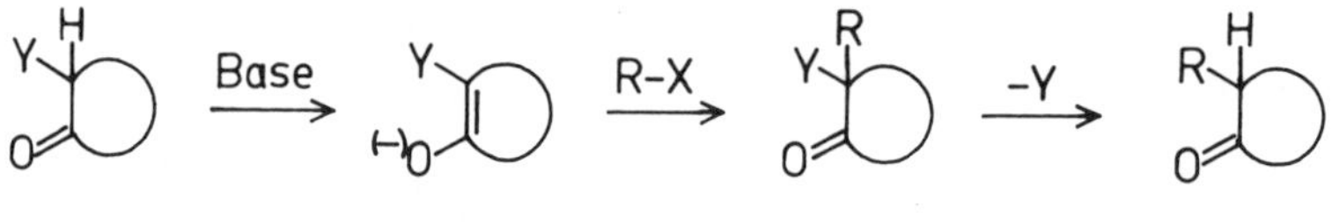

FIGURE 3.

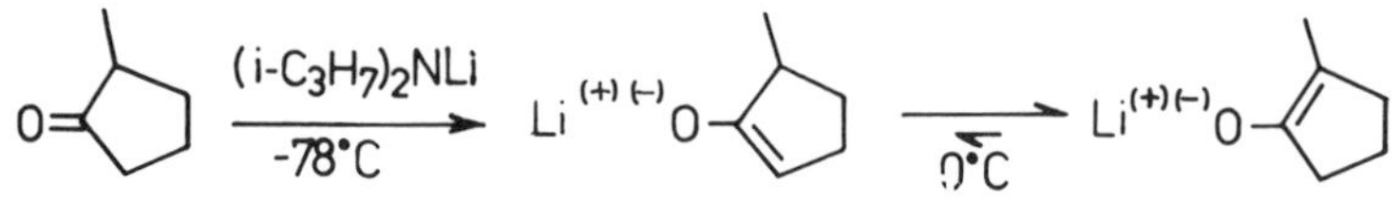

FIGURE 4.

addition,[1] etc.[2] The most frequently used methods in the second ring closing step are aldolization and related reactions, Wittig reaction, carbocation attack across a carbon-carbon double bond and others (*vide infra*).

1. Three-Carbon Polar Annulation (Cyclopentane Annulation)

In order to attach a three-carbon segment we usually need a synthon containing both the electrophilic and the nucleophilic centers (Figure 1). The suitable synthon can be a fragment bearing the carbonyl group (1), e.g. an α-halo ketone as its synthetic equivalent (Figure 2). The carbon atom bearing the halogen atom represents the electrophilic center, whereas the second (enolizable) position located α- to the carbonyl would serve as the nucleophile. The synthon 1 is generally suitable for a three-carbon annulation which, for an *ortho*-condensed system, would correspond to formation of a five-membered ring. The complementary dipolar synthon 2 is usually an enolate of a carbonyl compound,[3-6] or analogously an enamine,[7,8] enol ether,[9] etc.[10] The enolate can be prepared from the parent ketone under action of a strong base (Figure 3, Y = H). If the deprotonation is carried out at low temperature (−78°C) it often leads to a less stable enolate, a product of kinetic control, which isomerizes to the more stable form at elevated temperatures (Figure 4) (for a review see Reference 3). If, for some reason, we may not use a strong base, it is possible to increase the acidity of the enolizable hydrogens by introducing an electron-withdrawing group (Y in Figure 3). If Y = COOEt, the activating group can be then removed by simple hydrolysis-decarboxyl-

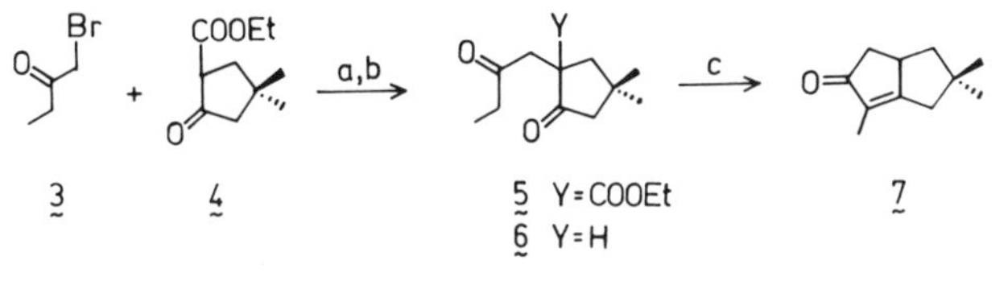

FIGURE 5. (a) NaH; (b) H_3O^+; (c) NaH, C_6H_6.

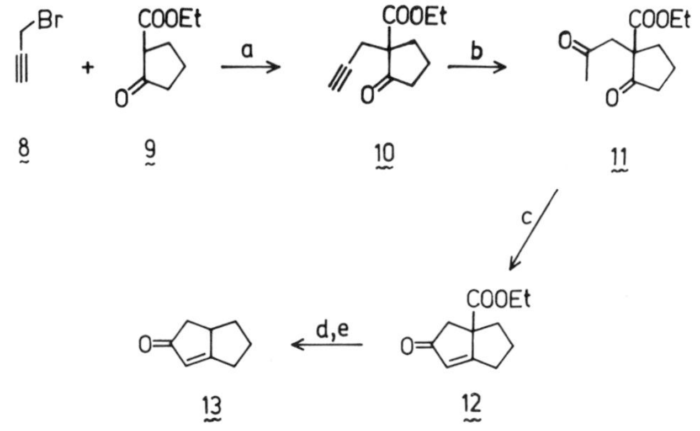

FIGURE 6. (a) t-BuOK, t-BuOH, reflux 45 min; (b) HgO, Dowex 50, H_2SO_4, H_2O, CH_3OH, 20°C, 60 hr; (c) NaH, toluene, reflux 18 hr, dist; (d) 1%-NaOH, 0—10°C, 1.5 hr; (e) 100°C 10 min.

ation. In addition to the activating effect, the auxiliary group can also direct the formation of the desired enolate in cases where simple enolization would give ambiguous results.[3]

Figure 5 shows an example of three-carbon polar annulation.[11] The acidity of the α-hydrogen in the parent ketone is boosted by the activating carbethoxy group. The synthon for the three-carbon annulation is here the bromo ketone 3. In the first step, the enolate of 4 is alkylated with the bromoketone 3 (a), the product 5 is then decarboxylated to 6, and in the next step the five-membered ring is closed in base-catalyzed aldolization (6 → 7). For other examples of this strategy see References 12 and 13.

The synthetic equivalent of the synthon 1 may also be a suitably substituted acetylene (the triple bond is a synthon for ketone), e.g. propargyl bromide (8) (Figure 6). After alkylation of the enolate of 9, the triple bond in the product 10 is hydrated to produce methylketone 11. The further synthetic procedure is quite analogous to that shown in Figure 5.[14]

Two other possible combinations are illustrated in Figure 7. The three-carbon dipolar synthon, e.g. 3-nitro-propionyl chloride (14) first generates the acylium cation which reacts as an electrophile with the enolate. The ring closure is accomplished by aldolization, whereby the nitro group is eliminated.[15]

In Trost's protocol,[16] derivative 15 serves as a synthetic equivalent of the dipolar synthon 1. Thus, mesylate 15 is coupled with an activated ketone and a five-membered ring is smoothly closed up by means of fluoride anion. It is pertinent to note that this method works very well in synthesis of [3.3.0] systems (for a review see References 17 and 18).

An alternative procedure for ring closure is that of using the Wittig reaction (Figure 8) which works even in cases where the aldolization fails (especially with [3.3.0] systems). In the synthesis depicted in Figure 8 the utilization of a chiral phosphine resulted in a 70% optical yield[19,19a] of 16. The α-position can also be activated by the carbethoxy group as in 17. Other three-carbon annulations may be found in References 2, 17, and 18.

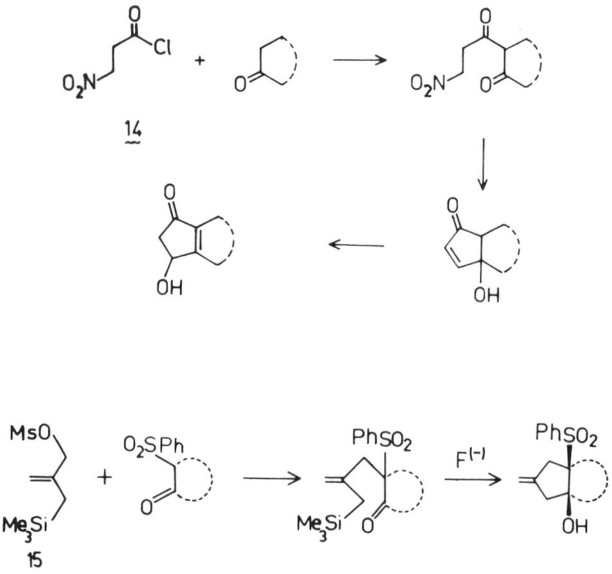

FIGURE 7.

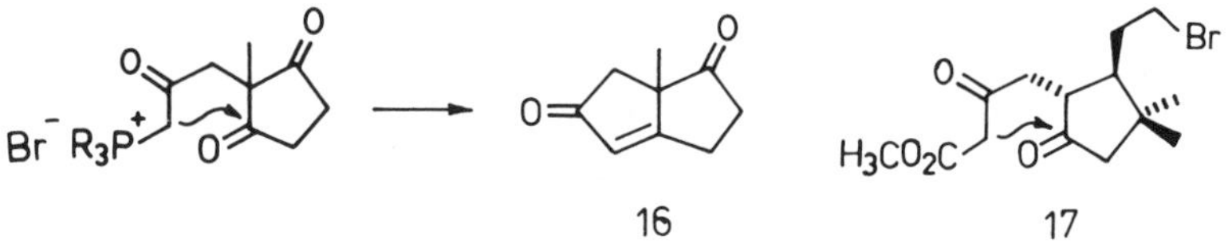

FIGURE 8.

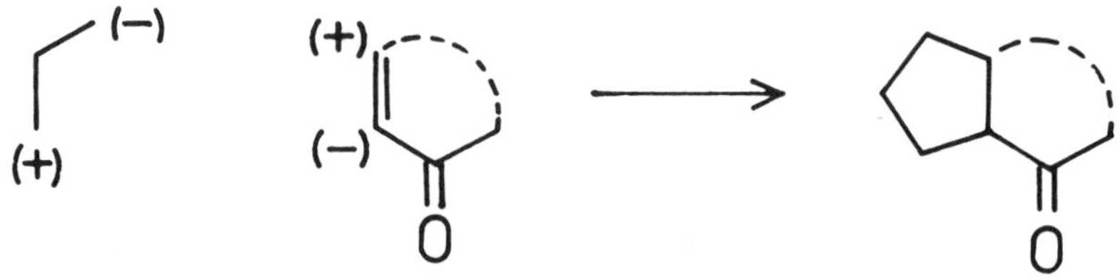

FIGURE 9.

A different choice of synthetic equivalents representing the same dipolar synthons is shown in Figure 9. The three-carbon building block is made up by the organometallic reagent (18 in Figure 10) carrying a masked aldehyde group as electrophile. The nucleophilic center is the carbon atom bearing the metal. This synthon is added to the enone double bond and the procedure is terminated by aldolization of the deprotected aldehyde group with the cyclohexanone moiety.[20] This approach has been recently used in the synthesis of ptilocaulin.[21] This general scheme can be applied to four-carbon annulation by using a homolog of 18.[22,23] Also silyl allenes (e.g. 19) can be used as fancy synthetic equivalents of 1,3-dipolar synthons in high yielding stereo- and regioselective three carbon annulations (Danheiser annulation).[24] For other synthetic equivalents and related methods see References 25 to 27.

So far we have described the methods based on two dipolar synthons, each containing one electrophilic and one nucleophilic center. Recently, Heathcock[28] reported a method for annulation of two dipolar synthons, one of which contains two nucleophilic centers while

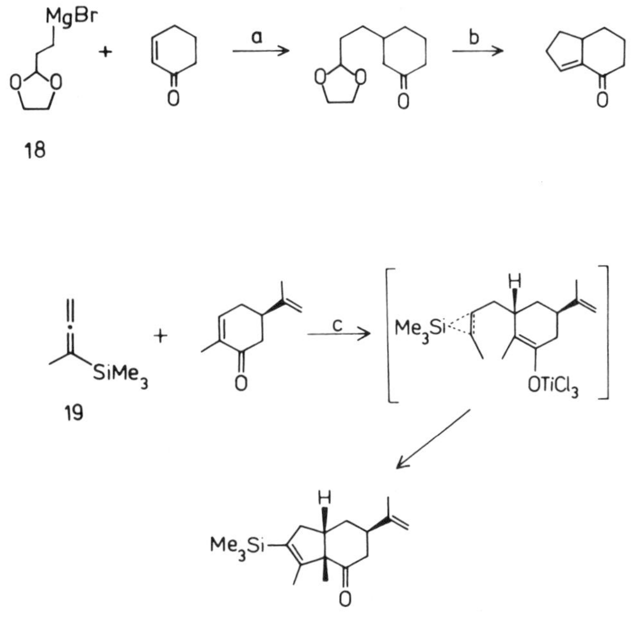

FIGURE 10. (a) CuBr, Me_2S, −78°C 4 hr, then 0°C 6 hr; (b) HCl; (c) $TiCl_4$, CH_2Cl_2, −78°C 1 hr.

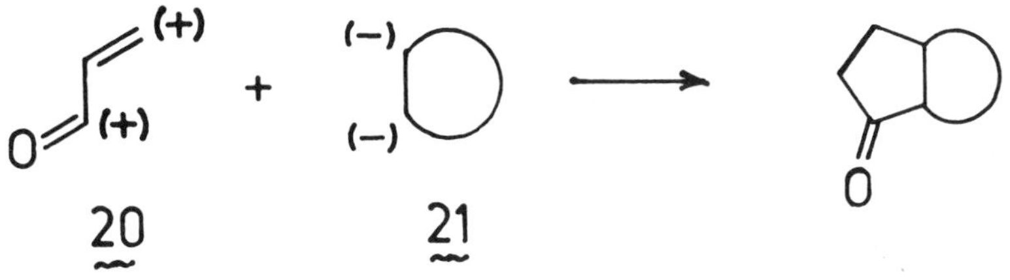

FIGURE 11.

the other carries two complementary electrophilic centers (Figure 6 in Chapter 2 and Figure 11). The synthetic equivalent of the doubly electrophilic synthon is represented by the α,β-unsaturated compound 20 (Figure 11). The nucleophilic center in 21 is the α-carbon with respect to the keto group. According to Figure 12, the second requisite nucleophilic center is created by introducing the double bond into the cyclopentanone ring, which activated the γ-position of the enone system (b, c in Figure 12). This procedure led to the stereoselective formation of a *cis*-annulated [3.3.0] system. A similar strategy has been used in the synthesis of coriolin.[29]

Other possibilities for building up a five-membered ring are exemplified by the ene reaction, tandem attachment of one- plus two-carbon synthons, epoxyannulation, or pyrolysis of vinylcyclopropanes. All these methods will be discussed in the following sections.

2. Four-Carbon Polar Annulation (Robinson Annulation)

As stated before, four-carbon annulation may be accomplished via monotopic or ditopic procedures. This section will deal with the first alternative.

Homologation of the dipolar synthon shown in Figure 1 produces the four-carbon dipolar synthon 22 with one electrophilic and one nucleophilic center (Figure 13). The corresponding synthetic equivalent can be realized as an α,β-enone 24 (Figure 14), or a β-haloketone, etc., which is able to alkylate the vicinal dipolar synthon 23, represented usually by the

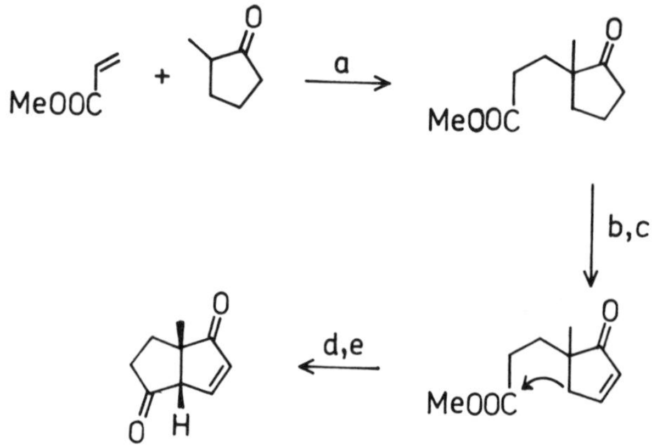

FIGURE 12. (a) CH_3ONa, THF; (b) Br_2, $CHCl_3$; (c) $CaCO_3$, $AcNMe_2$; (d) KOH, CH_3OH; (e) $BF_3{\cdot}Et_2O$, Ac_2O.

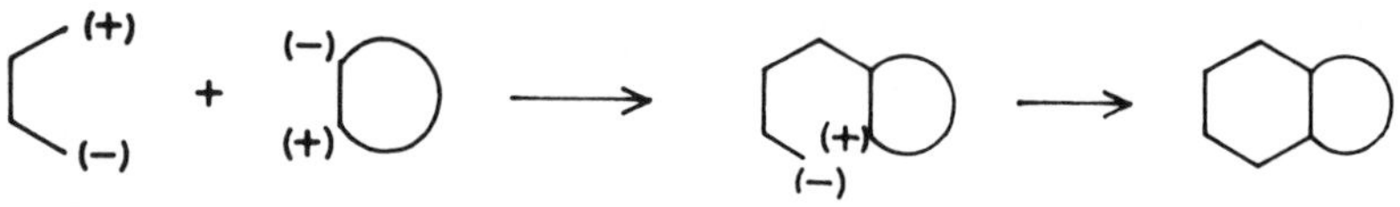

FIGURE 13.

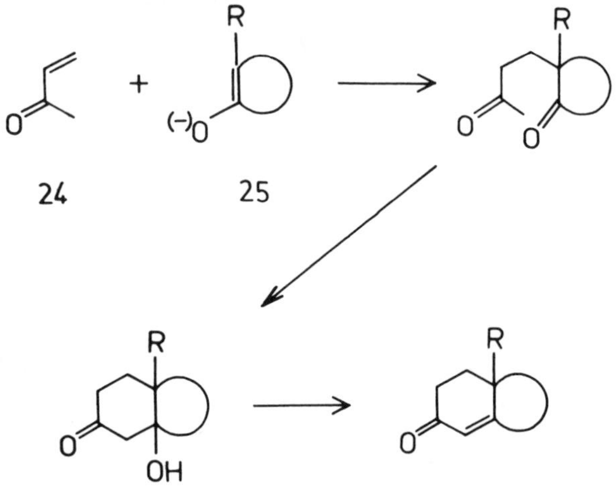

FIGURE 14.

enolate 25 (Figure 14) as its synthetic equivalent. The second step comprises aldolization followed by dehydration. This general procedure, referred to as Robinson annulation, represents one of the most frequently used modes of closing six-membered rings in decalin and hydrindane skeletons.[30,31] Numerous modifications of the above basic scheme have been employed in synthesis of isoprenoids and other natural products (for reviews see References 32 to 34).

Unfortunately, the key reagent in the Robinson annulation, 1-butene-3-one (24), undergoes facile polymerization under the strongly basic conditions that are necessary to generate

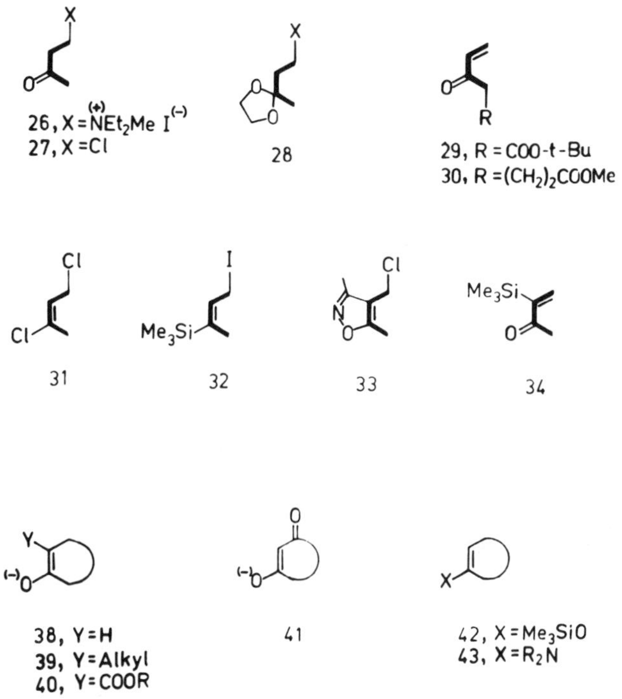

26, X = $\overset{(+)}{N}Et_2Me\ I^{(-)}$
27, X = Cl

28

29, R = COO-t-Bu
30, R = $(CH_2)_2COOMe$

31

32

33

34

38, Y = H
39, Y = Alkyl
40, Y = COOR

41

42, X = Me_3SiO
43, X = R_2N

FIGURE 15.

enolates from unactivated ketones (25). This obstacle gave an impetus to search for other synthetic equivalents of 22, which would survive the basic medium (Figure 15). As a suitable substitute of 24 it was suggested to use the quaternary ammonium salt 26,[35] β-chloroketone 27,[36-38] or hydroxyketal 28 (X = OH).[39] The modified methods have proved to be generally applicable, as documented especially by syntheses of nonaromatic steroids.[40-48] Another modification (the Nazarov reaction) makes use of enone activation by ester group (29).[49,50] Hindered enones may react with activated cycloalkanones at high pressure.[51]

In the steroid syntheses starting from 2-methylcyclopentan-1,3-dione in which the second oxo group works as an activator in enol formation, it was unnecessary to use a strong base. This opened new possibilities for using vinyl ketones 30 in which R served simultaneously as a synthon for the subsequent cyclization.[52-54] Bisannulation starting from such compounds became very useful for an easy preparation of tricyclic and tetracyclic skeletons.[55-64]

If unactivated ketones are used as dipolar synthons (23), the first step (Michael addition) may be troublesome because of slow formation of the enolate 25, especially if R = alkyl. One of the possibilities for overcoming this problem is to substitute the Michael addition by S_N alkylation. For this purpose 1,3-dichloro-2-butene (31) was introduced[65-69] (Wichterle reaction)[65,66] which has also found applications in steroid synthesis.[52,53,70-72] However, the hydrolysis of the vinylchloride group in the intermediate requires rather drastic conditions and although modifications of the reaction conditions had a beneficial effect,[73] it was shown by Stork that vinylsilane 32 (Figures 15 and 16)[74,75] or isoxazoline 33[56,76-78] are better synthetic equivalents generally leading to more favorable results. Eventually it was discovered that the Michael addition can be successfully performed with trimethylsilylenone 34. The trimethylsilyl group stabilizes through its vacant 3d orbitals both the enone function and the negative charge arising at the α-carbon after the Michael addition (Figure 16).[79-83] The trimethylsilyl group is then removed under basic conditions which simultaneously induce the aldolization. Another modification encompasses activation of 2-butenone by complex-

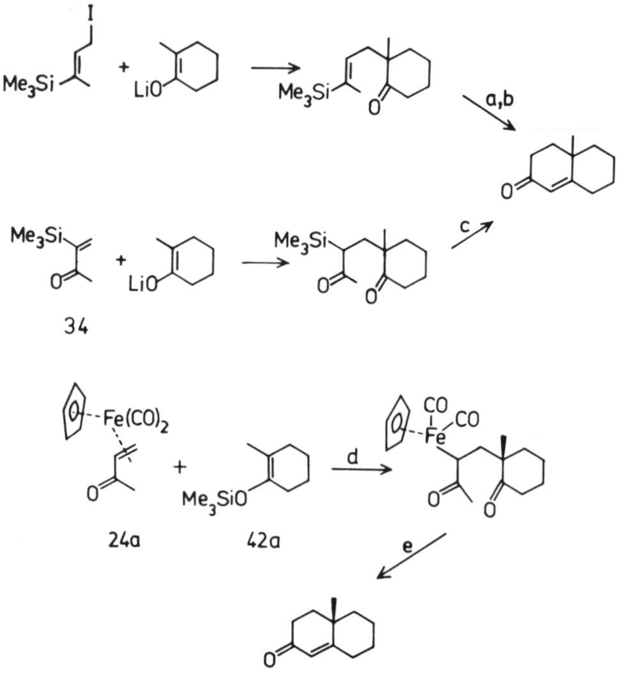

FIGURE 16. (a) m − Cl − $C_6H_4CO_3H$; (b) OH^-; (c) CH_3ONa; (d) 0°C 1 hr, CH_3CN; (e) Al_2O_3, CH_2Cl_2, reflux.

ation with transition metal (24a). The complex smoothly reacts with trimethylsilyl enol ether (42a) in the presence of fluoride anion to yield the corresponding adduct which is then cyclized by means of aluminum oxide.[84]

A modified sequence of Michael additions and alkylation has been recently devised by Danishefsky et al. for synthesis of cyclohexane rings in the tricyclic part of the molecule of aflavinine.[85] The sequence, referred to as [2 + 2 + 2] annulation, starts with Michael addition of the enolate 35 to the exocyclic double bond in the enone 36 (Figure 17). The intermediate enolate 37 undergoes the second, intramolecular Michael addition closing the six-membered ring and generating a new nucleophilic center. The latter attacks the carbonyl group of the original enolate which results in closure of the third ring.

As already mentioned, the formation of enolates from ketones (such as 38 and 39 in Figure 15) can be facilitated by introducing an activating group, e.g. ester (40). Enolates are easily produced from 1,3-diketones (41) as well. Beside ketone enolates, it is possible to employ trimethylsilyl enol ethers 42[2] or enamines 43.[86]

Although all the above-mentioned procedures furnish racemic products, application of chiral reagents[87-94] leads to compounds of high optical purity (Figure 18).[94] Michael addition of 2-methyl-1,3-cyclopentanedione (44) to 1-buten-2-one (24) proceeds under catalysis by a weak acid, so that polymerization of 24 is not a serious problem. The subsequent aldolization in the achiral triketone 45, if catalyzed by L-proline as a chiral base, affords the product 46 in practically quantitative yield and optical purity exceeding 90%! The hydrindane derivative 46, prepared in this way, involves the steroid C and D rings and was used for the synthesis of a pharmacologically important substance, (+)-19-nor-androst-4-en-3,17-dione (48).[88]

Another stereochemical problem may arise if the α,β-enone bears a substituent in the β-position, as in 49 (Figure 19).[95] On Robinson annulation the β-carbon atom becomes a new chirality center, and it has been found that the steric course of the reaction strongly depends

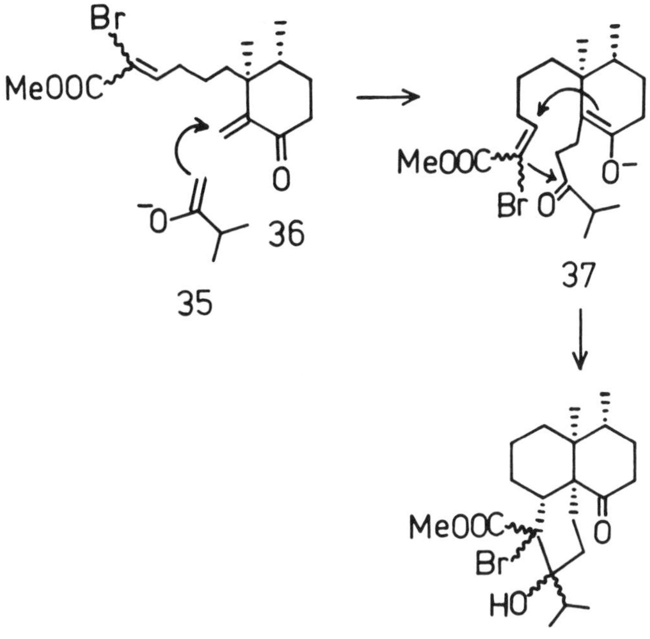

FIGURE 17.

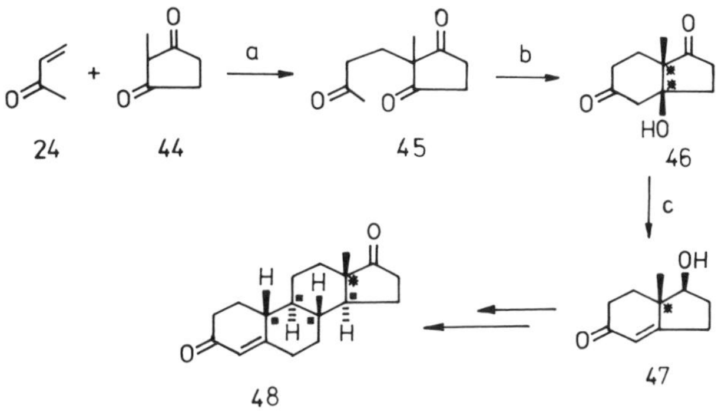

FIGURE 18. (a) AcOH, H_2O; (b) L-(−)-Proline; (c) 1. H_2SO_4, 2. $NaBH_4$.

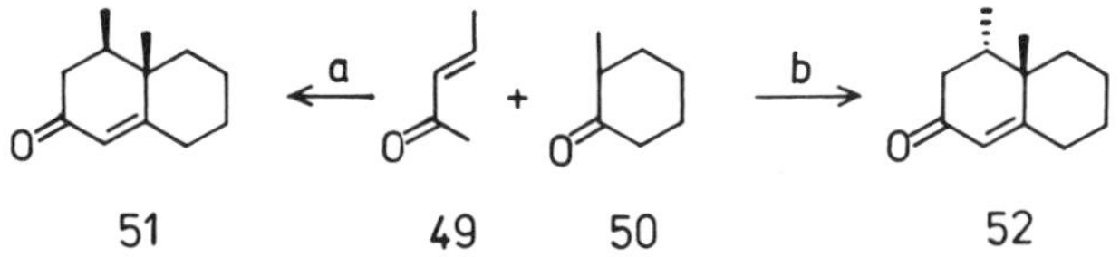

FIGURE 19. (a) NaH, dioxane, reflux 3 hr (→ enolate of 50), then 49 at r.t. for 100 hr; (b) NaH, Me_2SO (→ enolate of 50), then 49 at r.t. for 3 hr.

on the solvent used. For instance, the enolate derived from ketone 50, if generated with sodium hydride in dioxane and allowed to react with enone 49, affords the octalone derivative 51 with a *cis*-configuration of the methyl groups. If the same enolate is prepared by deprotonation with sodium hydride in dimethyl sulfoxide, the reaction with enone 42 leads to the octalone 52 with a *trans*-configuration of the methyls. Marshall et al.[96,97] and Scanio and Sarrett[95] have explored this dependence on solvent (Figure 20). In a nonpolar medium

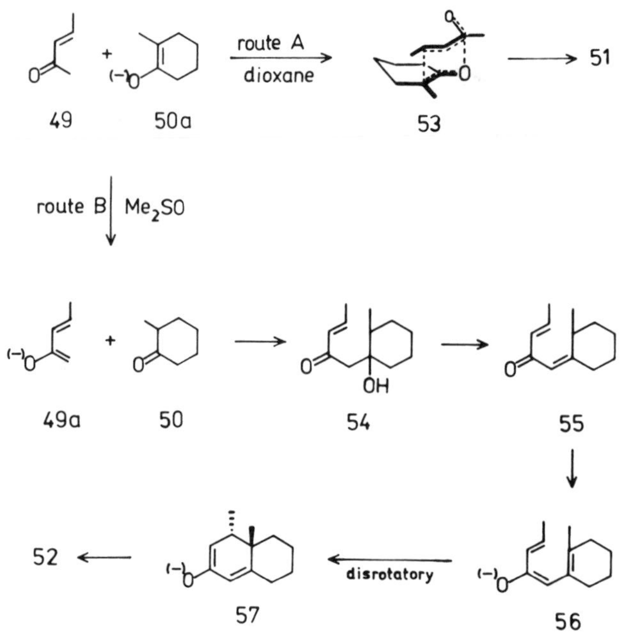

FIGURE 20.

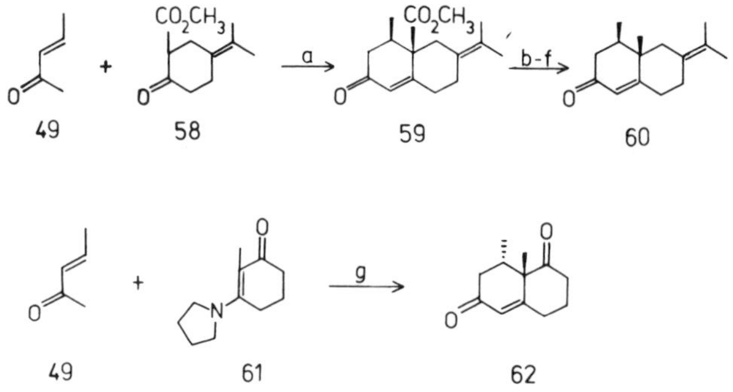

FIGURE 21. (a) CH_3ONa, CH_3OH, 0°C, then r.t. 11 hr; (b) $HOCH_2CH_2OH$, H^+; (c) $LiAlH_4$; (d) CH_3SO_2Cl; (e) Li, NH_3; (f) H_3O^+; (g) AcONa, AcOH, C_6H_6, reflux.

(dioxane) the reaction of 50a with 49 proceeds via a chair-like transition state 53 which is stabilized for steric reasons and which gives rise to the isomer 51 with *cis*-oriented methyl groups. In contrast, a strongly polar medium (dimethyl sulfoxide) facilitates rapid proton migration and hence it is possible to expect an equilibration of the enol forms 49 + 50a ⇄ 49a + 50 that would prefer the conjugated species 49a. The pair 49a + 50 then undergoes aldolization and the intermediate 54 is dehydrated to 55. The latter compound is converted to the enolate 56 which cyclizes to 57 in a disrotatory process. The keto-form of 57 corresponds to the *trans*-dimethyl derivative 52. This mechanism has been supported by further experimental evidence[98,99] (see, however, Reference 100).

In another case (Figure 21) the reaction between 49 and 58 results in the formation of the *cis*-disubstituted derivative 59. The latter compound probably arises via a transition state resembling 53 (path A in Figure 20). The alternative route B is disfavored due to stabilization of the enolate of 58 by the ester function and thus the equilibrium necessary for B does not

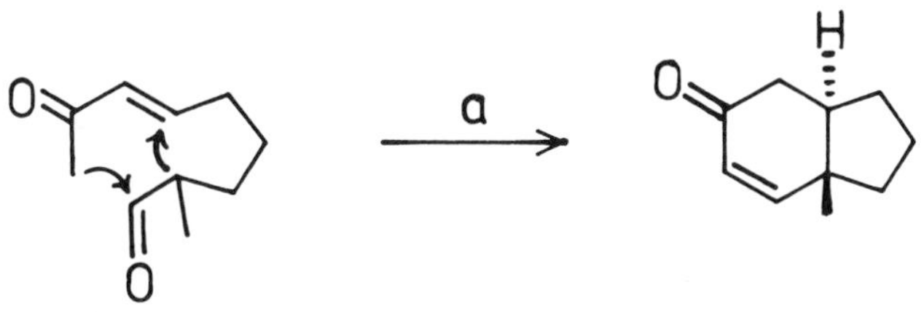

FIGURE 22. (a) $(PrO)_4Zr$, C_6H_6, LiOH, CH_3OH.

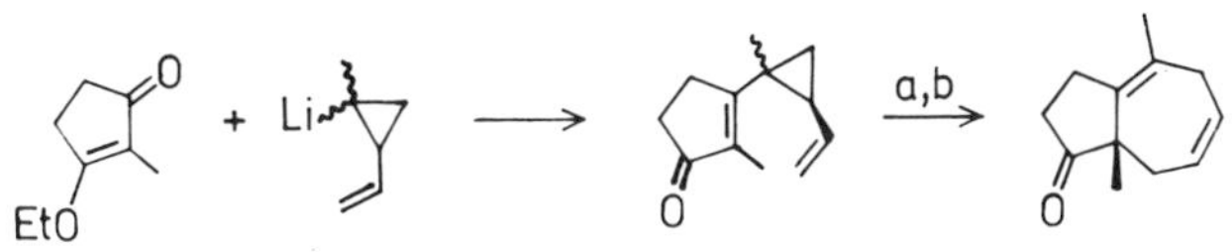

FIGURE 23. (a) hν; (b) 98°C.

evolve. On the contrary, the reaction of 49 with enamine 61 affords mainly the *trans*-dimethyl derivative 62. The stereochemistry of this reaction also depends on the solvent.[101]

Stork has recently suggested an intramolecular variant of the Robinson annulation (Figure 22)[102] which was applied to the synthesis of adrenosterone.[103] Other examples of the Robinson annulation and related reactions can be found in References 33 and 104 to 107. A retro-annulation approach has also been utilized in synthesis.[108]

3. Five-Carbon Polar Annulation

The synthon of a five-carbon fragment (Figure 23)[109] can be represented by the lithiated vinylcyclopropane 64, which reacts with the enol ether of 1,3-diketone 63 yielding a mixture of *cis*- and *trans*-isomers 65. Since only the *cis*-isomer can be further cyclized, it was necessary to improve the *cis:trans* ratio to ca. 4:1 by using a photochemical route. On pyrolysis of the mixture the *cis*-isomer affords 66 via opening the cyclopropane ring and forming subsequently the seven-membered ring. In this way it is possible to prepare compounds having the [5.3.0] bicyclo skeleton which constitutes or is incorporated in many sesquiterpenes. Other, more conventional ways of building up the [5.3.0] system will be treated later.

4. Carbocation Addition to the Double Bond

Nature herself uses this synthetic means in the biogenesis of isoprenoids and artificial synthetic strategies are tailored according to this general scheme. First, an electron-deficient center is generated at a carbon which then attacks the π-orbital of a suitably located double bond to form a new carbon-carbon bond, while regenerating a new electrophilic center. After one or more additional steps the reaction is terminated by nucleophile attachment or proton abstraction. The requisite electron-deficient center on carbon can be generated by solvolysis of a mesylate which proceeds with participation by the double bond (Figure 24).[110] This method was utilized in the synthesis of cycloartenol 67 (Figure 25).[111]

An alternative way of generating the electron-deficient center makes use of acid-catalyzed solvolysis of an allyl alcohol (Figure 26). If there is another double bond in the vicinity of the allylic cation, the cyclization can take place.[112] Tertiary alcohols can also be solvolyzed in this way. Figure 27 shows the fate of the cation 69 formed from the tertiary alcohol (+)-68: The cation first rearranges to 70a (which undergoes a partial racemization via 1,2-shift) and the latter cyclizes to produce the ketone (−)-71. The latter compound was transformed in two steps to (−)-albene 72. It is worth noting that this synthesis,[113,114] though not highly

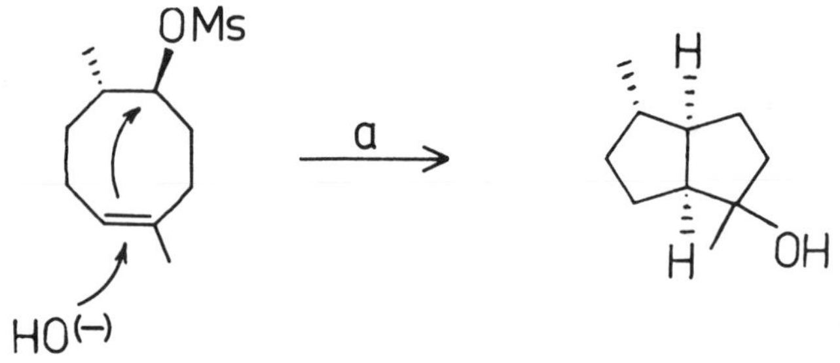

FIGURE 24. (a) Na_2CO_3, dioxane, H_2O, 60°C 12 hr.

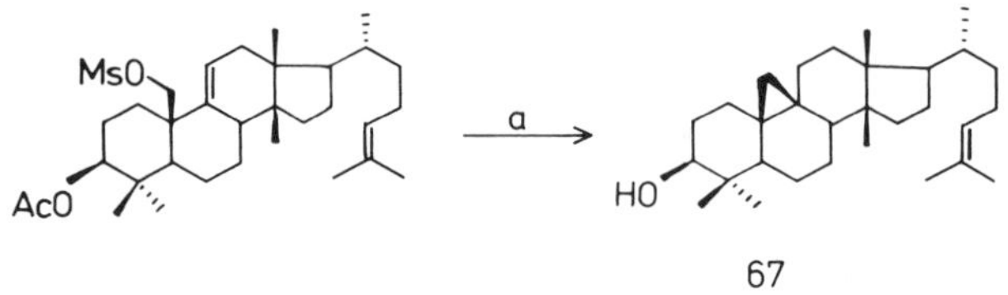

FIGURE 25. (a) $LiAlH_4$, Et_2O, reflux 1 hr.

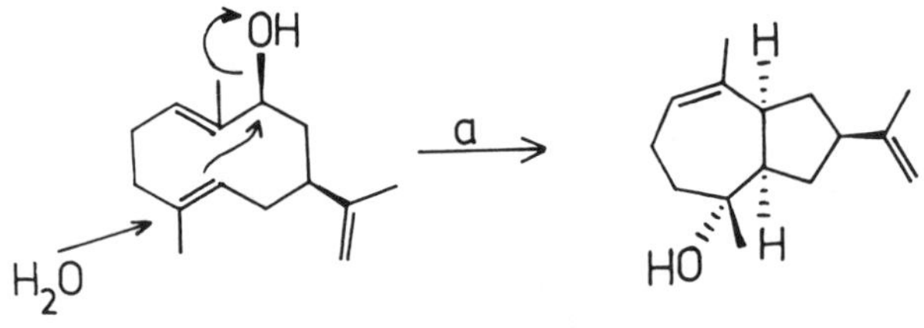

FIGURE 26. (a) 80% $AcOH/H_2O$.

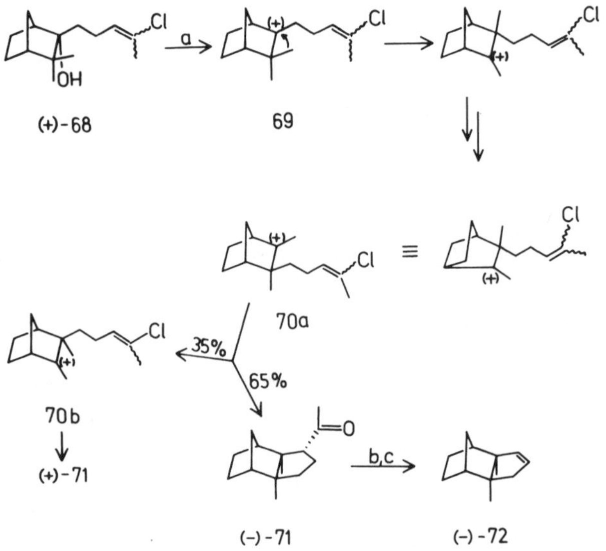

FIGURE 27. (a) HCO_2H; (b) H_2O_2, $(CF_3CO)_2O$, H_2SO_4; (c) $(AcO)_4Pb$.

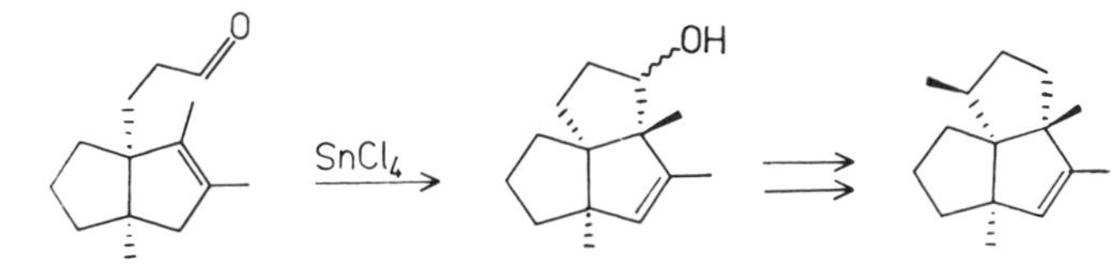

O

$SnCl_4$

OH

73

FIGURE 28.

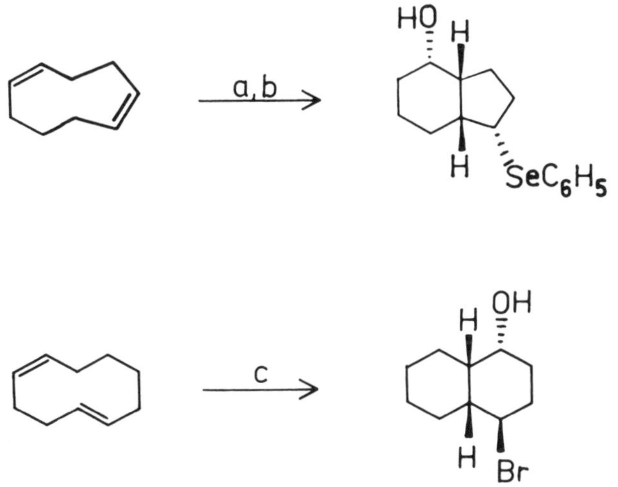

HO H

a,b

H SeC_6H_5

H OH

c

H Br

FIGURE 29. (a) PhSeCl, AcONa, AcOH; (b) K_2CO_3, CH_3OH; (c) NBS, dioxane, H_2O, cat. H_2SO_4, r.t.

stereoselective, definitely confirmed the absolute configuration of 72 and led to revision of the previously proposed structures and conclusions.[115-118] (For a detailed study of this reaction sequence and an exciting discussion see Reference 114.)

Another methodology relies upon the intramolecular Prins reaction catalyzed by Lewis acids (Figure 28).[119,120] Isocomene (73)[120] and pentalenene[121,122] have recently been synthesized in this way. The same strategy has been used in the synthesis of pentalenolactone.[123] A formally analogous reaction is the addition of an acid chloride to the double bond catalyzed by Lewis acids.[125]

In a different approach one may use the electrophilic addition across the double bond in order to create a transient electrophilic center which then interacts with another double bond present in the substrate molecule (Figure 29).[126,127] The configuration of the double bond in the reactant determines the relative configuration in the product. The reaction is useful with medium rings[124,127-132] and, under certain circumstances, with acyclic systems too.[128,134-137] Figure 29 shows a synthesis of an *ortho*-condensed bicyclic system, in which a cross-piece bond is formed between two positions of the larger ring. The original carbon-carbon bonds of the starting compound remain unchanged and constitute the peripheral framework of the target molecule.

By creating the transient carbocationic center it is sometimes possible to induce a biomimmetic cascade reaction involving several double bonds (Figure 30). It has proved advantageous to generate the cationic center from a tertiary allylic alcohol using a strong acid. This method has been applied to the so-called biomimetic polyene cyclization.[138-145] After being triggered by dehydration of the alcohol, the reaction proceeds as a multiple cyclization. The sequence is quenched by a triple bond,[138-142] a vinylfluoride moiety,[145] or an allylsilane group[146] which is eventually converted to a keto group. This scheme has been used in dozens

FIGURE 30.

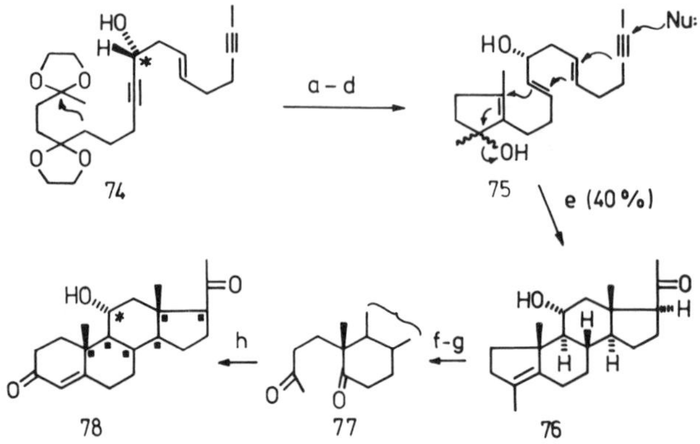

FIGURE 31. (a) Ac_2O, C_5H_5N; (b) TsOH, Me_2CO, H_2O, 25°C 76 hr; (c) 5%-NaOH, CH_3OH, 70°C 5 hr; (d) CH_3Li, Et_2O; (e) CF_3CO_2H, CH_2Cl_2, − 15°C; (f) O_3, CH_2Cl_2, CH_3OH, −78°C; (g) Zn, AcOH, 0 → 25°C 4 hr; (h) KOH, CH_3OH, H_2O, N_2, 25°C 1 hr.

of variations for total synthesis of a number of isoprenoids (for reviews see References 138 to 140). The elegant synthesis of the natural enantiomer of 11α-hydroxyprogesterone (78, Figure 31)[143,144] starts from alcohol 74 which was in turn prepared in high optical yield (R:S = 92:8) from the corresponding ketone by reduction with a chiral hydride. The key step forming the carbon-carbon bonds comprised dehydration of the allylic alcohol 75 followed by stereoselective, cascade polycyclization leading to the tricyclic compound 76 which was eventually converted in three steps to 78. The single chirality center (*) in the starting alcohol 74 induces the formation of other six centers (■) in the course of the synthesis.

Another fancy method which is based on olefin cyclization makes use of the photochemical [2 + 2] addition of allene to the double bond, followed by cleavage of one bond in the cyclobutene formed.[147] This method will be elaborated in Section II.B. Cyclizations of dienes catalyzed by transition metal complexes represent another elegant methodology (for reviews see References 148 to 152).

5. *Ene Reaction*

The intramolecular ene reaction (for a review see Reference 153) represents another means of annulation. It offers some advantages, especially in the synthesis of *ortho*-condensed systems which contain a quaternary carbon atom at the site of annulation, as shown by Oppolzer in his syntheses of isocomene[154] and modhephene.[155] Modhephene (81) was the first natural terpene of the propellane type.[156] The Oppolzer synthesis (Figure 32)[155] employs the ene-reaction 79 → 80 as the key step in building the quaternary center. In the reaction, only one allylic hydrogen atom of the endocyclic double bond is accessible for transfer, for the second allylic position is occupied by the keto group. That is why only the A transition state (Figure 32) can develop, leading specifically to compound 80 with a correct orientation of the methyl group. The resulting unsaturated ketone was transformed to modhephene (81)

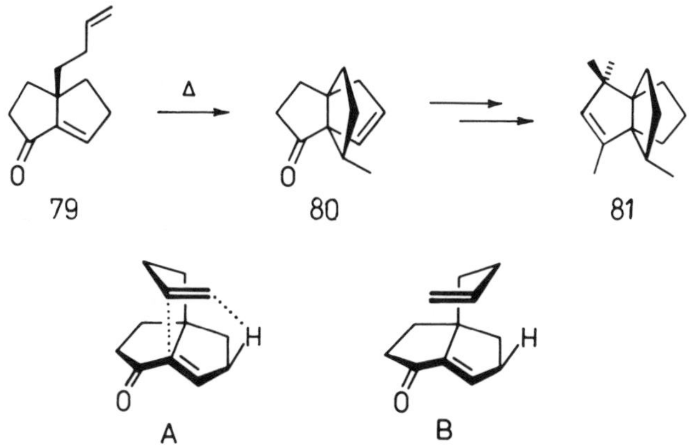

FIGURE 32.

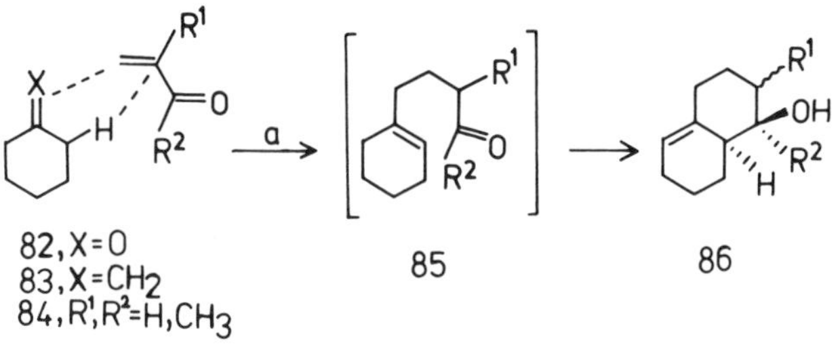

FIGURE 33. (a) $(CH_3)_2AlCl$.

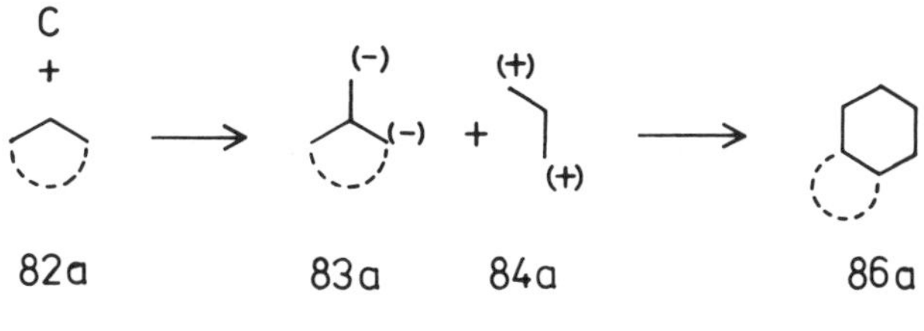

FIGURE 34.

in several steps. The described synthesis is simpler than Dreiding's variant.[156,157] Another elegant application of the ene reaction has been reported by Paquette.[11]

The ene reaction can be carried out in systems containing heteroatoms[158-160] and it can be catalyzed by Lewis acids[159-162] (for reviews see References 153, 158). Intramolecular "magnesium-ene" reactions have also found synthetic applications.[163,164]

An interesting annulation procedure, utilizing a bimolecular ene reaction in tandem with addition of a cationic center to a double bond, is depicted in Figures 33 and 34. The olefin 83 is first converted to 85 by ene reaction with the conjugated ketone 84. By action of Lewis acid the electron-deficient carbon atom of the carbonyl group in 85 attacks the double bond, and the reaction is terminated by proton elimination (85 → 86).[165,166] Formally, this scheme comprises a step-wise, four-carbon annulation (see Figure 34) with sequential attachment of one-carbon synthon (82a → 83a) (e.g. by Wittig methylenation of cyclohexanone) which produces the potentially binucleophilic synthon 83a. The latter is connected with the three-carbon, bielectrophilic synthon 84a to give rise to the six-membered ring in 86a.

The ene reaction may also be coupled in tandem with the Diels-Alder reaction in one pot, as has been recently shown by Trost et al. in the synthesis of verrucarol (Figure 35).[167]

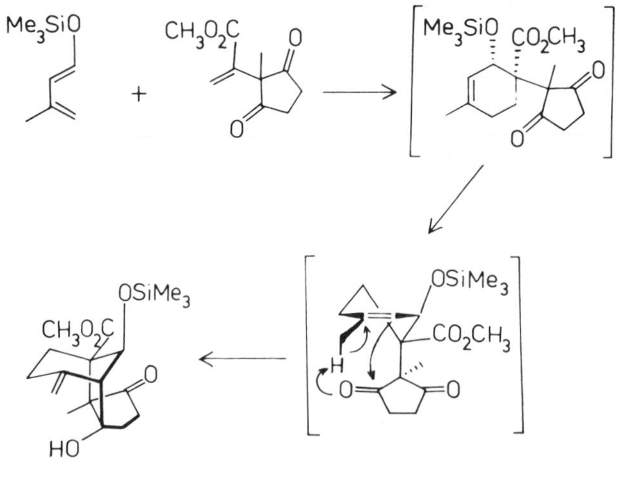

FIGURE 35.

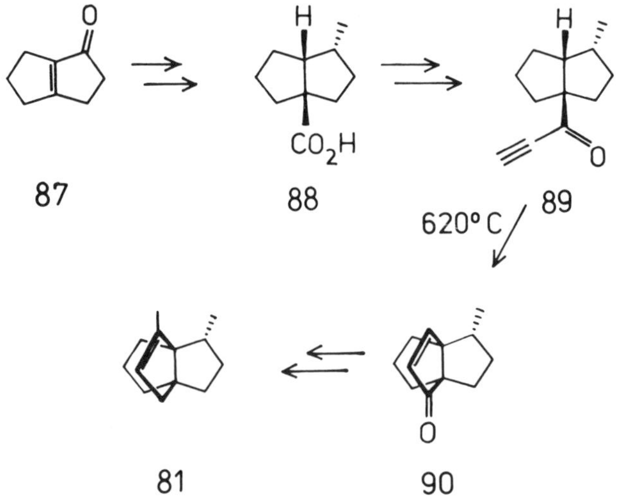

FIGURE 36.

6. Thermal Cyclization of α-Alkynones

A multistep, three-carbon annulation can be realized by applying the intramolecular cyclization of α-alkynones.[168] The synthesis of modhephene (81) will serve as an example (Figure 36). The sequence starts with the introduction of carboxyl group as a one-carbon synthon (87 → 88). After elongating the chain by ethinylation (two-carbon synthon, 88 → 89), the alkynone 89 is thermally cyclized to the propellane 90 which is then converted to modhephene (81) in several steps.[169] The thermal cyclization of α-alkynones played the key role in the synthesis of $\Delta^{9(12)}$-capnellene[170,171] and other natural substances.[172,173]

B. Ditopic Annulations

So far we have dealt with reactions in which only one carbon-carbon bond was formed in each step. In this Section we will focus on reactions of ditopic reagents, where two bonds are formed in one reaction step (Figure 1). It should be noted that the bond-forming reactions may or may not be concerted.[174,175]

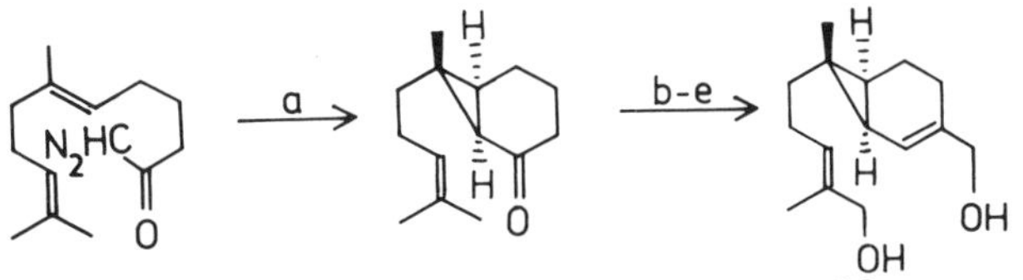

FIGURE 37. (a) $CuSO_4$, C_6H_{12}, reflux 2 hr; (b) $(CH_3O)_2CO$; (c) $NaBH_4$, t-BuCOCl, base; (d) SeO_2; (e) $LiAlH_4$, $AlCl_3$.

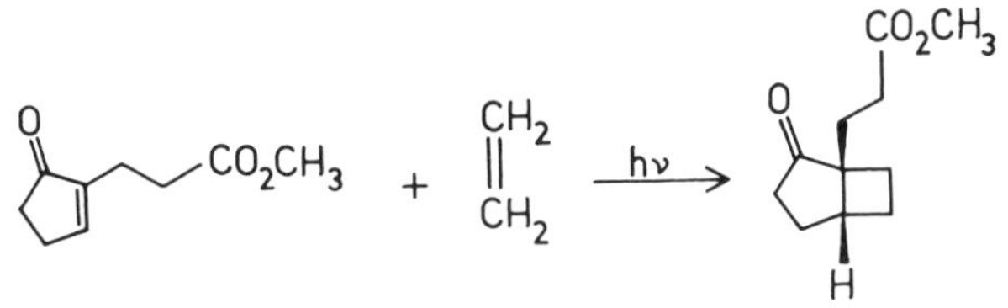

FIGURE 38.

1. [2 + 1] Cycloadditions

A number of natural products and their analogs contain the cyclopropane ring, e.g. the subsystem [3.1.0] or [4.1.0]. The synthetic approach to this type of annulation usually employs [2 + 1] addition of carbene or its equivalent, generated from a diazo compound, to a double bond, or the Simmons-Smith reaction. As a nice example of intramolecular addition of a diazoketone to the double bond we mention the synthesis of sirenine (93) (Figure 37).[176,177] The Simmons-Smith reaction has been employed in syntheses of analogs of steroid hormones.[178-180]

2. [2 + 2] Cycloadditions

According to the conservation of orbital symmetry, the supra-supra [2 + 2] cycloaddition is forbidden in the ground electronic state, but it can proceed in the excited state.[174,181] Concerted, stereospecific [2 + 2] cycloadditions are rare and they come into consideration only for synthesis using simple olefins (Figure 38).[182] On the other hand, the intramolecular variant, especially the photochemical [2 + 2] addition of an olefin to an enone proceeding by a radical mechanism, has recently found wide synthetic application. The reaction is markedly regioselective with respect to the length of the chain linking the olefinic double bond to the enone moiety. If possible, the major product arises by closing up a five-membered ring (the rule of five).[183-185] To illustrate this point, let us consider the photocyclization of heterodienones 94 and 95 (Figure 39).[186] The compound 94 is photocyclized to 96, forming the tetrahydrofuran ring. By contrast, the lower homolog 95 does not afford an analogous oxetane (98), but in accordance with the rule of five the photocyclization yields the cross-linked derivative 97. The rule of five may be violated either due to electronic effects (e.g. in ketene additions)[187] or because of geometric reasons. If the length of the spacer between the double bonds does not permit closing of the five-membered ring, the cyclization usually leads to formation of six-membered rings.

A spectacular application of the [2 + 2] addition to the synthesis of the strained terpene isocomene (102) is depicted in Figure 40.[188,189] Intramolecular photoaddition in 99 gives rise to the tricyclic compound 100 (a single product!) in which all three new chiral centers arose with correct relative configuration. The synthesis was completed by Wittig methylenation (100 → 101) followed by acid-catalyzed rearrangement yielding isocomene (102) in high yield. The excellent steric control in a nonconcerted reaction is due to the intramolecular

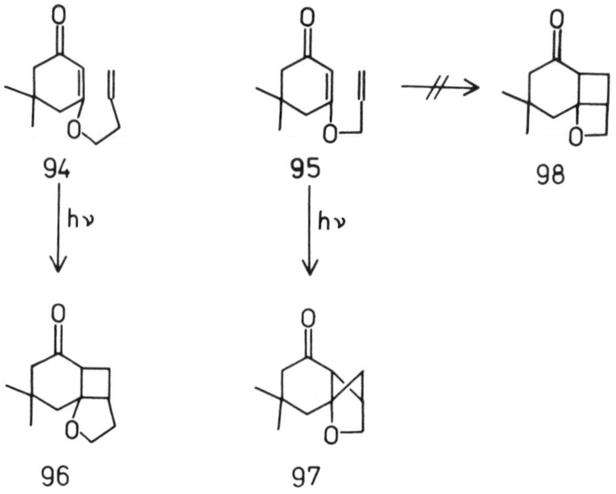

FIGURE 39.

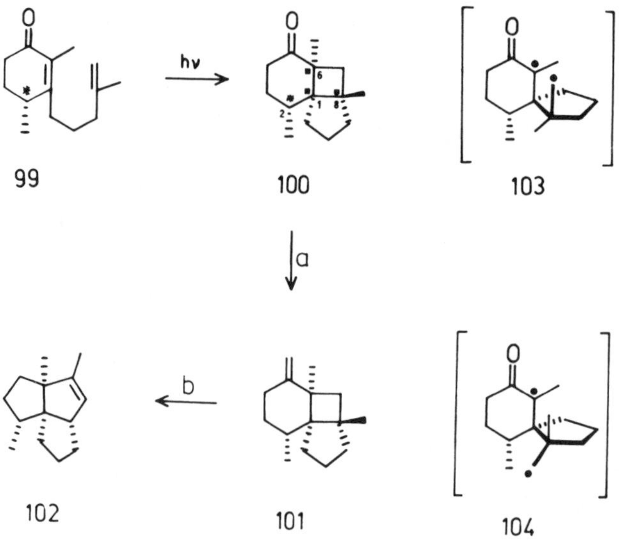

FIGURE 40. (a) $Ph_3P{=}CH_2$; (b) TsOH, C_6H_6, reflux.

character of the reversible initial step. The formation of the first carbon-carbon bond appears to be directed by the C-4 methyl, so that the olefinic part approaches the ring double bond from the side opposite to the methyl group. This dictates the configuration at C-2 after closing the five-membered ring (103). The subsequent closure of the cyclobutane ring from the biradical 103 results in *cis*-junction of the four- and five-membered rings. Conversely, the biradical 104 prefers reverting back to the dienone 99 instead of closing the strained *trans*-annulated system.

The intramolecular [2 + 2] cycloaddition has also been employed in the synthesis of α- and β-panasinensin[190] (107) (Figure 41) and other terpenoids[158,191-196] (for a review see Reference 197). The reverse fragmentation of the cyclobutane ring (de Mayo reaction) has also been applied to synthetic purposes[187,198-203] as will be discussed later.

The [2 + 2] photoaddition of allene to the double bond of an enone (Figure 42)[204] conforms to the Wiesner rule.[205] The rule postulates that the configuration of the product is determined

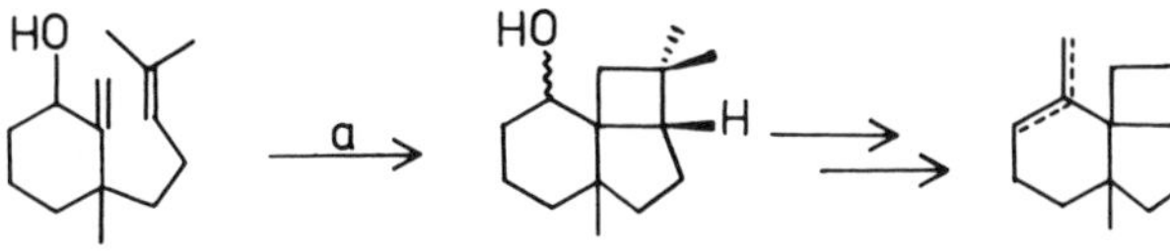

FIGURE 41. (a) hν, CF_3SO_3Cu.

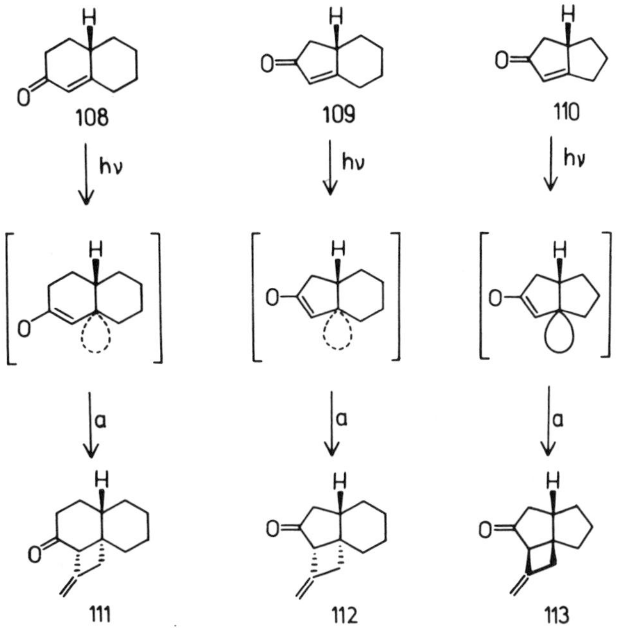

FIGURE 42. (a) $CH_2=C=CH_2$.

by the preferred steric arrangement of the excited state; the excited enone is considered to be isosteric with an anion-radical intermediate in the reduction of the enone with alkali metal in liquid ammonia. The approach of the allene molecule is directed by orbital geometry (for a detailed discussion see Reference 181 and the footnote in Reference 204). The *exo*-methylene group on the cyclobutane ring has considerable potential for further synthetic transformations (ozonolysis, epoxidation, etc.). In Figure 43 the intramolecular photoaddition of the allene system (114 → 115) serves as the key step in stereoselective annulation of the new six-membered ring. The steric course is due to the Wiesner rule. After modification of the *exo*-methylene group (115 → 116), one of the cyclobutane bonds is cleaved by retro-aldolization giving rise to the *ortho*-condensed product (116 → 117).[147,206-208]

[2 + 2] Cycloaddition without photochemical excitation can be accomplished by coupling acetylenecarboxylates with electron-rich olefins such as ketene acetals or enamines. The former reaction (Figure 44) led to the cyclobutene derivative 118 which served as a synthon in the synthesis of illudol 119.[209,210]

Photoexcitation is unnecessary in [2 + 2] cycloadditions of olefins to ketenes which yield *cis*-annulated products of the [n.2.0] type. The regioselectivity of this reaction has been discussed in detail by Fleming.[181] In this interpretation it is assumed that both the p_y and p_z orbitals of the ketene system develop overlap with the π-orbitals of the olefin (Figure 45).[181] The greater the electron deficiency at the central carbon atom, the higher the reactivity of the ketene and hence stems the advantageous use of halogenated ketenes (for reviews see

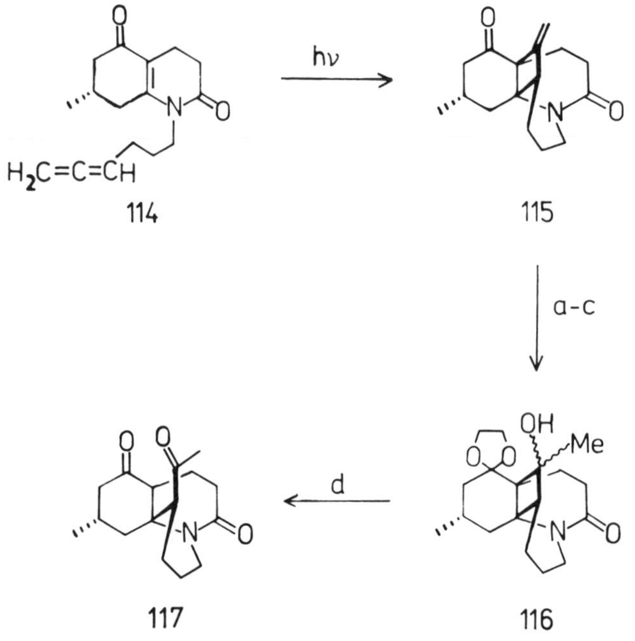

FIGURE 43. (a) Ketalization; (b) $m-Cl-C_6H_4CO_3H$, $CHCl_3$; (c) $LiAlH_4$, THF; (d) 1%-HCl, THF, H_2O, r.t. 45 min.

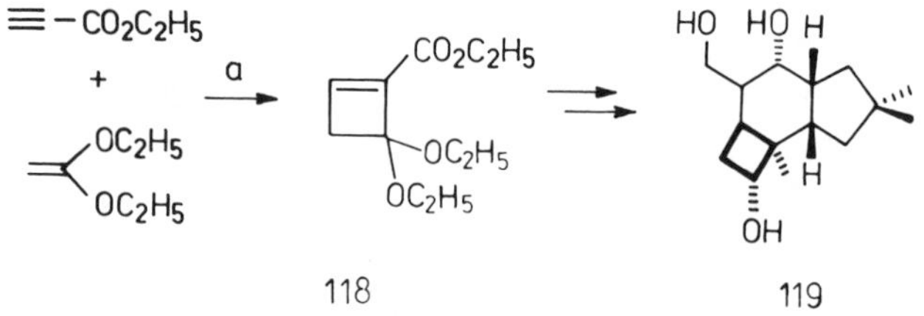

FIGURE 44. (a) Reflux 29 hr in CH_2Cl_2.

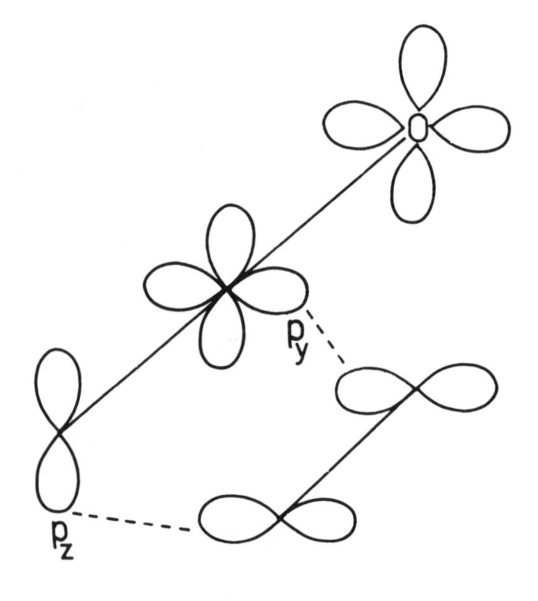

FIGURE 45.

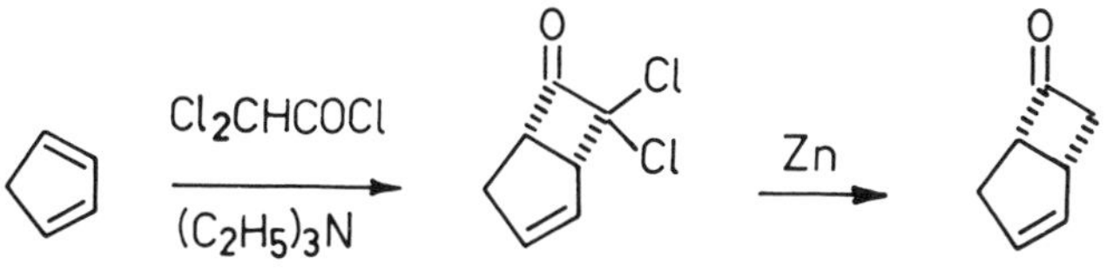

FIGURE 46.

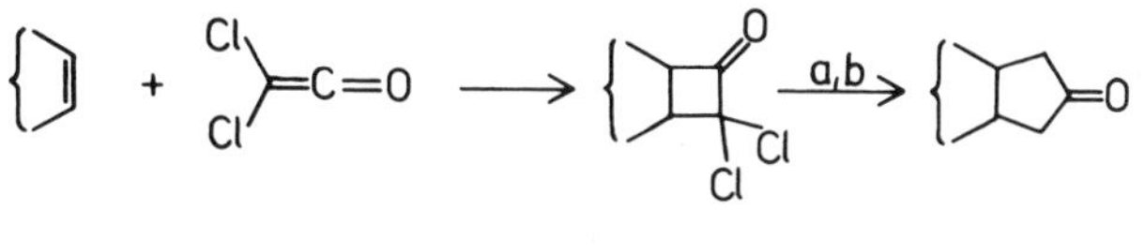

FIGURE 47. (a) CH_2N_2; (b) Zn.

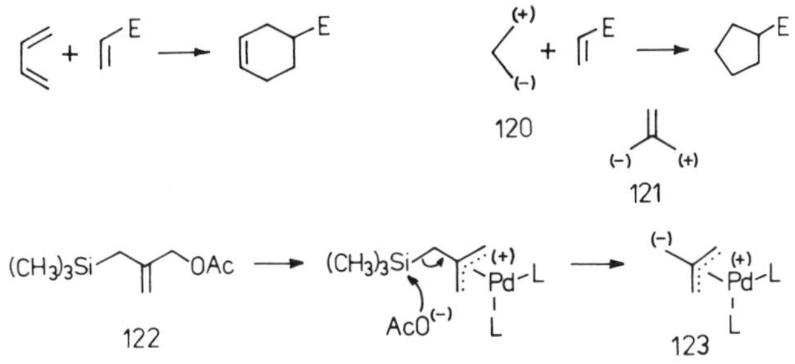

FIGURE 48. L = Ph_3P.

References 211 to 213). Dichloroketene has been especially popular as a $_{\pi}2_s$ component, as it can be conveniently generated by dehalogenation of trichloroacetyl chloride or by dehydrohalogenation of dichloroacetyl chloride (Figure 46).[214] It should be noted that the yield of the addition sometimes depends on the mode of ketene generation. The addition of dichloroketene to cyclopentadiene proceeds via a [2 + 2] mechanism, while the competing [4 + 2] reaction does not occur.

The cyclobutane ring obtained by [2 + 2] cycloaddition can be enlarged with diazomethane to cyclopentanone, while the chlorine atoms are easily removed on reduction with zinc (Figure 47).[214-217] This tandem reaction sequence may be regarded as an iterative three-carbon annulation.[218] In addition to ketenes, it is also possible to utilize keteneiminium salts; a high optical yield (55 to 90% ee)[219] was recorded in cycloaddition of optically active keteneimines.

3. *[3 + 2] Cycloadditions*

Reactions of this type are common in heterocyclic chemistry (see Volume II, Chapter 1). The carbocyclic modification, using the dipolar synthon 120 (or 121), is shown in Figure 48.[220] The synthetic equivalent of 121 is a stabilized palladium complex 123 prepared from the allylic silane 122. The cycloaddition has similar requirements as to the polarity of the components as has the Diels-Alder reaction, i.e. electron-withdrawing groups at the double bond of the dipolarophile are needed. A synthetic application[221] is shown in Figure 49. An intramolecular variant has also been reported,[222] and moreover, it was found that [3 + 2] reactions can be catalyzed by Lewis acids even without stabilizing the dipolar reagent in a Pd complex.[223]

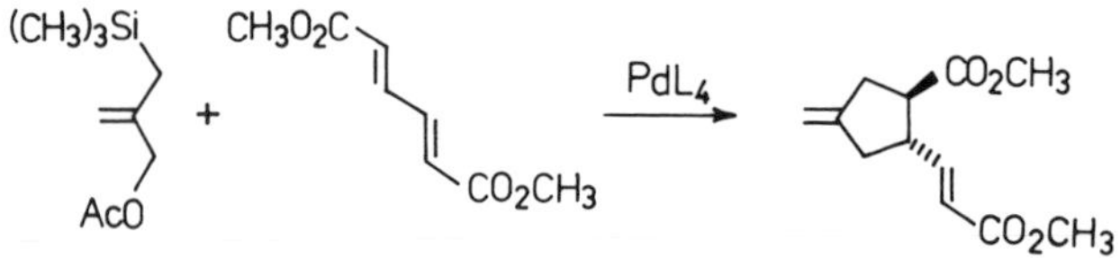

FIGURE 49. L = Ph_3P.

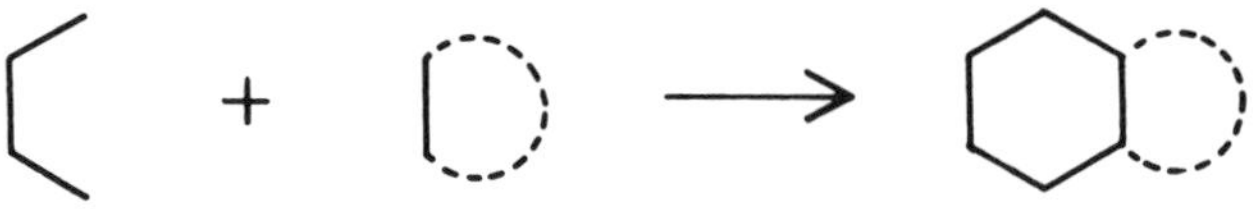

FIGURE 50.

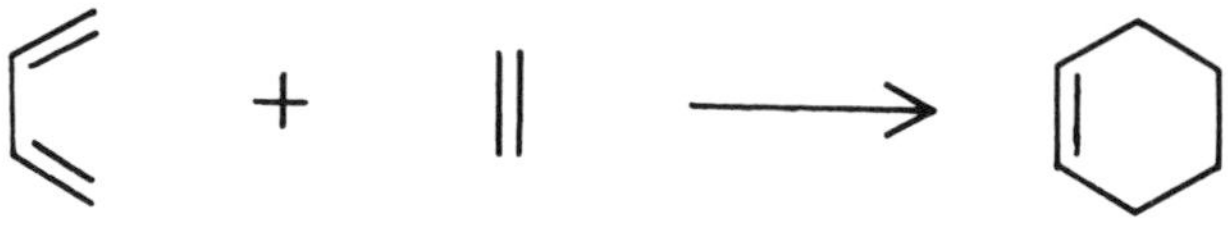

FIGURE 51.

4. [4 + 2] Cycloadditions

Four-carbon annulation (Figure 50) can be performed as the non-ionic, Diels-Alder reaction. The synthetic equivalents of the two synthons are the diene and the dienophile (Figure 51). The Woodward-Hoffmann scheme classifies the Diels-Alder reaction as a symmetry-allowed supra-supra process. The simplest pair of reactants, i.e. 1,3-butadiene and ethylene (Figure 51), undergoes the [4 + 2] cycloaddition only under extreme conditions. In order to facilitate the reaction it is necessary to activate one of the components. The commonest way of activation is to introduce electron-withdrawing substituents into the dienophile or to increase the electron density in the diene by electron-donating groups. Diels-Alder reaction with "reversed" polarity is also known and has been used for synthetic purposes (for discussion see Reference 181). The mutual orientation of the reactants in the transition state is controlled by secondary orbital interactions; these prefer the formation of *syn* (or *endo*) isomers over the *anti* (or *exo*) isomers.[181] Together with the Robinson annulation, the Diels-Alder reaction is the most frequently used method for constructing six-membered rings in synthesis of natural products. Due to its stereospecificity, using reactants with proper stereochemistry makes it possible to introduce into the target molecule substituents in defined and predictable positions and configuration (Figure 52). This contributes to the tremendous variability and general applicability of the [4 + 2] addition, not only in synthesis of ortho-condensed systems, but also in construction of bridged molecules and heterocycles. Briefly, if the target molecule contains a six-membered ring, the [4 + 2] cycloaddition and/or its intramolecular variant are the synthetic methods of choice. It is therefore no wonder that so much has been written about the various aspects of the [4 + 2] reaction, e.g. the selectivity rules,[181,224-237] mutual orientation of the reactants and its prediction,[174,227-229,238,239] catalysis,[174,237] and of course, synthetic applications.[240] Here we will present only a few recent illustrative examples which are intended to highlight the use of the Diels-Alder reaction in the strategy of stereoselective synthesis of natural products. For more details the reader should refer to excellent reviews.[181,229]

We have noted that the [4 + 2] addition is obviously the best way for preparation of a *cis*-annulated, six-membered ring (with *ortho*-condensed systems). Besides, if there is a carbonyl group that makes the annulation site enolizable, it is possible (especially in decalin

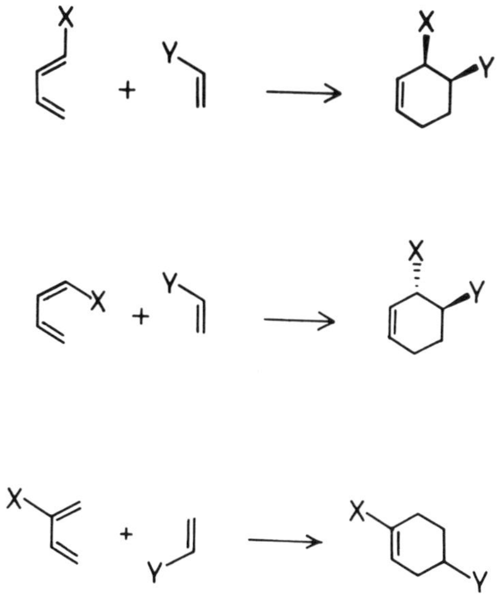

FIGURE 52.

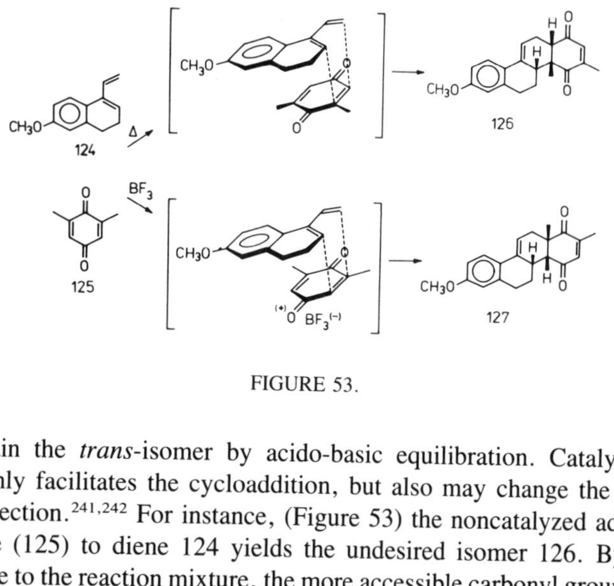

FIGURE 53.

systems) to obtain the *trans*-isomer by acido-basic equilibration. Catalysis with Lewis acids[181,229] not only facilitates the cycloaddition, but also may change the regioselectivity in the desired direction.[241,242] For instance, (Figure 53) the noncatalyzed addition of dimethylbenzoquinone (125) to diene 124 yields the undesired isomer 126. By adding boron trifluoride etherate to the reaction mixture, the more accessible carbonyl group of the quinone forms a complex with the Lewis acid; this reverses the polarity of the LUMO orbital in the quinone. As a consequence, the addition proceeds with reversed orientation giving the isomeric adduct 127. This provided an elegant solution to a long-standing problem in steroid synthesis of CD rings construction. For recent studies on Diels-Alder reactions of quinone see References 243 and 244.

The choice of suitable components for the [4 + 2] addition enables us to vary the position of the double bonds in the target molecule. If, for instance, a compound is needed in which the double bond should be located between the two carbon atoms constituting the original $_{\pi}2_s$ component (Figure 54),[245] we would use an acetylene derivative as the dienophile. Since the diene and the dienophile have a different polarity, the reactivity of the double bonds in the primary 1,4-cyclohexadiene differs and thus either double bond can be selectively removed or transformed. Conversely, if we demand introducing a 1,3-diene system into that

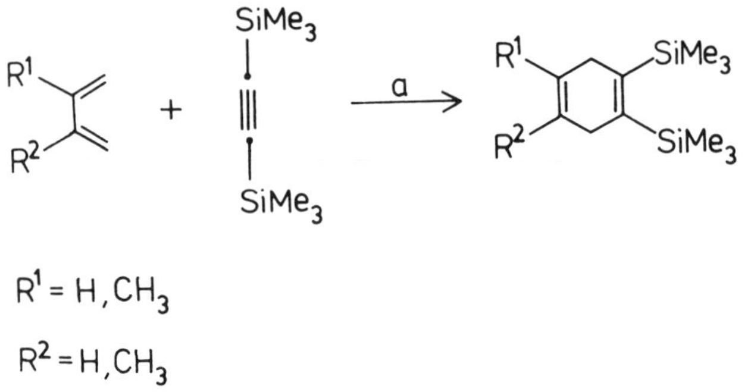
SiMe3
R1
R2
a
SiMe3
R1 = H,CH3
R2 = H,CH3

FIGURE 54. (a) $TiCl_4$, Et_2AlCl.

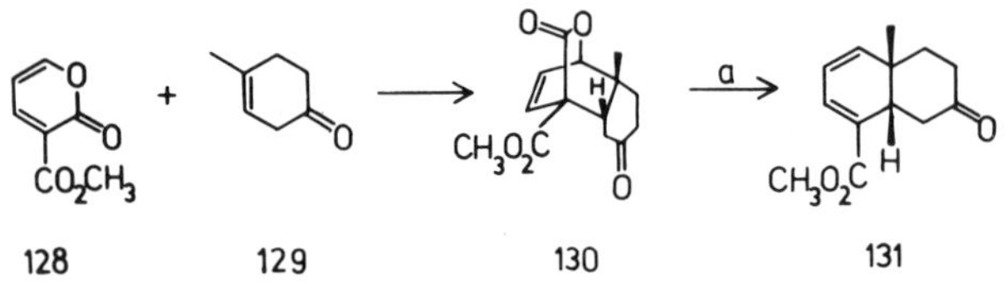
CO2CH3
CH3O2C
a
128
129
130
131

FIGURE 55. (a) $-CO_2$.

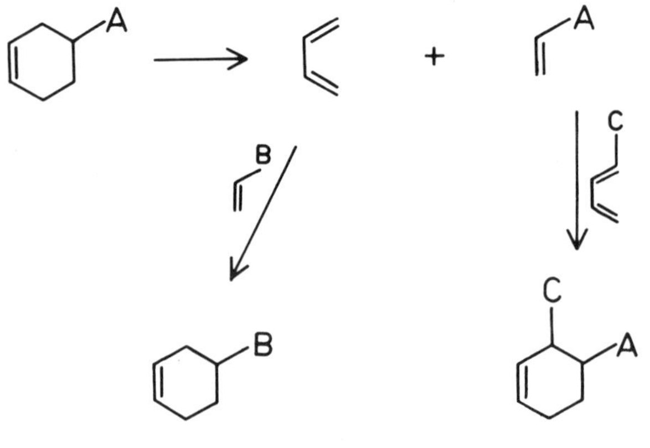
A
B
C

FIGURE 56.

part of the target molecule which corresponds to the original diene component (Figure 55)[246] we should start from the unsaturated lactone 128 and dienophile 129. The intermediate 130 undergoes a retro-Diels-Alder elimination of carbon dioxide, giving rise to the bicyclic diene 131.

In some cases it is advantageous or even necessary to introduce the diene or the dienophile in a latent form (for reasons of stability, selective protection, etc.) and then release it *in situ*. For this purpose one can use the retro-Diels-Alder reaction (Figure 56),[158,247] conrotatory opening of a cyclobutene ring,[248] or termolysis of sulfolenes (Figure 57).[158] It follows from Figure 56, however, that the components could enter competing [4 + 2] cycloaddition that would reform the starting compounds. This indirect procedure is therefore suitable, provided the desired [4 + 2] step proceeds as an intramolecular reaction (Figure 57) which rapidly consumes the components and drives the addition to completion (*vide infra*).

a. Heterosubstituted Dienes

In addition to the classical procedure in which both components bear substituents connected

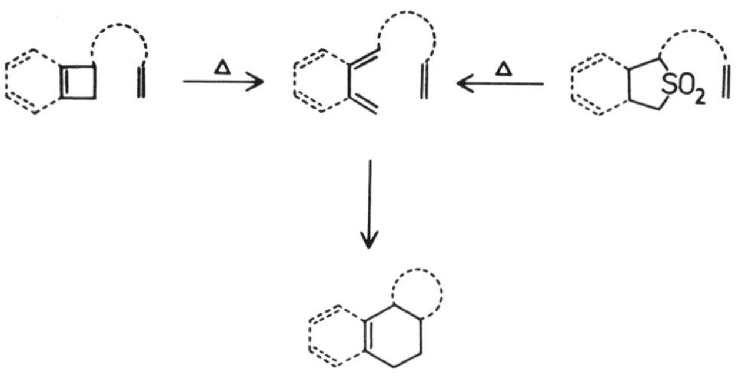

Δ Δ SO_2

FIGURE 57.

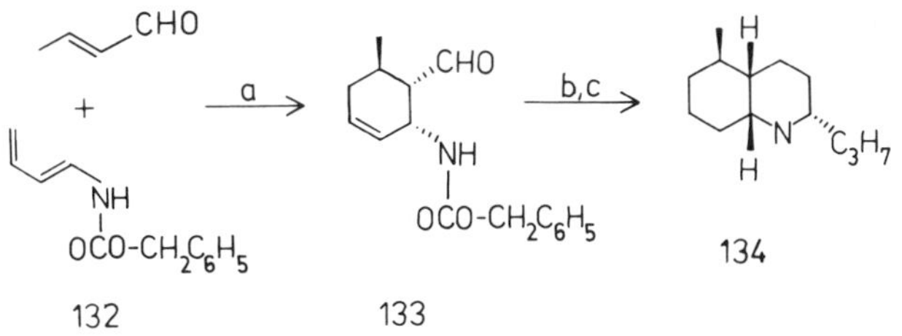

CHO + NH OCO-$CH_2C_6H_5$ 132 →(a) CHO NH OCO-$CH_2C_6H_5$ 133 →(b,c) H N C_3H_7 H 134

FIGURE 58. (a) 110°C; (b) $(CH_3O)_2P(O)-CH^{(-)}-CO-C_3H_7$; (c) H_2, Pd/C, HCl.

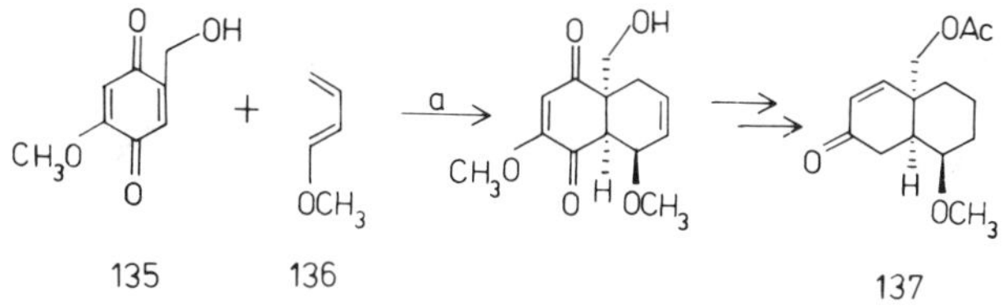

O OH CH_3O O 135 + OCH_3 136 →(a) O OH CH_3O O H OCH_3 → OAc O H OCH_3 137

FIGURE 59. (a) Reflux 3 hr in CH_3OH.

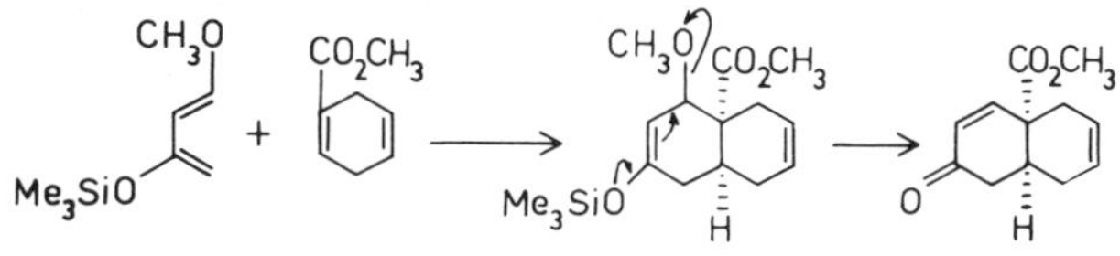

CH_3O Me_3SiO + CO_2CH_3 → CH_3O CO_2CH_3 Me_3SiO H → CO_2CH_3 O H

FIGURE 60.

to the diene or the dienophile by carbon-carbon bonds, it has proved advantageous to utilize various heterosubstituted compounds, namely, enol ethers, trialkylsilylenol ethers, enamines, enol esters (for review see References 249 and 250) or thiosubstituted dienes,[251] etc. This opened an easy way to a vast field of compounds in which the cyclohexane ring is substituted by heteroatoms. For instance, the [4 + 2] reaction of the enamine 132 with crotonaldehyde gives rise to the substituted cyclohexene 133 (Figure 58), which was converted in two steps to racemic pumiliotoxin (134).[252,253] An analogous procedure was used in the synthesis of (±)-perhydrogephyrotoxin.[254]

The synthesis of the substituted decalin system 137 started with addition of 1-methoxy-1,3-butadiene (136) to quinone 135 (Figure 59).[255-257] Trialkylsiloxydienes have broad and flexible applicability as diene components (for a review see Reference 250), as shown by Danishefsky et al. in their synthesis of vernolepine (Figure 60)[258-260] and other isoprenoids and natural products.[250,261,262]

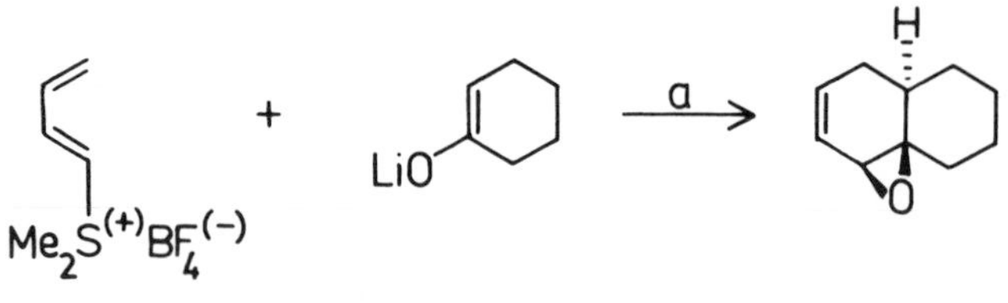

FIGURE 61. (a) −78°C, THF (58%).

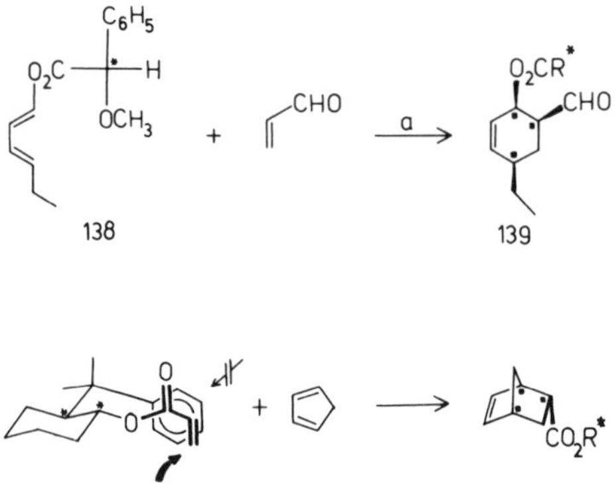

FIGURE 62. (a) BF_3, −10°C.

Beside the stable heterosubstituted dienes or enes, it is also possible to employ enolates (Figure 61). Here the cyclohexanone enolate serves as a dienophile, while the diene is activated by the dimethylsulfonimum group. This ingenious arrangement also solves the stereochemical problem in the next step in which it was necessary to close the β-oxirane ring[263] (see also References 264 to 267).

In connection with the heterosubstituted dienes and dienophiles it is worth mentioning the regioselective construction of the aromatic ring[225,226,268,269] using the [4 + 2] addition. In this way it is possible to prepare aromatic compounds with combinations of substituents otherwise hardly accessible. However interesting, this topic exceeds the framework of this book as it does not involve the problem of stereoselectivity.

b. Asymmetric Diels-Alder Reaction

An interesting approach in the area of Diels-Alder syntheses is represented by methods employing chiral components in order to prepare only one enantiomer of the target compound[250,270-272] (Figure 62). In the first case, acrolein is added to the enolester 138 containing a chirality center of the optically active acid. The stereochemistry of the [4 + 2] reaction secures the relative configuration of the three asymmetric centers in the product 139. Moreover, the product is formed in ca. 60% optical yield.[273] The auxiliary optically active acid was released by careful hydrolysis and the alcohol was used for synthesis of ibogamin.[273] In the second case (140),[274] one side of the dienophile double bond is hindered by the benzene ring and the Diels-Alder reaction with cyclopentadiene proceeds in practically quantitative optical yield! The parent optically active alcohol was regenerated by hydrolysis of the product 141. For related methods see References 160 and 275 to 277.

c. Intramolecular Diels-Alder Reaction

The fourth chapter on the Diels-Alder reaction involves its intramolecular variant (for reviews see References 248 and 278 to 281), which greatly extends its synthetic application. In the introduction to Section I.B.4 of this Chapter we have shown that unactivated com-

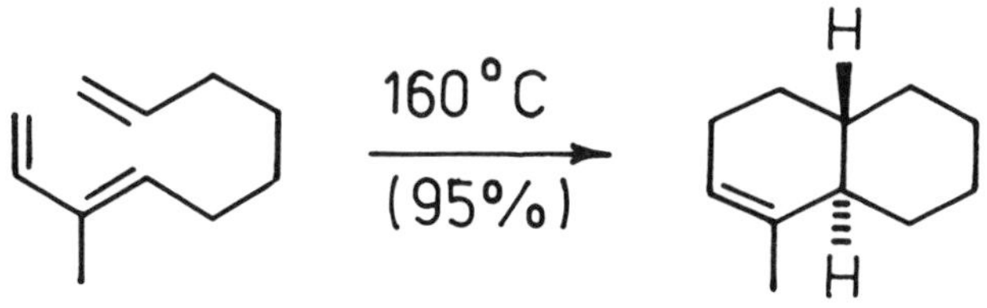

FIGURE 63.

FIGURE 64.

pounds, e.g. butadiene and ethylene, react only under drastic conditions and that activation with suitable polar substituents is necessary to secure successful results. If, however, the diene and the dienophile moieties are already present in the molecule of a synthetic precursor, the [4 + 2] addition can occur intramolecularly, even without activation by polar substituents (Figure 63).[282,283] This way of constructing the cyclic systems has gained considerable significance in the last few years, mainly because of its synthetic variability and flexibility.[281] The important feature is that the *cis* or *trans* geometry of double bond in the precursor largely determines the annulation of rings and configuration of substituents; this fact makes the intramolecular Diels-Alder reaction very popular[284-316] (for a recent review see Reference 317). On the other hand, it should be noted that the cyclization may be accompanied by E,Z isomerization, especially if the reaction is catalyzed by Lewis acids, and thus the stereochemistry of the product does not simply correspond to that of the starting compound. The actual situation of course depends on the system in question. The course of the intramolecular [4 + 2] addition can be demonstrated by the formation of a hydrindane (Figure 64).[282] For 1,3,8-nonatriene with *trans*-3-double bond one can anticipate two transition states (A and B) leading to *trans*- or *cis*-annulated products, respectively, the first being more favorable due to less steric strain;[287,288] however, cf. Section IV.D. This is also consistent with the preferential formation of the *trans*-annulated [4.4.0] product in Figure 63. Of course, the situation would be reversed with a 3-Z-triene.

By introducing a carboxymethyl group at the terminal double bond of the dienophile moiety, it is possible to prepare different isomers depending on the configuration of the ester group (Figure 65).[287-291] In both cases the reaction proceeds through the transition state A (see Figure 64) which leads to the *trans*-annulated hydrindane system. The orientation of the ester group is determined by the configuration of the dienophile part in the starting triene. In the first case, catalysis with Lewis acids enhances the stereoselectivity, whereas in the second case it has practically no effect. This approach was utilized in the synthesis of dendrobine.[279] Note that in the triene 143 the transition state does not obey the *syn*-rule as to the orientation of the dienophile substituent. An orientation obeying the *syn*-rule would require that the dienophile be rotated the other way, which would in turn lead to an unfavorable transition state B (see Figure 64). It appears that the secondary orbital interactions are insufficient to compensate for the energy requirements of the transition state B and thus the steric effects dominate (see, however, Section IV.D).

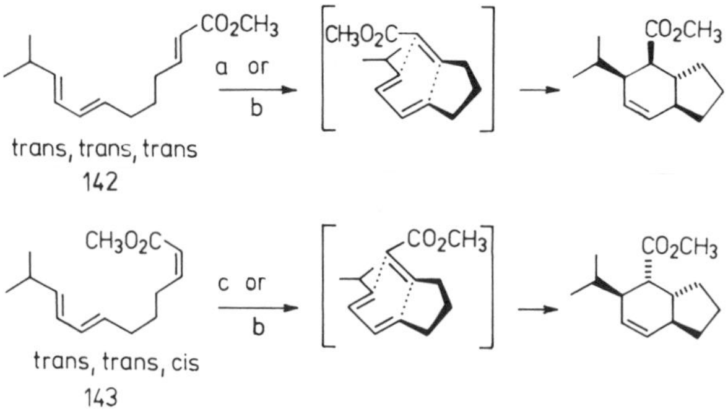

trans, trans, trans
142

trans, trans, cis
143

FIGURE 65. (a) 150°C 40 hr; (b) $AlCl_3$; (c) 180°C 6 hr.

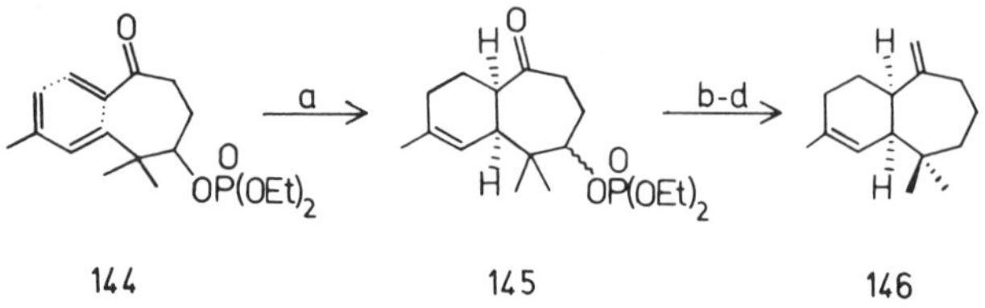

144 145 146

FIGURE 66. (a) 80°C 18 hr (90%); (b) CH_3Li; (c) Li, $EtNH_2$; (d) $SOCl_2$, C_5H_5N.

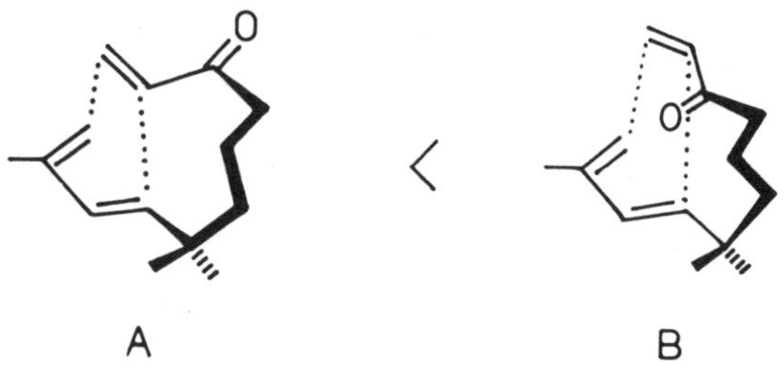

A B

FIGURE 67.

The [5.4.0] system of himachalene (146) (Figure 66)[296] was synthesized in an analogous manner, the only difference being that the Diels-Alder reaction afforded the *cis*-annulated skeleton 145. Due to the longer spacer chain in 144 connecting the diene and dienophile parts, the steric strain in the transition state was diminished and therefore the geometry B (Figure 67) was favored by secondary orbital interactions between the diene system and the carbonyl group. Since the *trans*-annulated isomer is thermodynamically more stable,[319] it follows that 145 is a product of a kinetically controlled reaction. The intramolecular [4 + 2] addition is therefore controlled by stereoelectronic factors, as is the bimolecular reaction.

If the middle double bond has a *cis*-configuration, the Diels-Alder reaction would afford a *cis*-annulated skeleton, provided the reaction is not accompanied by isomerization. This was demonstrated by Boeckman in the synthesis of marasmic acid (Figure 68).[301] *cis*-Triene 147 was cyclized less easily than the *trans*-isomer; nevertheless, this method provides a useful way to *cis*-annulated [4.3.0] system[299-301] (see also Section IV.D).

In designing stereoselective syntheses that make use of the intramolecular [4 + 2] addition it has to be kept in mind whether one of the double bonds of the triene system could shift to produce a conjugated system with the ester or keto groups (Figure 69).[302,303] The double bond shift may not proceed stereospecifically and hence the Diels-Alder reaction would not produce a configurationally homogeneous product.

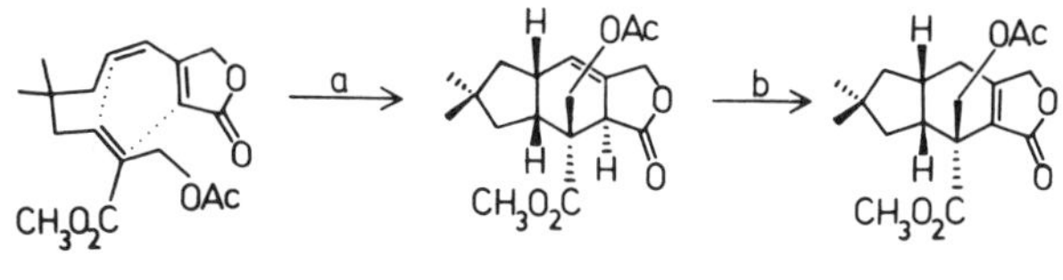

FIGURE 68. (a) 230°C; (b) t-BuOK.

cis : trans = 55 : 45

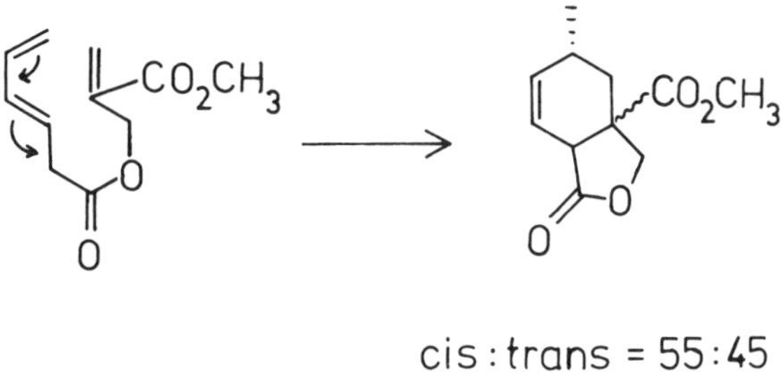

FIGURE 69.

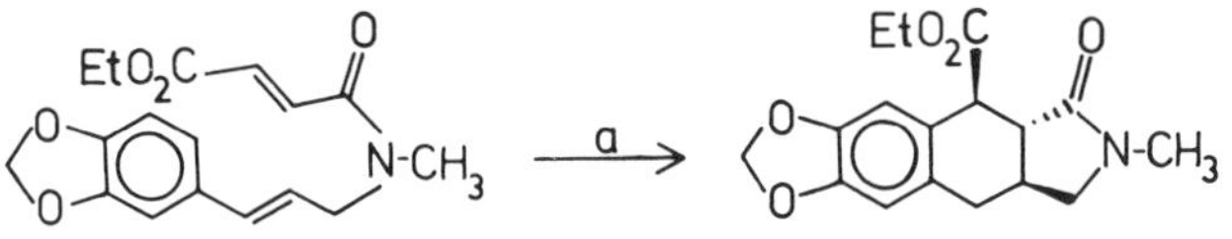

FIGURE 70. (a) 110°C, 8 hr (56%).

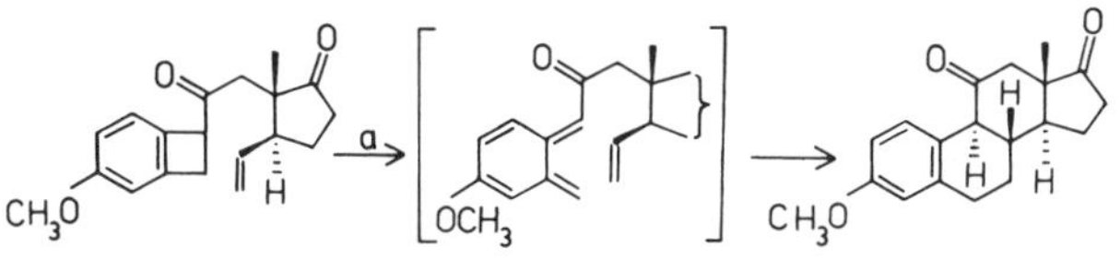

FIGURE 71. (a) 170°C (55%).

An interesting case is depicted in Figure 70, where one of the diene double bonds belongs to the aromatic ring.

So far, we have dealt with [4 + 2] additions that required a triene which had to be *a priori* synthesized by a stereoselective procedure. This, however, may be a complicated matter, especially if *cis* double bonds are needed. To circumvent this obstacle, methods have been developed which generate the diene or the dienophile parts *in situ,* as indicated in Section B.4.a. We now present some examples of this procedure, leading to construction of compounds containing an aromatic ring. The Kametani syntheses of A-aromatic steroids (for reviews see References 248, 279, and 320) employ the cyclobutene ring as a synthon of the diene (Figure 71),[321] other methods rely upon the cheletropic thermolysis the cyclic sulfone (Figure 72),[280,322-324] or elimination of trimethylsilyl group by means of the fluoride ion.[325,326] More examples of the *in situ* generation of diene or dienophile components will be given in Chapter 1 in Volume II.

A remarkable organometallics-mediated combination of tandem cycloaddition reactions has been devised by Vollhardt for the synthesis of the steroid skeleton.[148]

In this Chapter we have tried to present the principles and some aspects of the synthetic

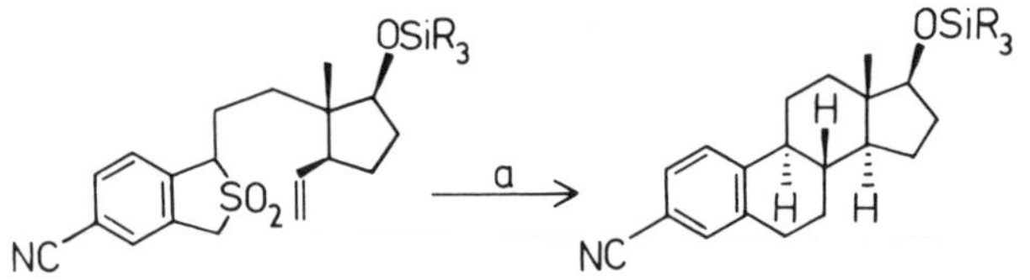

FIGURE 72. (a) 213°C.

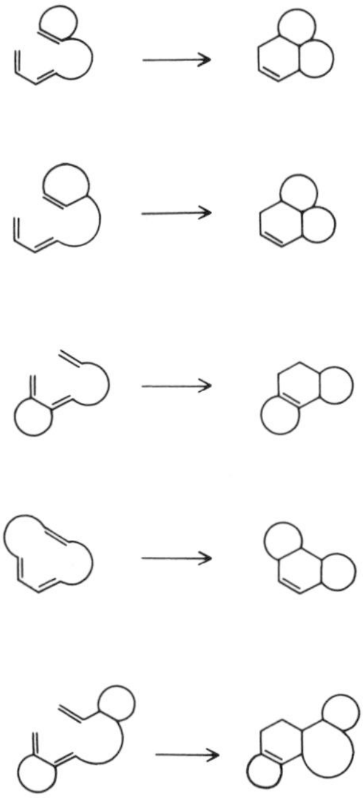

FIGURE 73.

wealth of the Diels-Alder reaction when applied to natural products. The examples were chosen to illustrate the topic; detailed information should be sought in specialized reviews. As a closing summary, Figure 73 depicts the strategies that the intramolecular Diels-Alder reaction provides for synthesis of polycyclic compounds. Other intramolecular cycloadditions, namely [6 + 4] and [8 + 2], have also been reported.[327,328]

II. CYCLIZATION OUTSIDE THE ANNULATION SITES

This alternative way of building up an *ortho*-condensed system requires that there exist an easily accessible cyclic structural unit furnished with two vicinal substituents. The substituents should have a defined relative configuration and should be equipped with functional groups enabling the closure of the new ring. This being the case, cyclization outside the annulation sites may be more advantageous than the methods described in the preceding Chapter. The problem of stereoselective annulation is then of course narrowed to specific introduction of the substituents on the first ring (see Chapter 7). The cyclization can be carried out by a variety of methods, for example, the Dieckmann, Thorpe, or acyloin condensations (Figure 74)[330-335] intramolecular aldolization,[336-340] McMurry reductive coupling,[341-343] electrolytic ring closure,[343] and other methods.[344,345,348,349] According to the number of bonds formed in the cyclization step we may distinguish monotopic and ditopic processes.

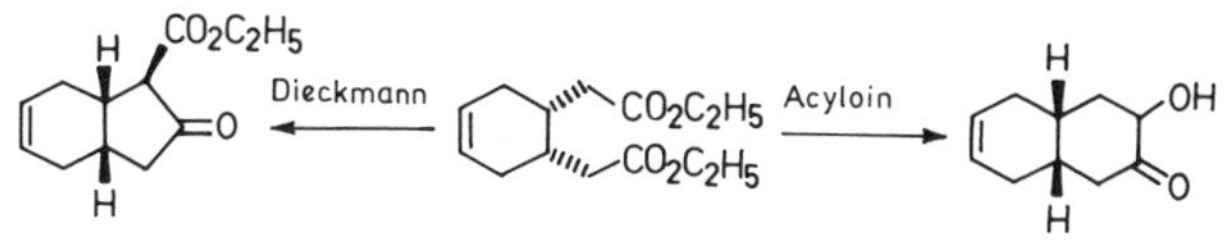

FIGURE 74.

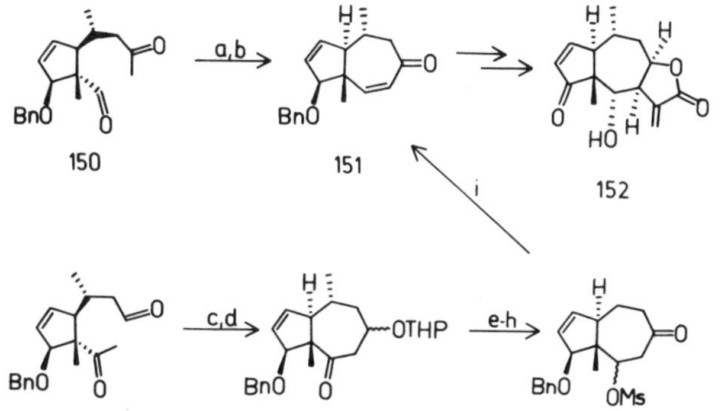

FIGURE 75. (a) 5%-KOH, CH_3OH, 20°C 30 min; (b) TsOH, C_6H_6, reflux 45 min; (c) KOH, CH_3OH; (d) Dihydropyran, TsOH, 0°C; (e) $NaBH_4$, EtOH, 0°C; (f) CH_3SO_2Cl, C_5H_5N, 0°C; (g) TsOH, CH_3OH, 0°C; (h) CrO_3, H_2SO_4, Me_2CO; (i) DBU, C_6H_6, 45 min.

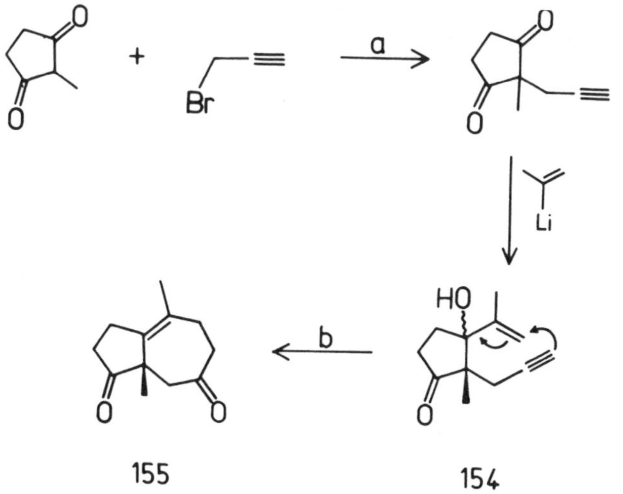

FIGURE 76. (a) $NaHCO_3$, H_2O; (b) 90%-HCO_2H.

Aldolization has played an important role in the synthesis of [5.3.0][336,337] and [3.3.0][340] systems. Figure 75 shows two ways which, starting from isomeric keto aldehydes (150 and 153), lead to the same [5.3.0] derivative which later served as an intermediate in the synthesis of helenanolide (152) and other sesquiterpenes.

The cationic π-cyclization can also be used for closing a ring outside the annulation site (Figure 76).[330] The electron-deficient center is generated as an allylic cation from the alcohol 154 by acid-catalyzed abstraction of the tertiary hydroxyl group. The acetylene triple bond functions as a quencher, as in biomimetic polyene cyclizations (see Figure 30). Instead of the carbon-carbon triple bond one may use a vinylchloride moiety.[332]

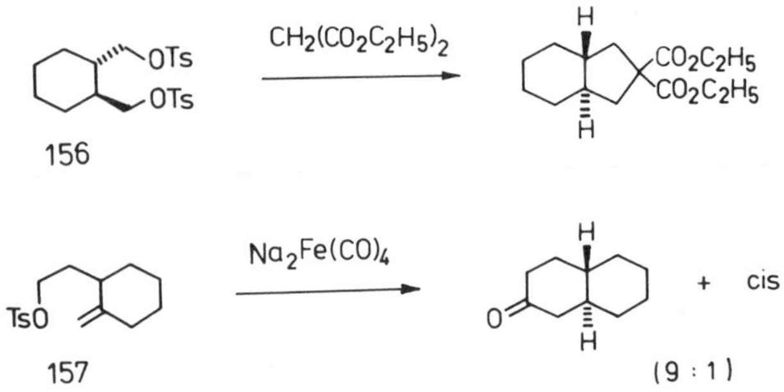

FIGURE 77.

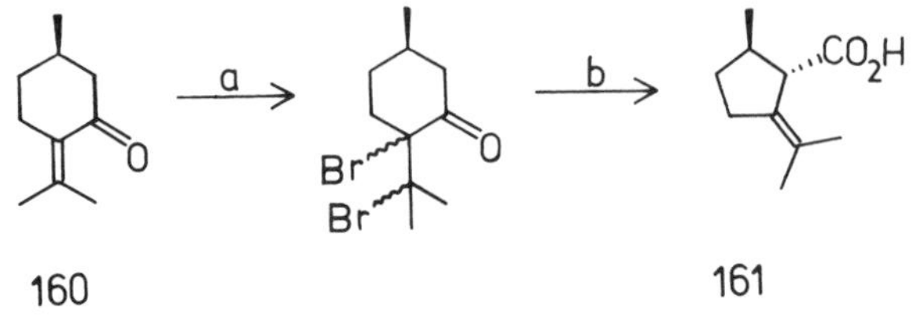

FIGURE 78. (a) Br_2; (b) OH^-.

The ditopic methods of cyclization are based on riveting the preformed side chains with a one-carbon or longer segment. For instance, the ditosylate 156 reacts with sodium diethyl malonate — a bidentate, one-carbon synthon to form the five-membered ring of the hydrindane skeleton (Figure 77).[335,350,351]

The reaction of unsaturated tosylates with sodium tetracarbonylferrate (157, Figure 77) results in a simultaneous cyclization and insertion of the carbonyl group.[352,353]

III. MODIFICATIONS OF CYCLIC SYSTEMS

Beside the *de novo* synthesis of cyclic systems that we have discussed in the previous Sections, it is often advantageous to modify an already existing skeleton. This Chapter will deal with modifications of cyclic skeletons which are provided by a bountiful nature or which are easily accessible by synthesis, usually via [4 + 2] or [2 + 2] cycloadditions. We shall focus our attention on methods of ring expansion and contraction (for review see Reference 354).

A. Ring Contractions

This subsection is devoted to different modes of conversion of n-membered rings to (n-1)-membered ones. This strategy has often been used for contracting cyclohexane derivatives to cyclopentanes.

1. Favorskii Rearrangement

Pulegone (160), available from natural material in an optically active form, is a suitable chiral synthon possessing a six-membered ring. Beside being used as a building block for compounds incorporating six-membered rings, it can be transformed to cyclopentane derivative 161 via a Favorskii rearrangement (Figure 78). The acid 161 served as a starting compound in the synthesis is actinidine.[355]

While the Favorskii rearrangement in 160 afforded a stereochemically homogenous product, ketones with seven-membered rings often yield mixtures of stereoisomers. This has been explained by greater conformational mobility in the transition state.[356]

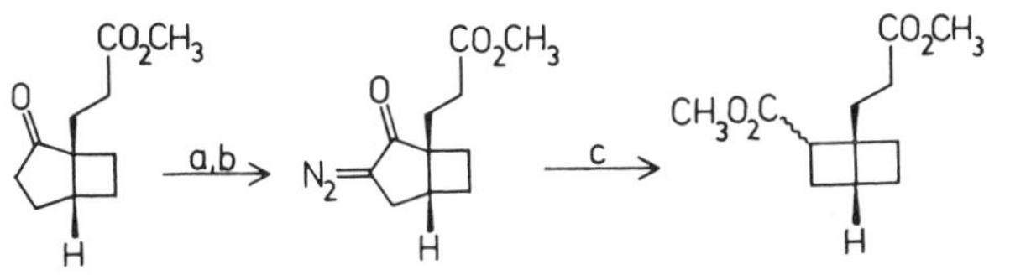

FIGURE 79. (a) HCO_2CH_3, CH_3O^-; (b) TsN_3; (c) $h\nu$,CH_3OH.

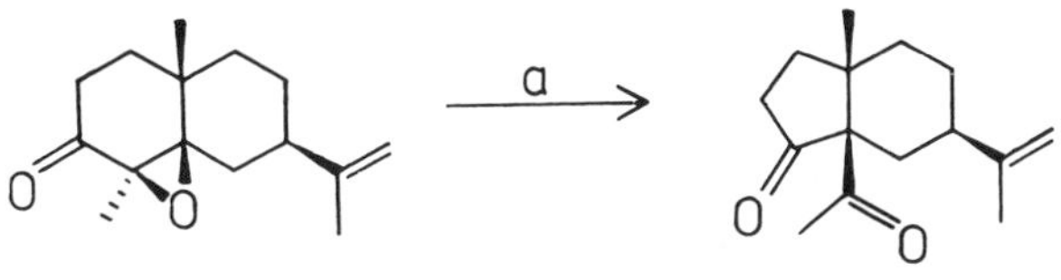

FIGURE 80. (a) $BF_3{\cdot}Et_2O$, C_6H_6.

The reaction of ketones[357] or endocyclic olefins[358] with thalium(III) nitrate is analogous to the Favorskii rearrangement, leading to esters or aldehydes with a contracted (n-1)-membered ring. In contrast, exocyclic olefins undergo ring expansion.[358] A ring containing the keto group can also be contracted with other methods.[359,360]

2. Wolf Rearrangement

The Wolf rearrangement of diazoketones[211] has been frequently used for ring contractions (Figure 79).[361] As a rule it does not proceed in a stereospecific manner and therefore is suitable only in cases where the reaction center is remote from the chirality centers, e.g. outside the annulation sites of the ring to be contracted.

3. Demyanov Rearrangement of Vicinal Aminoalcohols

In contrast to the Wolf reaction, the course of the Demyanov rearrangement is controlled by the mutual steric arrangement of the amino group and the hydroxyl.[362-364] A masterly application to contracting the cyclohexane ring to cyclopentane in the Woodward synthesis of prostaglandins[365] was highlighted in Chapter 1 (Figure 3). Mesylates can be rearranged in a similar way[366] (see Chapter 3, Section III.C).

4. Rearrangement of α,β-Epoxy Ketones

This well-known reaction, catalyzed by Lewis acids, has been employed by Japanese authors in the synthesis of cyperelone (162) (Figure 80).[367]

5. Tandem Oxidative Cleavage of the Double Bond Aldolization

Another possibility how to contract a ring in the starting compound consists of the oxidative cleavage of the double bond (especially in a cyclohexene ring) forming an intermediate dicarbonyl derivative which is subsequently cyclized by aldolization. The cleavage can be carried out by ozonolysis, hydroxylation, or epoxidation followed by oxidation of the intermediate vicinal diol, or by other methods. The regiochemistry of the aldolization step can be affected by the reagent used (Figure 81),[368,369] but sometimes the aldolization occurs spontaneously. This reaction sequence constituted the strategic step in the synthesis of various natural products.[370-373] A somewhat more complicated procedure was employed in the synthesis of B-norsteroids.[374,375]

B. Ring Expansions

This subsection will include expansions of n- to (n + 1)-membered rings. Most often,

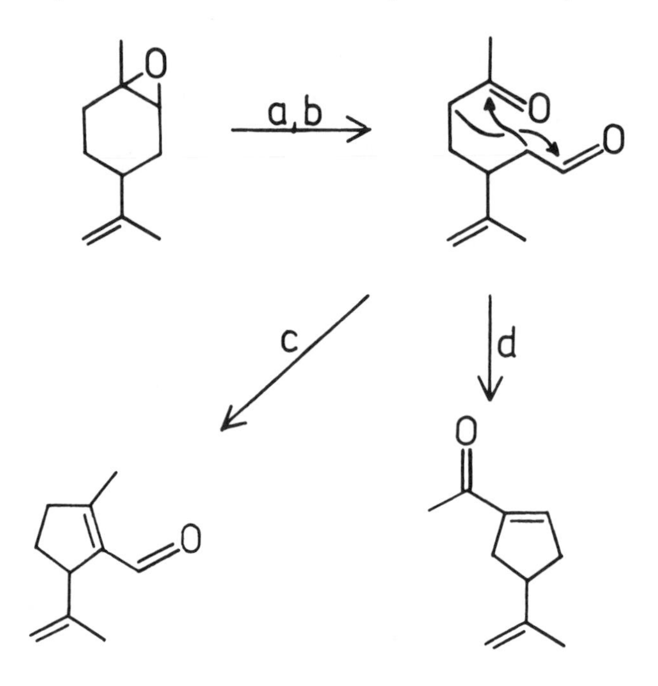

FIGURE 81. (a) H_3O^+; (b) $NaIO_4$; (c) Piperidinum acetate; (d) KOH.

such a procedure comes into consideration for expanding rings of n = 4, 5, and 6. The four-membered ring is accessible by [2 + 2] cycloaddition, while precursors with five- or six-membered rings can be frequently obtained from natural material. The pivotal role in this type of ring transformations again belongs to cyclic ketones or to their derivatives (for reviews see References 374, 376, and 377).

1. Reactions of Ketones with Diazocompounds

The reactions falling into this class are covered in every standard textbook on organic chemistry. Therefore, we will discuss only some selected topics here.

The ring expansion can be a part of a three-carbon annulation as shown in Figure 82.[379] [2 + 2] Cycloaddition of methylchloroketene to 163 yields the chloroketone 164 in which the cyclobutanone ring is enlarged with diazomethane (164 → 165). Note that the α-chloroketone grouping in 165 is ingeniously employed as a synthon of the double bond (165 → 166). It is pertinent to note that presence of halogen in the α-position to carbonyl controls the regioselectivity of ring expansion[380] in the way depicted by transformation 164 → 165.

If the reagent is used in large excess, the ring expansion may not stop in the (n + 1) step, and the subsequent insertions will lead to ketones with (n + 2), (n + 3), or larger rings.[376-378] These consequent reactions can be suppressed either by monitoring the reaction progress or, better, by using more specific reagents, e.g. trimethylsilyldiazomethane.[376] An intramolecular variant of ring expansion in a cyclic ketone with concomitant formation of a new ring is illustrated in Figure 83.[381] For another method see Reference 382.

2. Tiffeneu-Demyanov Ring Expansion

Diazotation of vicinal aminoalcohols is accompanied by rearrangement which, in cyclic compounds, results in ring contraction or expansion. The latter mode has been employed in a large-scale preparation of B-homosteroids (Figure 84).[383] Bromohydrins which are isosteric with the corresponding aminoalcohols also undergo ring expansion when treated with organometallic reagents.[384-387]

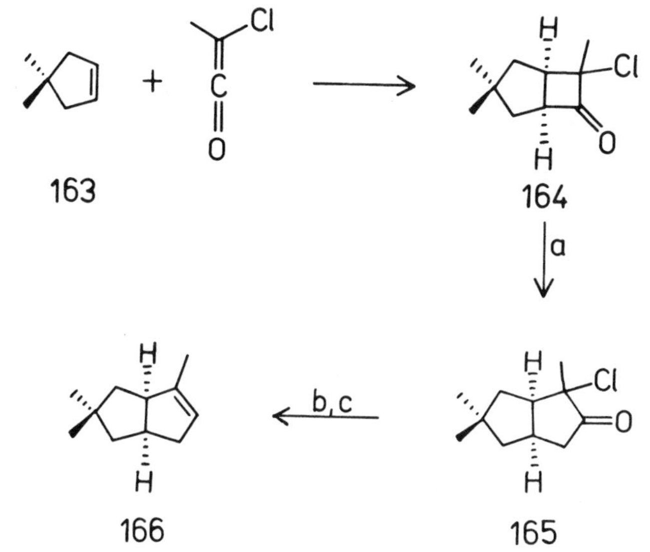

FIGURE 82. (a) CH_2N_2; (b) $NaBH_4$; (c) $Cr(ClO_4)_2$.

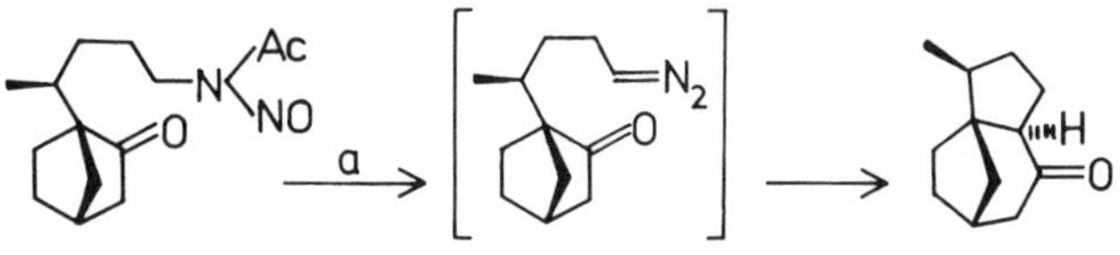

FIGURE 83. (a) CH_3ONa, CH_2Cl_2, reflux 15 min.

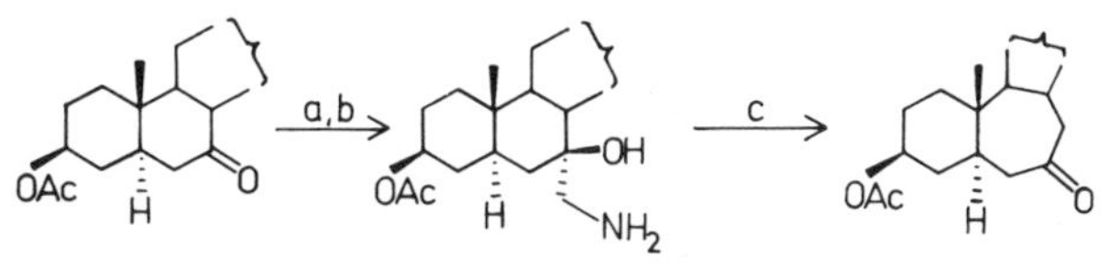

FIGURE 84. (a) $Me_2C(OH)CN$; (b) H_2, Pt, AcOH; (c) $NaNO_2$, AcOH.

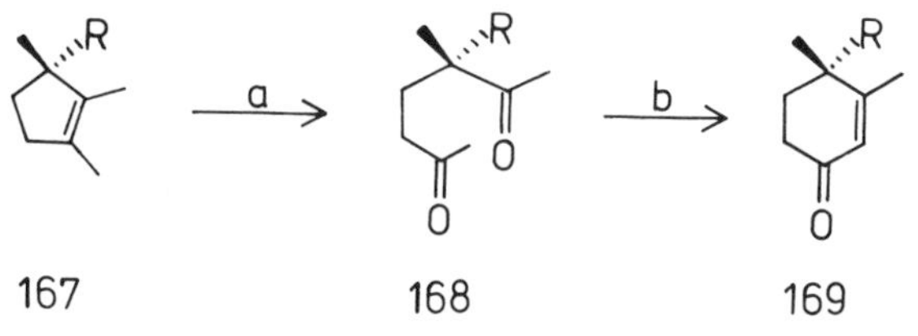

FIGURE 85. (a) Ozonization; (b) Piperidinium acetate.

3. Tandem Oxidative Cleavage of the Double Bond Aldolization

This way to ring enlargment is based on the same idea as described for ring contractions (Figure 81). By changing some structural features in the starting compound we can direct the aldolization step to form the desired expanded ring. The synthesis shown in Figure 85 starts with the dimethylcyclopentene derivative 167 which is converted by ozonolysis to diketone 168. Intramolecular aldolization in 168 results in closing the six-membered ring to give rise to 169.[388] A more involved reaction sequence which intricately uses ring enlargements and contractions has been reported by Schlessinger et al. (Figure 86).[389,390]

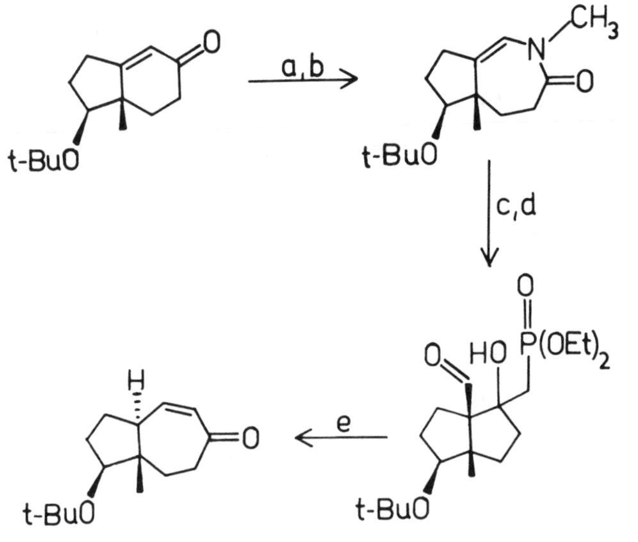

FIGURE 86. (a) $CH_3NHOH{\cdot}HCl$, C_5H_5N; (b) TsCl, C_5H_5N; (c) Li-$CH_2P(O){\cdot}(OEt)_2$; (d) AcONa, AcOH; (e) t-BuOK.

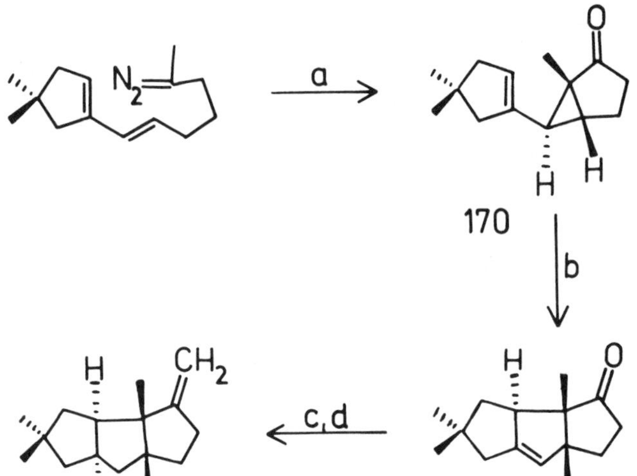

FIGURE 87. (a) $Cu(acac)_2$, heat; (b) 580°C, $PbCO_3$, glass; (c) H_2, Pt; (d) $Ph_3P{=}CH_2$.

4. Rearrangements of Vinylcyclopropane Derivatives

Vinylcyclopropanes undergo a thermolytic rearrangement whereby the cyclopropane ring expands to form a cyclopentane (170 → 171) by [1,3] sigmatropic shift.[391] This tactic has been utilized in building up the skeleton of hirsutene (172, Figure 87),[391] zizaene,[392] and other terpenes.[393] The reaction represents an expansion of the n → (n + 2) type, but simultaneously it may be regarded as a three-carbon annulation. It has been shown that both reaction rate and stereoselectivity of this rearrangement can be considerably enhanced by the presence of a hydroxy group (analogous to the Cope rearrangement).[394]

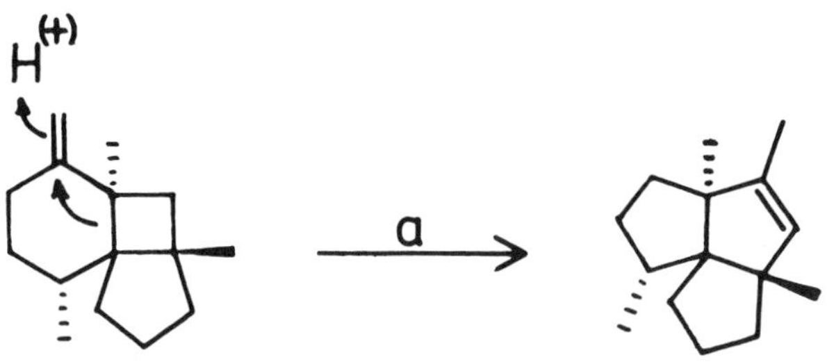

173 174

FIGURE 88. (a) TsOH, C_6H_6, reflux.

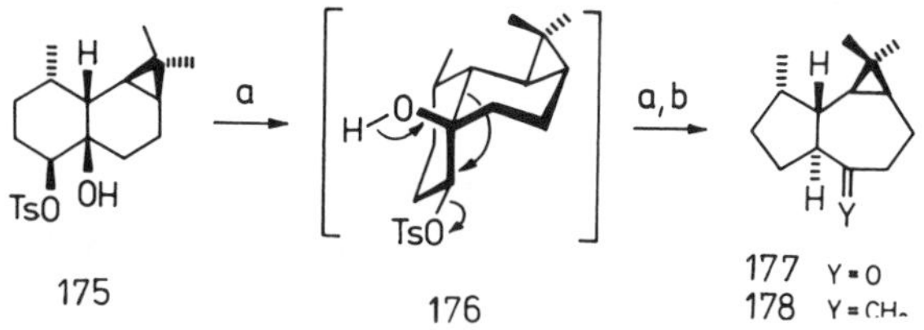

175 176 177 Y=O 178 Y=CH₂

FIGURE 89. (a) Al_2O_3, $CHCl_3$; (b) $Ph_3P=CH_2$.

C. Combined Ring Contraction and Expansion

If an electron-deficient center is generated in the neighborhood of a bridghead carbon atom in an *ortho*-condensed skeleton, the system can undergo a Wagner-Meerwein rearrangement forming a more stable carbocation which is eventually quenched by proton abstraction or nucleophilic attachment. In this process, the ring formerly bearing the electron-deficient center is contracted, whereas the neighboring ring is enlarged. In less strained cyclic systems the rearrangement exerts pronounced steric requirements; in highly strained systems, however, it can often proceed without meeting such requirements, the driving force being the relief of steric strain (*vide infra*).

1. Cationic Rearrangements in [x.y.0] Systems

In this paragraph we shall present some examples of the rearrangements converting [x.y.0] systems to [x − 1,y + 1,0] systems. A bond reorganization in a considerably strained system is illustrated by the acid-catalyzed rearrangement of the tricyclic olefin 173 to isocomene (174, Figure 88).[188,189] The requisite electron-deficient center arises upon protonation of the exomethylene group.

Vicinal diols possessing one tertiary hydroxyl group can be selectively tosylated on the other hydroxyl. On solvolysis of the tosylate the system rearranges,[395,396] whereby the carbon-carbon bond which is oriented antiperiplanar to the leaving tosyloxy group migrates by a 1,2-shift. The rearrangement is terminated by converting the free hydroxyl group to a ketone (Figure 89).[398]

The reaction has strict steric demands as to the arrangement of the bonds in the neighborhood of the electron-deficient center. It has often been applied to converting [4.4.0] systems to [5.3.0] ones. In conformationally mobile systems it is possible to control the course of the rearrangement by changing the molecular conformation.[397] The above-mentioned rearrangement of the tosylate 175 (Figure 89) played the key role in the Buchi synthesis of aromadendrene (178)[398] and other terpenes.[2] Numerous rearrangements of this type are known from the chemistry of steroids.[399]

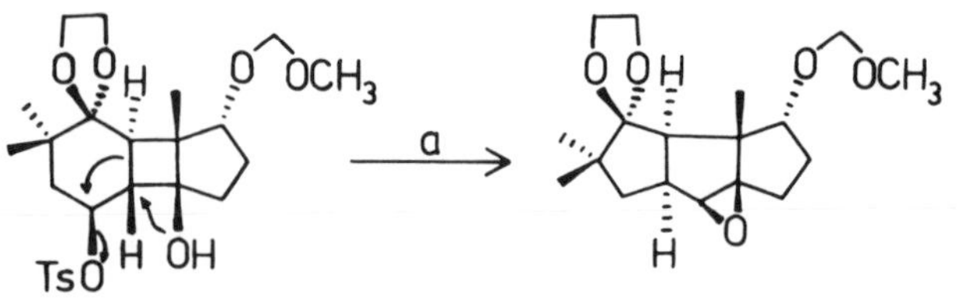

FIGURE 90. (a) K_2CO_3.

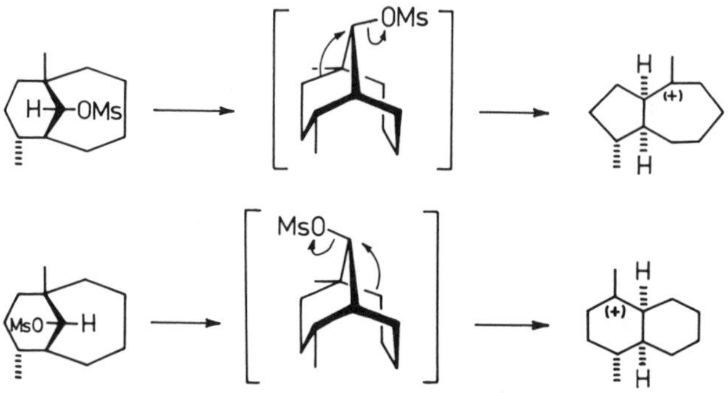

FIGURE 91.

It should be noted that the occurrence of this rearrangement is not restricted to monotosylates of vicinal diols. Figure 90 demonstrates the case in which the rearranged cation, formed by solvolysis of a 1,3-diol monotosylate (179), can be trapped by the hydroxyl group to form an epoxide ring (180).[400,401] The fragmentation of diol monotosylates will be discussed in more detail in Chapter 6.

Finally, we mention the photochemical rearrangements of a [4.4.0] → [5.3.0] type. For instance, santonin affords isophotosanthoninic lactone upon irradiation, and this rearrangement was employed in the synthesis of arborescin.[402] Other tactics using photorearrangements are described in References 403 and 404.

2. *Cationic Rearrangements in [x.y.1] Systems*

Sometimes it happens that a bridged system is more accessible (by synthesis or from the natural material) than an *ortho*-condensed one. Then it is advantageous to start from a compound with the bridged skeleton and proper configuration of substituents, and arrive at the desired *ortho*-condensed product by a rearrangement.

For instance the bridged [4.3.1] skeleton can be rearranged to an *ortho*-condensed one as shown in Figure 91.[2,405] Note that the configuration of the leaving mesyloxy group determines the direction of the rearrangement, i.e. whether it gives a [5.3.0] or a [4.4.0] *ortho*-condensed systems. The carbocation formed is usually stabilized by proton elimination, giving rise to the corresponding olefin with a tetrasubstituted double bond. The procedure was employed in the synthesis of bullnesol[406,407] and was also included in the synthetic strategy leading to pyrovallerolactone[408,409] and hirsutene.[410]

D. Fragmentation of a Cross-Piece Bond in the Ring

The methods based on fragmentation reactions are commonly used for construction of medium rings (Chapter 6), whereas applications to synthesis of smaller rings are rare.[411]

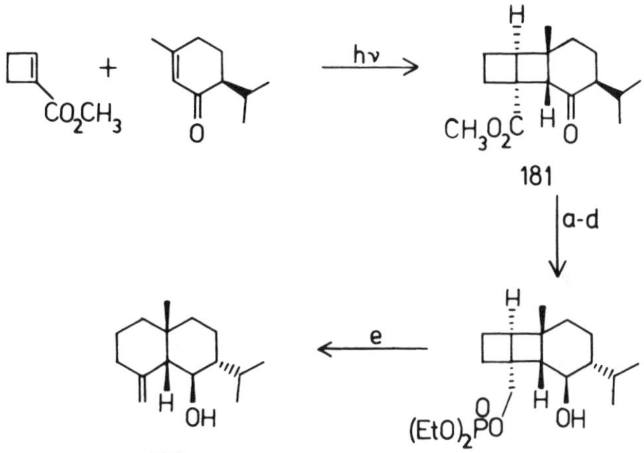

FIGURE 92. (a) $TsNHNH_2$; (b) NaH, toluene; (c) B_2H_6; (d) $(EtO)_2POCl$, BuLi; (e) $C_{10}H_7Li$, THF, 0°C.

FIGURE 93.

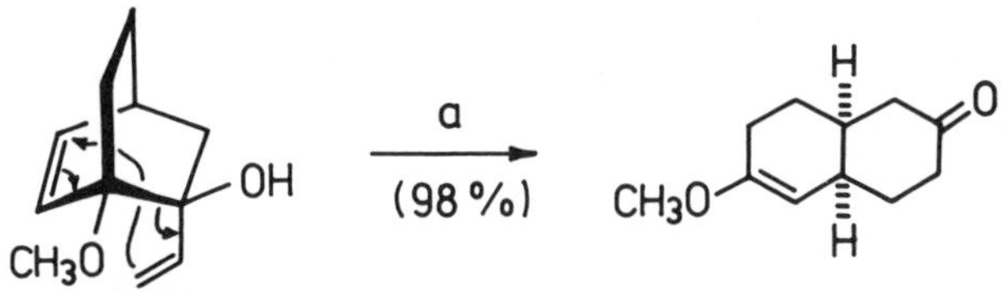

FIGURE 94. KH, THF, 66°C 10 min or 25°C 20 hr.

For instance, [2 + 2] cycloaddition of cyclobutene and cyclohexenone components (Figure 92) affords the tricyclic derivative 181 in which the [2.2.0] subsystem may be regarded as a cyclohexane spanned by a cross-piece bond. By removing this link, while preserving the peripheral bonds, one obtains the six-membered ring, as illustrated with the synthesis of 10-epijunenol (182).[412] Note the *cis*-annulation of the [4.4.0] system in the product 182.

Cyclopropane derivatives,[413] prepared by carbene addition to cyclic trimethylsilyl enol ethers, can be cleaved with Lewis acids in a similar way, preserving the peripheral ring.[414-417] Periodic acid cleavage of a suitably situated vicinal diol grouping has also been used for removing of a cross-piece bond.[414,418]

E. Cope Rearrangement

The Cope rearrangement has become a popular synthetic tool especially in the aliphatic chemistry or for attaching substituents to preformed rings. In contrast, its application to construction of cyclic skeletons is rather exceptional. One such application is depicted in Figure 93, where the [2.2.1] system is converted to *cis*-annulated [4.3.0] skeleton.[419,420] The rearrangement is remarkably accelerated in the presence of an alkoxide anion attached in the allylic position. This is referred to as the oxy-Cope rearrangement (Figure 94).[421] The

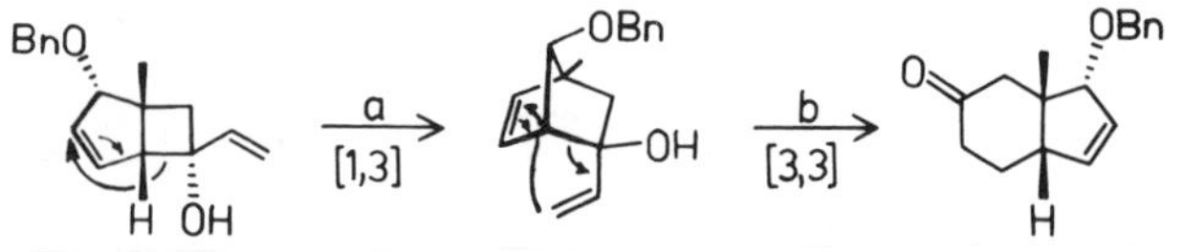

FIGURE 95. (a) NaH, THF, reflux 30 min; (b) ibid, reflux 1 hr.

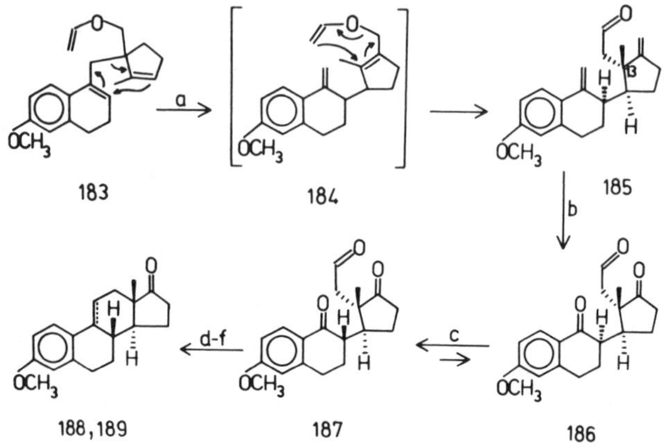

FIGURE 96. (a) 370°C, 20 sec; (b) O_3, CH_2Cl_2, −50°C; (c) CH_3ONa, CH_3OH, 25°C; (d) $TiCl_3$, Zn-Ag, reflux; (e) K, NH_3; (f) CrO_3.

reaction may be further facilitated by complexing the metal cation with a crown-ether[421] (for other examples see References 422 to 427). Tandem [1,3], [3,3] sigmatropic rearrangements have been applied to a conversion of a [3.2.0] system to a [4.3.0] via [2.2.1], as shown in Figure 95.[428] Photochemical methods for converting bridged systems to *ortho*-condensed ones can be found in Reference 429.

It is well documented that the Cope rearrangement proceeds preferentially through a chair-like transition state (*vide infra*). In cyclic systems, however, this preference may be lessened by other steric interactions. Because of the reversibility of the rearrangement, the formation of the products may be thermodynamically controlled, so that if a boat-like transition state leads to a more stable isomer, the latter will eventually prevail. In contrast to that, the Claisen rearrangement (a heteroatom analog of the Cope rearrangement) proceeds as an irreversible reaction. This fact can be utilized in the following manner. If we design the structure of the starting compound so as it be converted by Cope rearrangement to an intermediate which further undergoes Claisen rearrangement, we can trap the kinetic product of the former reaction. This sequence is referred to as tandem Cope-Claisen rearrangement. This approach elegantly solves two stereochemical problems: first, it controls the course of the Cope reaction and, second, it takes advantage of the stereospecificity of the Claisen rearrangement for other strategic purposes.

The synthetic application of tandem Cope-Claisen rearrangement is highlighted by a spectacular example of the synthesis of estrone methyl ether (189, Figure 96).[430] The Cope rearrangement in diene 183 affords 184 (a kinetic product) which is immediately consumed in Claisen rearrangement giving the stable compound 185. A small amount of the C-13 epimer also arises as an impurity. Ozonolysis of both exomethylene groups in 185 gave the triketone 186 which was isomerized to the more stable C-8 epimer 187. The latter already has the natural relative configuration of all chirality centers. The C ring was closed by the McMurry coupling and the synthesis was completed by reduction of the 9,10-double bond. The Cope-Claisen rearrangement will be treated in more detail in Chapter 7.

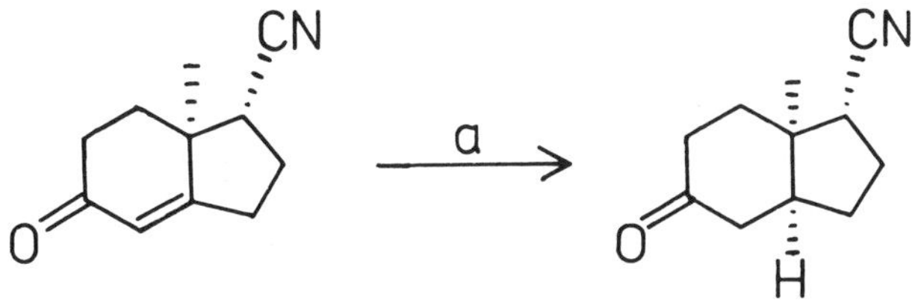

FIGURE 97. (a) H_2, Pd/$SrCO_3$.

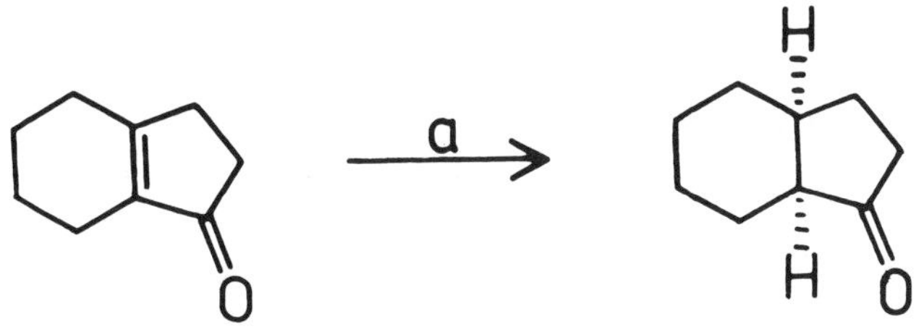

FIGURE 98. (a) H_2, Pd/C.

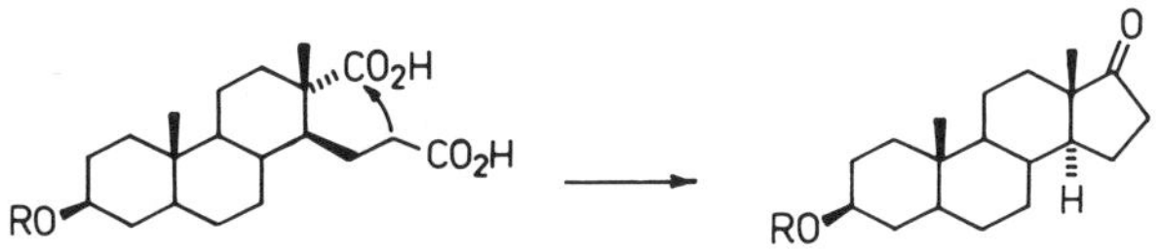

FIGURE 99.

IV. *cis,trans*-ANNULATION OF *ortho*-CONDENSED SYSTEMS

In the previous section we have outlined various possibilities of the stereoselective synthesis of *ortho*-condensed systems. In this Section we shall return to some special questions of synthesis of *cis*- and *trans*-annulated *ortho*-condensed systems, including stereoselective introduction of angular substituents. We shall focus our attention on [4.3.0], [4.4.0], and [5.3.0] systems, mainly due to their wide natural occurrence.

A. *cis*- and *trans*-Annulated Hydrindanes

Substituted *cis*-hydrindanes are often more stable than the corresponding *trans*-forms.[362,363] The *cis*-annulated compounds can be conveniently prepared by hydrogenation of precursors containing a double bond, e.g. from products of the Robinson annulation (Figure 97)[431] (see also Reference 432). The Diels-Alder reaction (cf. Figure 68) or cyclization of a suitable precursor outside the annulation sites also provide routes to *cis*-hydrindanes. Another synthetic approach makes use of the dichloroketene addition to cyclohexene, followed by expansion of the cyclobutanone ring[433] (Figures 47 and 82). 1(6)-Unsaturated bicyclic compounds (Figure 98) afford *cis*-annulated hydrindanes upon hydrogenation.[434] This method is suitable for preparation of other *cis*-annulated *ortho*-condensed systems, as well.[173]

trans-Annulated hydrindanes are accessible through intramolecular Diels-Alder reaction (cf. Figure 65) or by closing the five-membered ring outside the annulation sites[40] (Figure 99).

B. *cis*- and *trans*-Annulated Decalins

The decalin system is generally more stable in a *trans*-annulated form.[362,363] It is therefore

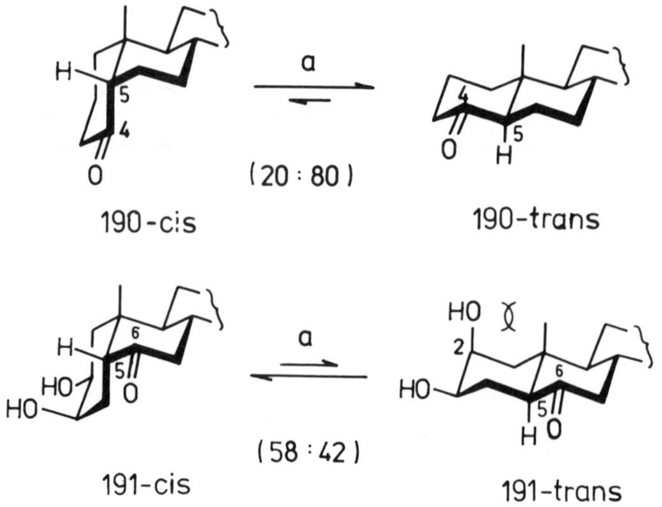

FIGURE 100. (a) KOH, CH_3OH, 20°C, 75 hr.

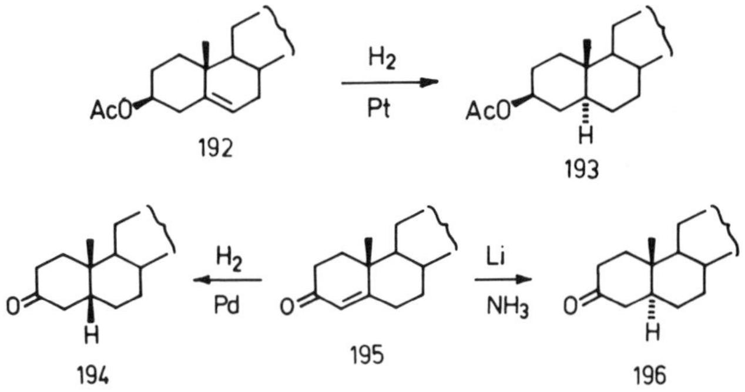

FIGURE 101.

possible to prepare a *trans*-isomer by equilibration of the corresponding *cis*-isomer, as exemplified with ketone 190 (Figure 100).[435,436] By constrast, a *trans*-decalin system with an unfavorable 1,3-diaxial interaction (191) is converted to a mixture in which the *cis*-annulated isomer slightly prevails (Figure 100).[437]

The *trans*-annulated decalin system can also be obtained by hydrogenation of the double bond located at the annulation site (192 → 193) (Figure 101).[399] Conversely, the hydrogenation of the conjugated ketone 195 affords predominantly the *cis*-annulated derivative 194, similarly as described for hydrindanes.[399,438] The yield of the *cis*-isomer can be enhanced by adding base to the reaction mixture.[438] The *trans*-isomer 196 is accessible by reduction of ketone 195 with lithium in liquid ammonia.[3,439] It is useful to compare here the course of the reduction of α,β-unsaturated ketones in [4.3.0], [4.4.0], and [3.3.0] systems. The steric course of the reduction depends on the relative stability of the corresponding intermediate (Figure 102)[439] which is protonated to produce the saturated ketone. In [4.3.0] and [4.4.0] systems the *trans*-annulated intermediates dominate, while in [3.3.0] systems the intermediate prefers *cis*-configuration, and thus protonation affords *cis*-annulated ketones (see also Figure 42). Conjugated ketones of the type 195 may also be reduced with Cr(II). For the stereoselectivity of this reduction see Reference 432.

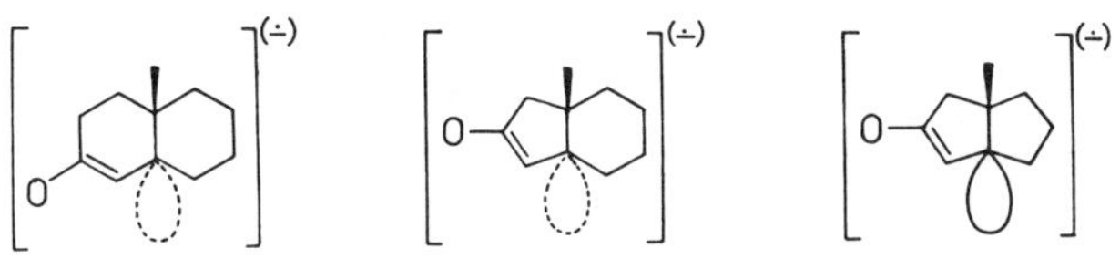

FIGURE 102.

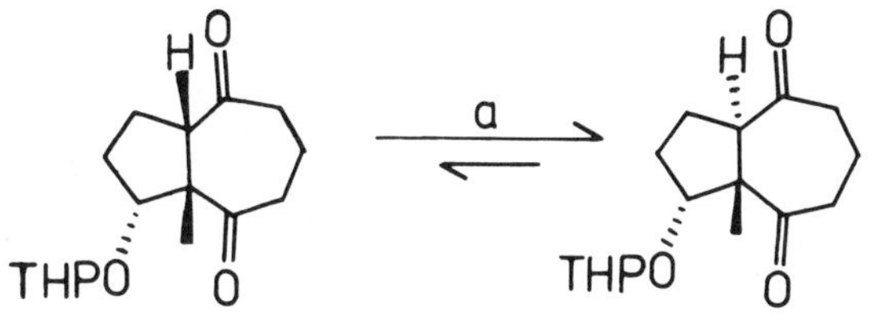

FIGURE 103. (a) CH_3ONa.

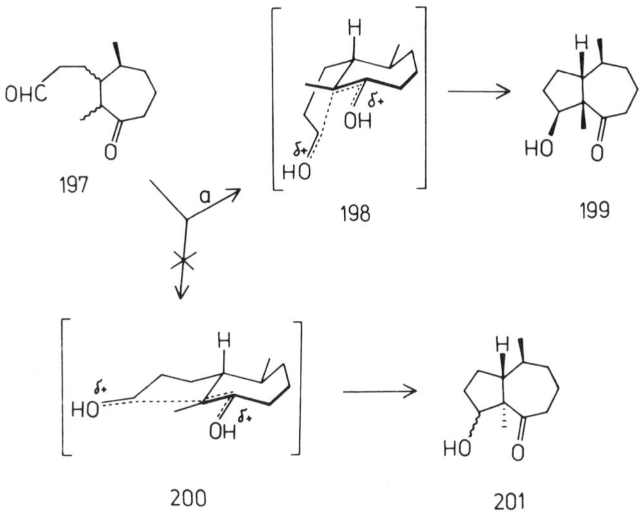

FIGURE 104. (a) 10%-HCl, AcOH, 30°C 48 hr.

C. *cis*- and *trans*-Annulated Hydroazulenes

The third subsystem occurring widely in natural products (namely, isoprenoids) is the hydroazulene skeleton. Its *trans*-annulated form appears to be somewhat more stable than the *cis*-form (cf. Figure 103).[414,440,441] Nevertheless, the equilibrium position cannot be unambiguously predicted, as it depends on the presence and nature of substituents, just as with some decalins (see Figure 100). In the previous text we have dealt with syntheses of the *trans*-annulated [5.3.0] system which made use of cyclization outside the annulation sites (cf. Figure 75), rearrangements (Figure 89), or ring expansion (Figures 83 and 86). In order to make the picture complete we present here formation of a *cis*-annulated isomer by cyclization at the annulation site (Figure 104).[414]

The acid-catalyzed aldolization of the keto aldehyde 197 proceeds preferentially via the transition state 198 with a more favorable geometry, leading to the *cis*-annulated compound 199. Transition state 200 which would have led to the *trans*-isomer 201 is not populated. *cis*-Annulated [5.4.0] systems can be made by using intramolecular Diels-Alder reaction as has been shown in Figure 66.

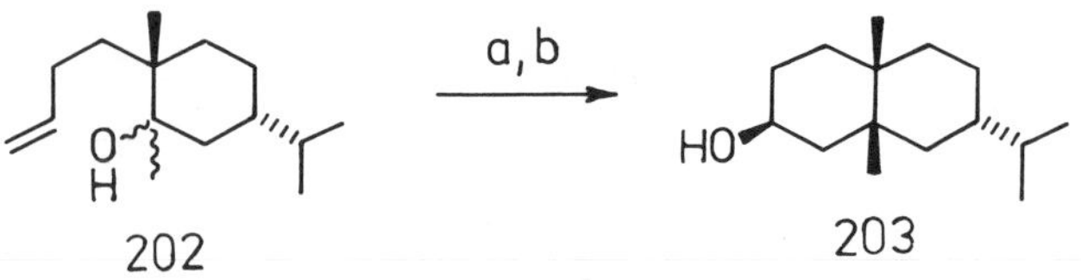

FIGURE 105. (a) HCO_2H; (b) OH^-.

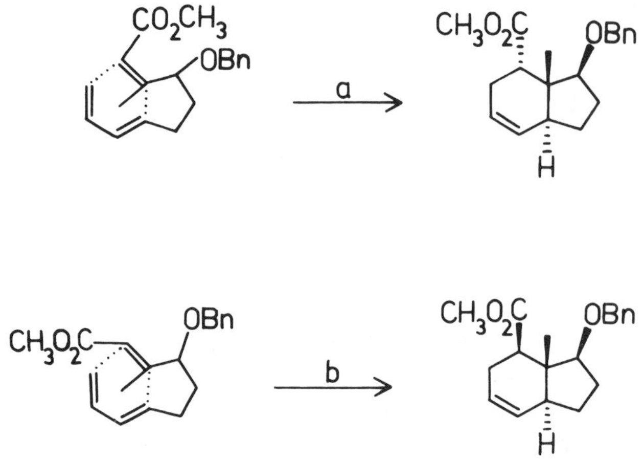

FIGURE 106. (a) $AcN(SiMe_3)_2$, toluene, 210°C 11 hr (82%); (b) $AcN(SiMe_3)_2$, toluene, 240°C 11 hr (42%).

In the last three sections we have summarized the cyclization methods that have made it possible to build up the [4.3.0], [4.4.0], and [5.3.0] systems in a stereospecific way. Having constructed the basic skeleton, it is often necessary to complete it with substituents which are to be introduced by stereoselective reactions. Some aspects of this methodology are discussed in the next Section and in Chapter 7.

D. Introduction of the Angular Methyl Group

The presence of the angular methyl group is a frequent structural feature of naturally occurring *ortho*-condensed systems. This is why the introduction of the angular methyl (or other angular groups) is given a special paragraph.

In principle, an angular alkyl group can be introduced in two ways, i.e. either by cyclization of a precursor that already contains the alkyl, or by subsequent introduction of the alkyl into the completed ring system. Beside these artificial methods, nature sometimes provides us with compounds containing angular groups in proper positions and configuration. To illustrate the first synthetic means we have chosen two examples. In Figure 105, the cyclic system with angular methyl groups (203) arises from the unsaturated alcohol 202 by cationic π-cyclization.[442]

Another possible alternative is the intramolecular Diels-Alder reaction of a substituted triene (Figure 106).[443] Here the stereochemical course can be affected by subtle variations of the structure of the starting diene (cf. References 2, and 444 to 446). Thus, for instance, in contrast to Figures 63 to 65 and 106, trienone 204 affords on intramolecular Diels-Alder reaction preferentially *cis*-annulated product 205[446] (Figure 107) whereas the corresponding ketal 206 gives rise mostly to the *trans*-product 207.[444,445] Again, as in Figure 64, we can assume two transition states A and B. In 204 the energetic preference of A is not as high as in other cases (probably due to the presence of the methyl group) and this results in

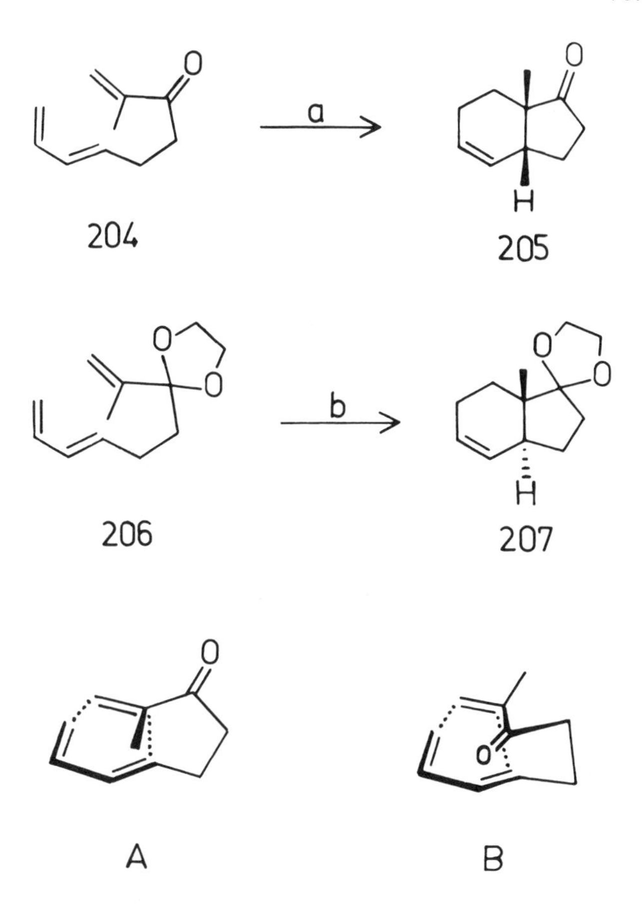

FIGURE 107. (a) 190°C, benzene; (b) 200°C, 15 hr, toluene.

FIGURE 108. (a) Me_2CuLi.

adherence of the system to the *endo* rule, i.e., it favors B. However, if the carbonyl group is transformed to ketal (206), the latter effect is removed and, also, steric interactions of the bulky ketal group favor A over B (Figure 107).

A posteriori introduction of an angular alkyl group can be accomplished by several methods. The procedure which is apparently most often used is the 1,4-addition of organocuprates to enones or unsaturated esters (Figure 108).[2,447,448] The reaction obeys well-defined and predictable stereochemical rules, and affords mainly *cis*-adducts with both [4.3.0] and [4.4.0] bicyclo systems (Figure 109).[2]

In addition to organocuprates there are other nucleophiles that may be added to the enone system, namely, the cyanide ion,[449] allyltrimethylsilanes,[450] etc.[451,452] The addition usually affords *cis*-annulated hydrindanes and decalins (Figure 109).

Another possibility for introducing the methyl group into the desired position is based on migration of the methyl via a Wagner-Meerwein type of rearrangement. For instance, the

FIGURE 109.

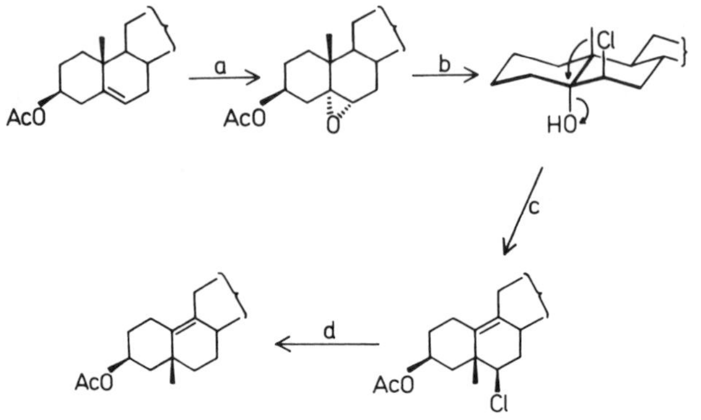

FIGURE 110. (a) m – Cl – $C_6H_4CO_3H$, $CHCl_3$, r.t. 2 hr; (b) HCl, Et_2O, H_2O, r.t. 15 min; (c) $KHSO_4$, Ac_2O, AcOH, 90°C 30 min; (d) Bu_3SnH, AIBN, C_6H_6, reflux 2 hr.

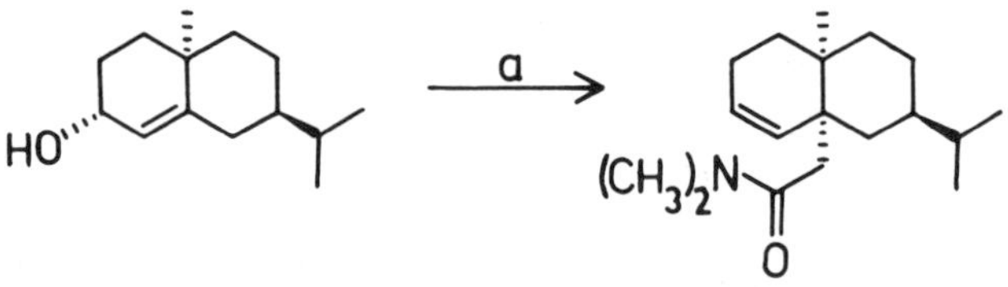

FIGURE 111. (a) $CH_3C(OMe)_2NMe_2$.

Westphalen rearrangement[399,453-455] was employed in syntheses of analogs of steroid hormones[455] and cardiotonics (Figure 110).[456,457] More extensive backbone rearrangements have been reviewed.[458]

Beside the simple alkyl groups it may be necessary to introduce a functional group into the target molecule. As discussed with the introduction of alkyl groups, this can be achieved in several ways: (1) cyclization of a precursor already containing the group in a proper position and with correct stereochemistry; (2) subsequent planting of the substituent by a stereoselective reaction. A nice example of this point is the introduction of a two-carbon segment via the Claisen rearrangement (Figure 111);[451] and (3) an inactive alkyl group can be functionalized through a remote reaction center, usually a radical. The last approach has found wide use in steroid and terpenoid chemistry. The functionalization requires the presence of an auxiliary polar group which is located close to the alkyl to be functionalized. For instance, the radical cyclization 209 → 210 results in functionalization of the 10β-methyl in the steroid skeleton (Figure 112).[435,438,459,460] Other routes to functionalized angular alkyl groups, e.g. the Barton reaction, photolysis of hypochlorites and chloramines, etc., are based on similar principles (for reviews see References 399 and 438).

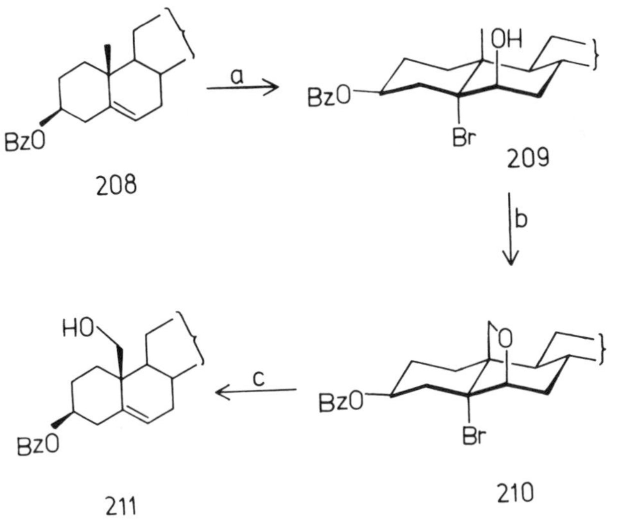

208 → (a) 209 → (b) 210 → (c) 211

FIGURE 112. (a) $CH_3CONH \cdot Br$, $HClO_4$, dioxane, H_2O; (b) $(AcO)_4Pb$, cat. I_2,C_6H_6, reflux 30 min; (c) Zn, AcOH, reflux 10 min.

V. NOTES ADDED IN PROOF

New dipolar synthons for the (3 + 2) annulation have been developed[461,462] and a second generation of the Danheiser annulation has been announced.[463] Isoxazoline routes to natural products have been reviewed.[464] Several new cyclopentane[465,466] and cyclohexane[467] annulations have been published including new modifications of the Robinson annulation.[468,469] Stereoselectivity of two competing strategies, namely the Michael and Diels-Alder methods, has been compared.[470] Further examples of the (2 + 2 + 2) polar annulation appeared[471,472] and the synthetic arsenal has been extended by an improved (3 + 2 + 1) methodology.[473] Spanning an 11-membered ring to create a [5.4.0] subsystem was implemented in a synthetic study aiming at taxane skeleton.[474] A further application of the cationic polyene cyclization enabled an efficient synthesis of inhibitor K-76.[475] The analogous tandem radical cyclization proved to be highly stereoselective and flourished in elegant syntheses of hirsutene,[476] $\Delta^{9(11)}$-capnellene,[477] and podocarpic acid.[478] This topic has been reviewed.[479] A detailed study of the scope and mechanism of the intramolecular Prins reaction has been published.[480] Transition-metal catalyzed intramolecular cyclization of aliphatic enynes[481,482] has evolved into a versatile synthetic method.[483-486] The asymmetric ene-reaction has been reviewed.[487]

The Weisner rule was used to predict correctly the stereochemistry of allene addition in a new synthesis of isocomene.[488] A nonconcerted intramolecular [2 + 2] addition of a ketene unit across an olefinic double bond served as the key step in a synthesis of retigeranic acid[482] (see also References 490 to 492). [2 + 2] addition of dichloroketene to menthyl enol ether of cyclopentanone gives 67% ee.[493] A further theoretical study of stereoselectivity of the Diels-Alder reaction has appeared.[494-496] Photo-induced Diels-Alder reaction has been proposed as a novel route to trans-fused [5.3.0] and [5.4.0] systems.[497] Regeneration of the diene system by CO_2 extrusion after intramolecular [4 + 2] addition (see Figure 55) has been used in a novel synthesis of reserpine.[498] Acetylene equivalents in cycloadditions[499] and asymmetric Diels-Alder reactions[487] have been reviewed. Two new comprehensive reviews on intramolecular Diels-Alder reactions have appeared.[500,501] A 1:1 complex of $TiCl_4$ with a chiral acrylate ester gives the Diels-Alder product with cyclopentadiene in a high R/S ratio (93:7).[502] The furane ring has been proposed as a masked dienophile in intramolecular Diels-Alder additions.[503] 1-Nitro-1,6,8-trienes that can be readily prepared by the

Henry reaction smoothly afford products of intramolecular [4 + 2] addition.[504] It has been found that hydrogen fluoride catalyzed intramolecular [4 + 2] addition of unsaturated glycol esters gives a higher endo/exo selectivity than ordinary Lewis acid catalysts.[505] Excellent diastereofacial selectivity (up to 97:3) has been observed in intramolecular Diels-Alder reactions of chiral triene-N-acyloxazolidones.[506] For further progress in this field see References 507 and 508. Intramolecular triple bond addition to a diene system was a part of the approach to forskolin[509] (see also Reference 510). Intramolecular imino Diels-Alder reactions have been reviewed[511] as well as new synthetic applications of [4 + 2] cycloreversions.[512]

Lewis acid catalyzed rearrangement of α,β-epoxyketones may lead either to ring contraction or expansion depending on the substrate structure.[513] A novel palladium-catalyzed rearrangement of 1-vinyl-1-cyclobutanols leads to methylcyclopentenones.[513] An excellent review appeared on thermolytic [1,3] rearrangements of vinylcyclopropanes.[515] A study of Wagner-Meerwein rearrangements in propellane skeletons[516] led to a short, enantioselective synthesis of quadrone.[517] This synthesis proved that the absolute configuration previously assigned to quadrone from its CD spectrum was wrong. A novel stereoselective approach to cis-fused [5.3.0] systems relies on the use of organo-iron complexes.[518]

REFERENCES

1. **Corey, E. J., Mitra, R. B., and Uda, H.,** Total synthesis of d,l-caryophyllene and d,l-isocaryophyllene, *J. Am. Chem. Soc.,* 86, 485, 1964.
2. **Apsimon, J.,** *The Total Synthesis of Natural Products,* Vols. 1—5, Wiley-Interscience, New York, 1973—1983.
3. **D'Angelo, J.,** Ketone enolates. Regiospecific preparation and synthetic uses, *Tetrahedron,* 32, 2979, 1976.
4. **Jackman, L. M. and Lange, B. C.,** Structure and reactivity of alkali metal enolates, *Tetrahedron,* 33, 2737, 1977.
5. **Kuwajima, I. and Nakamura, E.,** Quatenary ammonium enolates as synthetic intermediates. Regiospecific alkylation reaction of ketones, *J. Am. Chem. Soc.,* 97, 3257, 1975.
6. **Kuwajima, I., Nakamura, E., and Shimizu, M.,** Fluoride mediated reaction of enol silyl ethers. Regiospecific monoalkylation of ketones, *J. Am. Chem. Soc.,* 104, 1025, 1982.
7. **Hickmott, P. W.,** Enamines. Recent advances in synthetic, spectroscopic, mechanistic and stereochemical aspects, *Tetrahedron,* 38, 1975, 1982.
8. **Cook, A. G.,** *Enamines: Synthesis, Structure and Reactions,* Marcel Dekker, New York, 1969.
9. **Rasmussen, J. K.,** O-Silylated enolates — versatile intermediates for organic synthesis, *Synthesis,* 91, 1977.
10. **Carey, F. A. and Sundberg, R. J.,** *Advanced Organic Chemistry,* Part B, Plenum Press, New York, 1983.
11. **Kagawa, S., Matsumoto, S., Nishida, S., Yu, S., Morita, J., Ichihara, A., Shirahama, H., and Matsumoto, T.,** Synthesis of illudol. 1. Proto illudane skeleton, *Tetrahed. Lett.,* 3913, 1969.
12. **Paquette, L. A., Galemmo, R. A., Jr., and Springer, J. P.,** Synthesis of the alleged structure of senoxydene, the triquinane sesquiterpene derived from Senecio oxyodontus, *J. Am. Chem. Soc.,* 105, 6975, 1983.
13. **Paquette, L. A. and Leone-Bay, A.,** Triquinane sesquiterpenes. An alternative, highly stereocontrolled synthesis of (±)-silphinene, *J. Am. Chem. Soc.,* 105, 7352, 1983.
14. **Klipa, D. K. and Hart, H.,** Synthesis of bicycle [3.3.0] oct-1(2)-en-3-one, *J. Org. Chem.,* 46, 2815, 1981.
15. **Seebach, D., Hoekstra, M. S., and Protschuk, G.,** 4-Hydroxy-2-cyclopenten-1-on aus Ketonen und 3-Nitropropionylchlorid. Eine einfache Methode zum Aufbau von Fünfringen, *Angew. Chem.,* 89, 334, 1977.
16. **Trost, B. M. and Vincent, J. E.,** A three-carbon condensative expansion. Application to muscone, *J. Am. Chem. Soc.,* 102, 5680, 1980.
17. **Paquette, L. A.,** Recent synthetic developments in polyquinane chemistry, *Topics Curr. Chem.,* 119, 1, 1984.
18. **Paquette, L. A.,** The development of polyquinane chemistry, *Topics Curr. Chem.,* 79, 41, 1979.

19. **Trost, B. M. and Curran, D. P.,** An enantiodirected cyclopentenone annulation. Synthesis of useful building block for condensed cyclopentanoid natural products, *J. Am. Chem. Soc.*, 102, 5699, 1980.
19a. **Piers, E. and Abeysekera, B.,** Alkylation of enolate anions with dimethyl 3-bromo-2-ethoxypropenylphosphonate. A convergent cyclopentenone annulation method, *Can. J. Chem.*, 60, 1114, 1982.
20. **Marfat, S. A. and Helquist, P.,** Copper-catalyzed conjugate addition of an acetal-containing Grignard reagent. A method for cyclopentene annulation, *Tetrahed. Lett.*, 4217, 1978.
21. **Snider, N. N. and Faith, W. C.,** Total synthesis of (±) and (−)-ptilocaulin, *J. Am. Chem. Soc.*, 106, 1443, 1984.
22. **Bal, S. A., Marfat, A., and Helquist, P.,** Cyclopentene and cyclohexene annulation *via* copper-catalyzed conjugate addition of acetal-containing Grignard reagents, *J. Org. Chem.*, 47, 5046, 1982.
23. **Posner, G. H., Whitten, C. E., Sterling, J. J., and Brunelle, D. J.,** Alkylation of enolate ions generated regiospecifically via organocopper conjugate addition reactions. Synthesis of decalin sesquiterpene valerane and of a prostaglandin model system, *Tetrahed. Lett.*, 2591, 1974.
24. **Danheiser, R. L., Carini, D. J., and Basak, A.,** (Trimethylsilyl)-cyclopentene annulation: A regiocontrolled approach to the synthesis of five-membered rings, *J. Am. Chem. Soc.*, 103, 1604, 1981.
25. **Ley, S. V., Simpkins, N. S., and Whittle, A. J.,** The total synthesis of the clerodane diterpene insect antifeedant ajugarin I, *J. Chem. Soc. Chem. Commun.*, 503, 1983.
26. **Piers, E. and Karunarante, V.,** Conjugate addition of lithium phenylthio- and cyano-[2-(4-chlorobut-1-enyl)]cuprate to cyclic enones. An efficient methylenecyclopentane annulation process, *J. Chem. Soc. Chem. Commun.*, 935, 1983.
27. **Roush, W. R. and Walts, A. E.,** Total synthesis of (−)ptilocaulin, *J. Am. Chem. Soc.*, 106, 721, 1984.
28. **Tice, C. M. and Heathcock, C. M.,** Synthesis of sesquiterpene antitumor lactones. An approach to the synthesis of pseudoguaianolides based on oxy-Cope rearrangement, *J. Org. Chem.*, 46, 9, 1981.
29. **Korreda, M. and Mislankar, S. G.,** Chemistry of the dianions of 3-heteroatom-substituted cyclopent-2-en-1-ones: An expedient route to dl-coriolin, *J. Am. Chem. Soc.*, 105, 7203, 1983.
30. **Miller, S. A. and Robinson, R.,** Condensation of phenols with unsaturated ketones of aldehydes I. β-Naphtol and vinyl methyl ketone, *J. Chem. Soc.*, 1535, 1934.
31. **Rapson, W. S. and Robinson, R.,** Experiments on the synthesis of substances related to sterols II. A new general method for the synthesis of substituted cyclohexanones, *J. Chem. Soc.*, 1285, 1935.
32. **Gawley, R. E.,** Robinson annelation and related reactions, *Synthesis*, 777, 1976.
33. **Jung, M. E.,** A review of annulation, *Tetrahedron*, 32, 3, 1976.
34. **Ramage, R.,** Synthesis of sesquiterpenoids of biogenetic importance, in *Further Perspectives in Organic Chemistry*, Ciba Foundation Symposium 53, Kenner, G. W., Ed., Elsevier, Amsterdam, 1978, 67.
35. **Du Feu, E. C., McQuillin, F. J., and Robinson, R.,** Experiments on the synthesis of substances related to sterols. XIV. A simple synthesis of certain octalones and ketotetrahydrohydrindenes which may be of angle-methyl-substituted type. A theory of the biogenesis of sterols, *J. Chem. Soc.*, 53, 1937.
36. **Walker, J.,** A synthesis of dl-piperitone (dl-Δ^1-p-menthen-3-one), *J. Chem. Soc.*, 1585, 1935.
37. **Zoretic, P. A., Bendiksen, B., and Branchaud, B.,** Robinson annelation by reaction of 2-methyl 1,3-diketones with a β-chloro ketone, *J. Org. Chem.*, 41, 3767, 1976.
38. **Zoretic, P. A., Branchaud, P., and Maestrone, T.,** Robinson annelations with a β-chloroketone in the presence of an acid, *Tetrahed. Lett.*, 527, 1975.
39. **Zoretic, P. A., Branchaud, B., and Maestrone, T.,** Robinson annelation with a β-hydroxyketal in the presence of an acid, *Org. Prep. Proc. Int.*, 7, 51, 1975.
40. **Woodward, R. B., Sondeimer, F., Taub, D., Heusler, K., and McLamore, W. M.,** Total synthesis of a steroid, *J. Am. Chem. Soc.*, 73, 2403, 1951.
41. **Woodward, R. B., Sondeimer, F., Taub, D., Heusler, K., and McLamore, W. M.,** The total synthesis of steroids, *J. Am. Chem. Soc.*, 74, 4243, 1952.
42. **Cardwell, H. M. E., Cornforth, J. W., Duff, S. R., Holtermann, H., and Robinson, R.,** Experiments on the synthesis of substances related to the sterols LI. Completion of the synthesis of androgenic hormones and of cholesterol group of sterols, *J. Chem. Soc.*, 361, 1953.
43. **Sarett, L. H., Lukes, R. M., Beyler, R. E., Poos, G. I., Johns, W. F., and Constantin, J. M.,** Approaches to the total synthesis of adrenal steroids, *J. Am. Chem. Soc.*, 75, 422, 1707 and 2112, 1953.
44. **Wilds, A. L., Ralls, J. W., Tyner, D. A., Daniels, R., Kraychy, S., and Harnick, M.,** Total synthesis of racemic methyl-3-oxo-etiocholanate, *J. Am. Chem. Soc.*, 75, 4878, 1953.
45. **Wieland, P., Ueberwasser, H., Anger, G., and Miescher, K.,** Steroids. Preparation of 8,10a-dimethyl-1,7-dioxo-Δ^8-dodecahydrophenanthrene, *Helv. Chim. Acta*, 36, 376, 1953.
46. **Wieland, P., Anner, G., and Miescher, K.,** Steroids. The steric relationship of $\Delta^{8(8a)}$-1,7-dioxo-8,10a-dimethyldodecahydrophenanthrene with sterols. Total synthesis in the sterol series II., *Helv. Chim. Acta*, 36, 646, 1953.
47. **Wieland, P., Ueberwasser, H., Anner, G., and Miescher, K.,** Steroids, Total synthesis of D-homosteroids, *Helv. Chim. Acta*, 36, 1231, 1953.

48. **Johnson, W. S., Bannister, B., Pappo, R., Rogier, E. R., and Smuszkovicz, J.,** Steroid total synthesis — hydrochryrsene approach. Metal-in-ammonia reduction of the aromatic nucleus. dl-Epiandrosterone and the lumi-epimer, *J. Am. Chem. Soc.*, 78, 6331, 1956.
49. **Ohta, S., Shimabayashi, A., Hatano, S., and Okamoto, M.,** Preparation of t-butyl 3-oxopent-4-enoate and its use as a Nazarov reagent, *Synthesis,* 715, 1983.
50. **Santelli-Rouvier, C. and Santelli, M.,** The Nazarov Cyclization, *Synthesis,* 429, 1983.
51. **Dauben, W. G. and Bunce, R. A.,** Organic reactions at high pressure. A Robinson Annulation sequence initiated by Michael addition of activated cycloalkanones with hindered enones, *J. Org. Chem.*, 48, 4643, 1983.
52. **Velluz, L., Nominé, G., Mathieu, J.,** Neuere Ergebnisse bei der Totalsynthese von Steroiden, *Angew. Chem.*, 72, 725, 1960.
53. **Velluz, L., Nominé, G., Mathieu, J., Toromanoff, E., Bertin, D., Tessier, J., and Pierdet, A.,** Sur l'accès stéreospécifique, par synthèse totale, à la série 19-nor-steroide. La 19-nor-testostérone de synthese, *Compt. Rend. Acad. Sci. Paris,* 250, 1084, 1960.
54. **Danishefsky, S., and Migdalof, B. H.,** β-Chloroethyl vinyl ketone, a useful reagent for the facile construction of fused ring systems, *J. Am. Chem. Soc.*, 91, 2806, 1969.
55. **Eschenmoser, A., Schreiber, J., and Julia, A. S.,** Steroids and sex hormones. Synthesis of 8,10a-dimethyl-1,7-dioxo-$D^{4a,8}$-decahydrophenanthrene, *Helv. Chim. Acta,* 36, 482, 1953.
56. **Stork, G. and McMurry, J. E.,** Stereospecific synthesis of steroids *via* isoxazole annelation. dl-D-Homotestosterone and dl-progesterone, *J. Am. Chem. Soc.*, 89, 5464, 1967.
57. **Stork, G. and Ganem, B.,** α-Silylated vinyl ketones. A new class of reagents for the annelation of ketones, *J. Am. Chem. Soc.*, 95, 6152, 1973.
58. **Ireland, R. E., Dawson, M. I., Kowalski, C. J., Lipinski, C. A., Marshall, D. R., Tilley, J. W., Bordner, J., and Trus, B. L.,** Experiments directed towards the total synthesis of terpenoids. Synthesis of 8-methoxy-4aβ,10aβ,12aα-trimethyl-3,4,4a,4bβ,5,6,10b,11,12,12a-decahydrochrysen-1(2H)-one, a key intermediate in the total synthesis of (+)-shionone, *J. Org. Chem.*, 40, 973, 1975.
59. **Danishefsky, S., Cain, P., and Nagel, A.,** Bis annelations via 6-methyl-2-vinylpyridine. An efficient synthesis of dl-D-homoestrone, *J. Am. Chem. Soc.*, 97, 380, 1975.
60. **Pelletier, S. W., Chappell, R. L., and Prabhakar, S.,** A stereoselective synthesis of racemic androgrypholide lactone, *J. Am. Chem. Soc.*, 90, 2889, 1968.
61. **Stork, G. and Guthikonda, R. N.,** Stereoselective total synthesis of (±)-yohimbine, ψ-yohimbine and (±)-β-yohimbine, *J. Am. Chem. Soc.*, 94, 5109, 1972.
62. **Trost, B. M. and Kunz, R. A.,** New synthetic reactions. A convenient approach to methyl 3-oxo-4-pentenoate, *J. Org. Chem.*, 39, 2648, 1974.
63. **Wenkert, E. and Berges, D. A.,** The stereospecific introduction of a vicinally functionalized angular methyl group. A synthesis of l-valeranone, *J. Am. Chem. Soc.*, 89, 2507, 1967.
64. **Ireland, R. E., Marshall, D. R., and Tilley, J. W.,** A convenient stereoselective synthesis of 9,10-dimethyl-*trans*-1-decalones through the photolysis of fused methoxycyclopropanes, *J. Am. Chem. Soc.*, 92, 4754, 1970.
65. **Wichterle, O.,** Transformation des chlorures du type vinylique en cétones, *Collect. Czech. Chem. Commun.*, 12, 93, 1947.
66. **Wichterle, O., Procházka, J., Hofman, J.,** L'acetylacétate γ-chlorocrotylé et sa cyclisation par l'acide, *Collect. Czech. Chem. Commun.*, 13, 300, 1948.
67. **House, O.,** *Modern Synthetic Reactions,* 2nd ed., W. Benjamin, San Francisco, 611, 1972.
68. **Ireland, R. E. and Kierstead, R. C.,** Experiments directed towards the total synthesis of terpenes. A stereoselective scheme for diterpenoid resin acid synthesis, *J. Org. Chem.*, 31, 2543, 1966.
69. **Caine, D. and Tuller, F. N.,** An alternative synthesis of *trans* -8,10-dimethyl-1(9)-octal-2-one, *J. Org. Chem.*, 34, 222, 1969.
70. **Velluz, L., Nominé, G., Burcourt, R., Pierdet, A., and Dufay, P.,** Synthèse stereospecifique totale d'un homologue angulaire de l'hormone folliculinique naturelle. Le 13-propyl nor-estradiol, *Tetrahed. Lett.*, 127, 1961.
71. **Velluz, L., Nominé, G., Burcourt, R., Pierdet, A., and Tessier, J-.,** L'extension de la synthèse totale dans le groupe des nor-testostérones, *Compt. Rend. Acad. Sci. Paris,* 252, 3903, 1961.
72. **Burcourt, R., Tessier, J., and Nominé, G.,** Extension de la synthese totale steroide à la nor-19progesterone, *Bull. Soc. Chim. Fr.*, 1923, 1963.
73. **Kobayashi, M. and Matsumoto, T.,** Modifizierte Wichterle reaktion. Ein neuer Weg zur Synthese von Δ^2-Cyclohexenones, *Chem. Lett.*, 957, 1973.
74. **Stork, G., Jung, M. E., Colvin, E., and Noel, Y.,** Synthetic route to halomethyl vinylsilanes, *J. Am. Chem. Soc.*, 96, 3684, 1974.
75. **Stork, G. and Jung, M. E.,** Vinylsilanes as carbonyl precursors. Use in annelation reactions, *J. Am. Chem. Soc.*, 96, 3682, 1974.

76. **Stork, G. and McMurry, J. E.,** The mechanism of isoxazole annelation, *J. Am. Chem. Soc.,* 89, 5463, 1967.
77. **Stork, G., Danishefsky, S., and Ohashi, M.,** The isoxazole annelation reaction. A method for the construction of cyclohexenone rings in polycyclic system, *J. Am. Chem. Soc.,* 89, 5459, 1967.
78. **Stork, G., Ohashi, M., Kamachi, H., and Kakisawa, H.,** A new pyridine synthesis via 4-(3-oxoalkyl)isoxazoles, *J. Org. Chem.,* 36, 2784, 1971.
79. **Stork, G. and Ganem, B.,** α-Silylated vinyl ketones. A new class of reagents for the annelation of ketones, *J. Am. Chem. Soc.,* 95, 6152, 1973.
80. **Stork, G. and Singh, J.,** Regiospecific Michael reactions in aprotic solvents with α-silylated electrophilic olefins. Application to annelation reactions, *J. Am. Chem. Soc.,* 96, 6181, 1974.
81. **Boeckman, R. K.,** Conjugate addition-annelation. A highly regiospecific and stereospecific synthesis of polycyclic ketones, *J. Am. Chem. Soc.,* 95, 6867, 1973.
82. **Boekman, R. K.,** Regiospecificity in enolate reactions with α-silyl vinyl ketones. An application to steroid total synthesis, *J. Am. Chem. Soc.,* 96, 6179, 1974.
83. **Boeckman, R. K., Jr.,** The utility of silicon in organic synthesis. Annulation methodology and applications employing α-trimethylsilyl vinyl ketones, *Tetrahedron,* 39, 925, 1983.
84. **Rosan, A. and Rosenblum, M.,** Metal assisted C-C bond formation. Use of a methyl vinyl ketone complex in Michael condensations, *J. Org. Chem.,* 40, 3621, 1975.
85. **Danishefsky, S., Chackalamannil, S., Silvestri, M., and Springer, J.,** A stereospecific 2 + 2 + 2 annulation, *J. Org. Chem.,* 48, 3615, 1983.
86. **Stork, G. and Sherman, D. H.,** Efficient de novo construction of the indanpropionic acid precursor of 11-keto steroids. An improved internal Diels-Alder sequence, *J. Am. Chem. Soc.,* 104, 3758, 1982.
87. **Hajos, Z. G., Parrish, D. R., and Oliveto, E. P.,** Total synthesis of optically active (−)-17β-hydroxy-$\Delta^{9,10}$-des-A-androsten-5-one, *Tetrahedron,* 24, 2039, 1968.
88. **Saucy, G., Borer, R., and Fürst,** A Total Synthese von Steroiden. Rac-17-Hydroxy-des-A-androst-9-en-5-one, *Helv. Chim. Acta,* 54, 2034, 1971.
89. **Saucy, G. and Borer, R.,** Steroid total synthesis. (−)-17-Hydroxy-des-A-androst-9-en-5-one, *Helv. Chim. Acta,* 54, 2121, 1971.
90. **Saucy, G. and Borer, R.,** Steroid total synthesis. 3. 9,10-Testosterone, *Helv. Chim. Acta,* 54, 2517, 1971.
91. **Rosenberger, M., Duggan, A. J., Borer, R., Muller, R., and Saucy, G.,** Steroid total synthesis. (+)-Estr-4-ene-3,17-dione, *Helv. Chim. Acta,* 55, 2663, 1972.
92. **Eder, U., Sauer, G., and Wiechert, R.,** Neuartige asymmetrische Cyclisierung zu optische aktiven Steroid-CD-Teilstücken, *Angew. Chem.,* 83, 492, 1971.
93. **Hajos, Z. G. and Parrish, D. R.,** Asymmetric synthesis of bicyclic intermediates of natural product chemistry, *J. Org. Chem.,* 39, 1615, 1974.
94. **Micheli, R. A., Hajos, Z. G., Cohen, N., Parrish, D. R., Portland, L. A., Sciamanna, W., Scott, M. A., and Wehrli, P. A.,** Total syntheses of optically active 19-norsteroids. (+)-Estr-4-ene-3,17-dione and (+)-13β-ethylgon-4-ene-3,17-dione, *J. Org. Chem.,* 40, 675, 1975.
95. **Scanio, C. J. V. and Sarrett, R. M.,** A remarkably stereoselective Robinson annelation reaction, *J. Am. Chem. Soc.,* 93, 1539, 1971.
96. **Marshall, J. A., Faubl, H., and Warne, T. M., Jr.,** The total synthesis of *cis-* and *trans-*4,4a-dimethyl-2-octalone derivatives, *Chem. Commun.,* 47, 1967.
97. **Marshall, J. A. and Warne, T. M., Jr.,** The total synthesis of (±)-isonootkanone. Stereochemical studies of the Robinson annelation reaction with 3-penten-2-one, *J. Org. Chem.,* 36, 178, 1971.
98. **Ramage, R. and Sattar, A.,** Thermal isomerisation of a hexatriene system: synthesis and rearrangement of 2-methyl-3-(cis-,cis-penta-1,3-dienyl)cyclohex-2-en-1-one, *J. Chem. Soc. Chem. Commun.,* 173, 1970.
99. **Zoretic, P. A., Ferrari, J. L., Bhakta, C., Barcelos, F., and Branchaud, B.,** Sesquiterpene synthesis. Studies relating to the synthesis of (±)-dugesialactone, *J. Org. Chem.,* 47, 1327, 1982.
100. **Kikuchi, M. and Yoshikoshi, A.,** A reexamination of Robinson annelation of 2-methylcyclohexanone with 3-penten-2-one and 4-phenyl-3-buten-2-one, *Bull. Chem. Soc. Jpn.,* 54, 3420, 1981.
101. **Coates, R. M. and Shaw, J. E.,** Stereoselectivity in the synthesis of *cis-* and *trans-*4,4a-dimethyl-2-octalone derivatives, *Chem. Commun.,* 47, 1968.
102. **Stork, G., Shiner, C. S., and Winkler, J. D.,** Stereochemical control of the internal Michael reaction. A new construction of *trans-*hydrindane system, *J. Am. Chem. Soc.,* 104, 310, 1982.
103. **Stork, G., Winkler, J. D., and Shiner, C. S.,** Stereochemical control of intramolecular conjugate addition. A short, highly stereoselective synthesis of adrenosterone, *J. Am. Chem. Soc.,* 104, 3767, 1982.
104. **Irie, H., Katakawa, J., Mizuno, Y., Udaka, S., Taga, T., and Osaki, K.,** New stereoselective synthesis of 9-methyl-*cis-*decalin derivatives by double Michael Reaction of 3,5-dimethyl-4-methylenecyclohex-2-enone- and dimethyl 3-oxoglutarate; X-ray crystal and molecular structure of two of the products, *J. Chem. Soc. Chem. Commun.,* 717, 1978.

105. **Hajos, Z. G. and Parrish, D. R.,** Stereocontrolled total synthesis of 19-nor steroids, *J. Org. Chem.*, 38, 3244, 1973.
106. **Stork, G. and D'Angelo, J.,** Condensation of formaldehyde with regiospecifically generated anions, *J. Am. Chem. Soc.*, 96, 7114, 1974.
107. **Sauer, G., Eder, U., Haffer, G., Neef, G., Wiechert, R., and Rosenberg, D.,** Darstellung eines 7α-Methylostratriens durch stereoselektive methylierung von Ostratrien-6-on, *Justus Liebigs Ann. Chem.*, 459, 1982.
108. **Newman, M. S. and Mekler, A. B.,** The synthesis of 8-hydroxy-1-keto-1,2,3,5,6,7-hexahydronaphthalene, *J. Am. Chem. Soc.*, 82, 4039, 1960.
109. **Wender, P. A., Eissenstadt, M. A., and Filosa, M. P.,** A general methodology for pseudoguaiane synthesis: Total synthesis of (±)-damsinic acid and (±)-Confertin, *J. Am. Chem. Soc.*, 101, 2196, 1979.
110. **Matthews, R. S. and Whitesell, J. K.,** Transannular cyclizations. A stereoselective synthesis of the cyclopentanoid monoterpenes, *J. Org. Chem.*, 40, 3312, 1975.
111. **Boar, R. B. and Copsay, D. B.,** Synthesis of cycloartenol *via* 19-oxygenated lanostanes, *J. Chem. Soc. Perkin Trans.*, 1, 563, 1979.
112. **Bellesia, F., Pagnoni, U. M., and Trave, R.,** Stereospecific cyclisation of agerol to an isovetivane carbon framework, *J. Chem. Soc. Chem. Commun.*, 34, 1976.
113. **Baldwin, J. E. and Barden, T. C.,** Absolute stereochemistry of (−)-albene, *J. Org. Chem.*, 48, 625, 1983.
114. **Baldwin, J. E. and Barden, T. C.,** Discrimination between exo- and endo-3,2-methyl shifts in substituted 2-norbornyl cations on the (+)-camphenilone route to (−)-albene, *J. Am. Chem. Soc.*, 105, 6656, 1983.
115. **Kreiser, W. and Janitschke, L.,** The real structure of "albene" and its total synthesis, *Tetrahed. Lett.*, 601, 1978.
116. **Kreiser, W., Janitschke, L., and Sheldrick, W. S.,** The putative structure of albene; X-ray structure of an analogue, *J. Chem. Soc. Chem. Commun.*, 269, 1977.
117. **Lansbury, P. T. and Boden, R. M.,** On the structure of albene, *Tetrahed. Lett.*, 5017, 1973.
118. **Vokáč, K., Samek, Z., Herout, V., Šorm, F.,** The structure of albene a hydrocarbon from the plants of the genera *Petasites* and *Adenostyles*, *Tetrahed. Lett.*, 1665, 1972.
119. **Stetter, H.,** Die katalysierte Addition von Aldehyden an aktivierte Doppelbindungen — ein neues Syntheseprinzip, *Angew. Chem.*, 88, 695, 1976.
120. **Paquette, L. A. and Han, Y. K.,** Stereospecific total synthesis of (±)-isocomene (berkheyaradulene), *J. Org. Chem.*, 44, 4014, 1979.
121. **Annis, G. D. and Paquette, L. A.,** Total synthesis of (±)-pentalenene, *J. Am. Chem. Soc.*, 104, 4505, 1982.
122. **Paquette, L. A. and Annis, G. D.,** Total synthesis of (+)-pentalenene, the least oxidized neutral triquinane metabolite of streptomyces griseochromogenes, *J. Am. Chem. Soc.*, 105, 7358, 1983.
123. **Ohtsuka, T., Shirahama, H., and Matsumoto, T.,** Synthesis of pentalenolactone E and F through biogenetic like cyclization of humulene, *Tetrahed. Lett.*, 14, 3851, 1983.
124. **Geetha, K. Y., Rajagopalan, K., and Swaminathan, S.,** Transannular reaction during the base cytalysed rearrangement of bicyclo(4.3.0)-2β-hydroxy-2α-vinyl-1β-methyl-8-oxo-$\Delta^{6,7}$-nonene, *Tetrahedron*, 34, 2201, 1978.
125. **Miyashita, M., Makino, N., Singh, M., and Yoshikoshi, A.,** Synthesis of the enantiomers of irones from (+)-citronellal, *J. Chem. Soc., Perkin Trans.*, 1, 1303, 1982.
126. **Clive, D. L. J., Chittattu, G., and Wong, C. K.,** A new use of cyclofunctionalisation with selenyl reagents. An example of carbon-carbon bond formation, *J. Chem. Soc. Chem. Commun.*, 441, 1978.
127. **Haufe, G., Mählstädt, M., and Graefe, J.,** Darstellung von 4^t-brom-1^c,9^c10^c-decalol und 4^t-brom-1^c-mehtoxy-9,$^c10^c$-decalin, *Monatsh., Chemie*, 108, 199, 1977.
128. **Renold, W., Ohloff, G., and Norin, T.,** New olefinic cyclizations by oxymetallation. Conversion of (−)-elemol to (−)-selina-4α,11-diol (cryptomeridiol) and (−)-Guai-1(10)-ene-4α,11-diol, *Helv. Chim. Acta*, 62, 985, 1979.
129. **Niwa, M., Iguchi, M., and Yamamura, S.,** Regio- and stereospecific cyclizations of germacrones, *Bull. Chem. Soc. Jpn.*, 49, 3148, 1976.
130. **Tsankova, E., Ognyanov, I., and Norin, T.,** Transannular cyclization of germacrone and isogermacrone via oxymercuration — demercuration, *Tetrahedron*, 36, 669, 1980.
131. **Clark, A. M. and Hufford, C. D.,** Microbial transformations of the sesquiterpene lactone costunolide, *J. Chem. Soc. Perkin Trans.*, 1, 3022, 1979.
132. **Rastetter, W. H., Richard, T. J., Bordner, J., and Hennessee, G. L. A.,** X-ray crystal and molecular structure of a bridgehead diene, *J. Org. Chem.*, 44, 999, 1979.
133. **Tolstikov, G. A., Dzhemilev, U. M., and Shavanov, S. S.,** Isomerisation of *cis,trans*-1,5-cyclodecadiene by means of triisobutylaluminum, *Izv. Akad. Nauk USSR, Ser. Khim.*, 1207, 1974 (in Russian).
134. **Jain, T. C., Banks, C. M., and McCloskey, J. E.,** Novel cyclization of *trans*-1,2-divinylcyclohexane-3,4-*trans*-γ-lactone unit, *Tetrahed. Lett.*, 2387, 1970.

135. **Hoye, T. R. and Kurth, M. J.,** Mercuric trifluoroacetate mediated cyclizations of dienes. Total synthesis of *dl*-3β-bromo-8-epicaparrapi oxide, *J. Org. Chem.*, 44, 3461, 1979.
136. **Toshimitsu, A., Uemura, S., and Okano, M.,** Carbon-carbon bond formation in diolefins using a new reagent benzylselenyl iodide, *J. Chem. Soc. Chem. Commun.*, 87, 1982.
137. **Brunke, E. J., Hammerschmidt, F.-J., and Struwe, H.,** Cyclization of 1,3-disubstituted 2-(3-butenyl)-2-cyclohexen-1-ols, *Tetrahedron*, 37, 1033, 1981.
138. **Johnson, W. S.,** Biomimetische cyclisierung von Polyene, *Angew. Chem.*, 88, 33, 1976.
139. **Johnson, W. S.,** Biomimetic polyene cyclizations, *Bioorg. Chem.*, 5, 51, 1976.
140. **van Tamelen, E. E.,** Bioorganic chemistry: Total synthesis of tetra and pentacyclic triterpenoids, *Accounts Chem. Res.*, 8, 152, 1975.
141. **Kametani, T., Kurobe, H., Nemoto, H., and Fukumoto, K.,** Stereoselective olefin cyclization mediated by the selenyl group; direct formation of a selenyl caparrapi oxide, *J. Chem. Soc. Perkin Trans.*, 1, 1085, 1982.
142. **Schmidt, C., Chsati, N. H., and Breining, T.,** An efficient annulation of the sterically hindered 2,2,6-trimethylcyclohexanone, *Synthesis*, 391, 1982.
143. **Johnson, W. S., Frei, B., and Gopalan, A. S.,** Improved asymmetric total synthesis of corticoids *via* biomimetic polyene cyclization methodology, *J. Org. Chem.*, 46, 1512, 1981.
144. **Johnson, W. S., Brinkmeyer, R. S., Kapoor, V. M., and Yarnell, T. M.,** Asymmetric total synthesis of 11α-hydroxyprogesterone via a biomimetic polyene cyclization, *J. Am. Chem. Soc.*, 99, 8341, 1977.
145. **Johnson, W. S., Lyle, T. A., and Daub, G. W.,** Corticoid synthesis via vinylic fluoride terminated biomimetic polyene cyclizations, *J. Org. Chem.*, 47, 161, 1982.
146. **Johnson, W. S., Chen, Y.-Q., and Kellogg, M. S.,** Termination of biomimetic cyclizations by the allylsilane function. Formation of the steroid nucleus in one step from an acyclic polyenic chain, *J. Am. Chem. Soc.*, 105, 6653, 1983.
147. **Wiesner, K., Musil, V., and Wiesner, K. J.,** Synthesis in the series of lycopodium alkaloids. Two simple stereospecific syntheses of 12-epi-lycopodine, *Tetrahed. Lett.*, 5643, 1968.
148. **Funk, R. L. and Vollhardt, K. P. C.,** A cobalt-catalyzed steroid synthesis, *J. Am. Chem. Soc.*, 99, 5483, 1977.
149. **Collman, J. P. and Hegedus, L. S.,** *Principles and Applications of Organotransition Metal Chemistry*, University Science Books, Mill Valley, California, 1980.
150. **Scheffold, R., Ed.,** *Modern Synthetic Methods*, Vol. 3, J. Wiley & Sons, Frankfurt, 1983.
151. **Tsuji, J.,** *Organic Synthesis with Palladium Compounds*, Springer-Verlag, Berlin, 1980.
152. **Davies, S. G.,** *Organotransition Metal Chemistry: Application to Organic Synthesis*, Pergamon Press, Oxford, 1982.
153. **Oppolzer, W. and Snieckus, V.,** Intramolecular En-Reactionen in der organischen Synthese, *Angew. Chem.*, 90, 506, 1978.
154. **Oppolzer, W., Bättig, K., and Hudlický, T.,** The total synthesis of (±)-isocomene by an intramolecular ene reaction, *Helv. Chim. Acta*, 62, 1493, 1979.
155. **Oppolzer, W. and Marazza, F.,** A new, stereoselective approach to the [3,3,3] propellane system: Synthesis of (±)-modhephene, *Helv. Chim. Acta*, 64, 1575, 1981.
156. **Zalkow, L. H., Harris III, R. N., and van Derveer, D.,** Modhephene: a sesquiterpenoid carbocyclic [3.3.3] propellane. X-ray crystal structure of the corresponding diol, *J. Chem. Soc. Chem. Commun.*, 420, 1978.
157. **Karpf, M. and Dreiding, A. S.,** Anwendung der α-Alkinon-Cyclisierung. Synthese von rac-Modhephene, *Helv. Chim. Acta*, 64, 1123, 1981.
158. **Oppolzer, W.,** Regio- and stereo-selective synthesis of cyclic natural products by intramolecular cycloaddition and ene-reaction, *Pure Appl. Chem.*, 53, 1181, 1981.
159. **Lindner, D. L., Doherty, J. B., Shoham, G., and Woodward, R. B.,** Intramolecular ene reaction of glyoxylate esters: an anisatin model study, *Tetrahed. Lett.*, 23, 5111, 1982.
160. **Bussas, R., Münster, H., and Kresze, G.,** Ene reaction mechanisms. 1. Chirality transfer to the enophile 4-methyl-N-sulfinylbenzensulfonamide, *J. Org. Chem.*, 48, 2828, 1983.
161. **Snider, B. B. and Roush, D. M.,** Lewis acid induced cyclizations of ethylenetricarboxylates, *J. Org. Chem.*, 44, 4229, 1979.
162. **Snider, B. B., Roush, D. M., and Killinger, T. A.,** Intramolecular reactions of 1-allylic 2,2-dimethyl ethylenedicarboxylates, *J. Am. Chem. Soc.*, 101, 6023, 1979.
163. **Oppolzer, W. and Battig, K.,** Total synthesis of (±)-$\Delta^{9(12)}$-capnellene via iterative intramolecular type-I-"magnesium-ene" reactions, *Tetrahed. Lett.*, 23, 4669, 1982.
164. **Oppolzer, W., Strauss, H. F., and Simmons, D. P.,** Stereoselective total syntheses of (±)-sinularene and of (±)-5-epi-sinularene via intramolecular type-I-"magnesium-ene" reaction, *Tetrahed. Lett.*, 23, 4673, 1982.
165. **Snider, B. B. and Deutsch, E. A.,** Sequential ene reaction. A new annelation procedure, *J. Org. Chem.*, 47, 745, 1982.

166. **Snider, B. B. and Deutsch, E. A.,** Sequential ene reaction. A new annelation procedure, *J. Org. Chem.*, 48, 1822, 1983.
167. **Trost, B. M., McDougal, P. G., and Haller, K. J.,** A tandem cycloaddition-ene strategy for the synthesis of (±)-verrucarol and (±)-4,11-diepi-12,13-deoxyverrucarol, *J. Am. Chem. Soc.*, 106, 383, 1984.
168. **Kaneti, J., Karpf, M., and Dreiding, A. S.,** On the mechanism of α-alkynone cyclization, *Helv. Chim. Acta*, 65, 2517, 1982.
169. **Karpf, M. and Dreiding, A. S.,** Application of the α-alkynone cyclization. Synthesis of rac-modhephene, *Tetrahed. Lett.*, 21, 4569, 1980.
170. **Huguet, J., Karpf, M., and Dreiding, A. S.,** Consecutive application of the α-alkynone cyclization: Total synthesis of (±)-$\Delta^{9(12)}$-capnellene, *Helv. Chim. Acta*, 65, 2413, 1982.
171. **Little, R. D. and Carrol, G. L.,** Intramolecular 1,3-diyl trapping reactions: Total synthesis of the marine natural product (*d,l*)-$\Delta^{9(12)}$-capnellene, *Tetrahed. Lett.*, 4389, 1981.
172. **Stevens, K. E. and Paquette, L. A.,** Stereocontrolled total synthesis of (±)-$\Delta^{9(12)}$-capnellene, *Tetrahed. Lett.*, 4393, 1981.
173. **Paquette, L. A.,** Recent synthetic developments in polyquinane chemistry, *Topics Curr. Chem.*, 119, 1, 1984.
174. **Woodward, R. B. and Hoffmann, R.,** *The Conservation of Orbital Symmetry*, Verlag Chemie, Weinheim, 1970.
175. **Pearson, R. G.,** *Symmetry Rules for Chemical Reactions*, Wiley-Interscience, New York, 1976.
176. **Plattner, J. J., Bhalearo, U. T., and Rapoport, H.,** Synthesis of dl-sirenin, *J. Am. Chem. Soc.*, 91, 4933, 1969.
177. **Bhalearo, U. T., Plattner, J. J., and Rapoport, H.,** Synthesis of *dl*-sirenin and *dl*-isosirenin, *J. Am. Chem. Soc.*, 92, 3429, 1970.
178. **Joska, J., Fajkoš, J., and Šorm, F.,** 3,5-Cyclosteroid androgen analogues, *Collect. Czech. Chem. Commun.*, 33, 3342, 1968.
179. **Mironowicz, A., Kohout, L., and Fajkoš, J.,** 5,7-Cyclo-B-homopregnane derivatives with an oxygen function in position 21, *Collect. Czech. Chem. Commun.*, 39, 1780, 1974.
180. **Kočovaký, P., Kohout, L., and Fajkoš, J.,** 5,7-Cyclo-B-homo-pregnane derivatives with an oxygen function in position 17α, *Collect. Czech. Chem. Commun.*, 40, 468, 1975.
181. **Fleming, I.,** *Frontier Orbitals and Organic Reactions*, Wiley-Interscience, Chichester, 1976.
182. **Wiberg, K. B., Olli, L. K., Golembski, N., and Adams, R. D.,** Tricyclo [4.2.0.$0^{1,4}$] octane, *J. Am. Chem. Soc.*, 102, 7467, 1980.
183. **Srinivasan, R. and Carlough, H. K.,** Mercury ($^3P_1$) photosensitized internal cycloaddition reaction in 1,4,1,5, and 1,6-dienes, *J. Am. Chem. Soc.*, 89, 4939, 1967.
184. **Liu, R. S. H. and Hammond, G. S.,** Photosensitized internal addition of dienes to olefins, *J. Am. Chem. Soc.*, 89, 4936, 1967.
185. **Agosta, W. C. and Wolff, S.,** Cyclization of substituted 5-hexenyl radicals as a model for photocyclization of 1,5-hexadien-3-ones, *J. Org. Chem.*, 45, 3139, 1980.
186. **Tamura, Y., Ishibashi, H., Hirai, M., Kita, Y., and Ikeda, M.,** Photochemical syntheses of 2-aza- and 2-oxabicyclo [2.1.1] hexane ring system, *J. Org. Chem.*, 40, 2702, 1975.
187. **Becker, D. and Birnbaum, D.,** Intramolecular photo-addition of ketenes to conjugated cycloalkenones, *J. Org. Chem.*, 45, 570, 1980.
188. **Pirrung, M. C.,** Total synthesis of (±)-isocomene, *J. Am. Chem. Soc.*, 101, 7130, 1979.
189. **Pirrung, M. C.,** Total synthesis of (±)-isocomene and related studies, *J. Am. Chem. Soc.*, 103, 82, 1981.
190. **McMurry, J. E. and Choy, W.,** Total synthesis of α- and β-panasinsene, *Tetrahed. Lett.*, 21, 2477, 1980.
191. **Hoye, T. R., Martin, S. J., and Peck, D. R.,** Intramolecular photochemical cycloaddition reactions of 3-(1,5-dimethyl-hex-4-enyl)cyclohex-2-enone: regio and stereochemical aspects, *J. Org. Chem.*, 47, 331, 1982.
192. **Baker, A. J. and Pattenden, G.,** Zizaane sesquiterpenes. Synthesis of the Coates-Sowerby tricyclic ketone, *Tetrahed. Lett.*, 22, 2599, 1981.
193. **Oppolzer, W., Gorrichon, L., and Bird, T. G. C.,** A stereoselective approach to the spiro[4,5]decane system via intramolecular photocycloaddition and reductive fragmentation, *Helv. Chim. Acta*, 64, 186, 1981.
194. **Oppolzer, W. and Bird, T. G. C.,** Intramolecular de Mayo reactions of 3-acetoxy-2-alkenyl-2-cyclohexenones, *Helv. Chim. Acta*, 62, 1199, 1979.
195. **Oppolzer, W. and Godel, T.,** A new and efficient total synthesis of (±)-longifolene, *J. Am. Chem. Soc.*, 100, 2583, 1978.
196. **Cookson, R. C., Hudec, J., Szabo, A., and Usher, G. E.,** The optical and photochemical properties of methylisopulegone, *Tetrahedron*, 24, 4353, 1968.
197. **de Mayo, P.,** Enone photoannelation, *Acc. Chem. Res.*, 4, 41, 1971.

198. **Corey, E. J., Ohno, M., Mitra, R. B., and Vatakencherry, P. A.,** Total synthesis of longifolene, *J. Am. Chem. Soc.*, 86, 478, 1964.
199. **McMurry, J. E. and Isser, S. J.,** Total synthesis of longifolene, *J. Am. Chem. Soc.*, 94, 7132, 1972.
200. **Volkmann, R. A., Andrews, G. C., and Johnson, W. S.,** A novel synthesis of longifolene, *J. Am. Chem. Soc.*, 97, 4777, 1975.
201. **Oppolzer, W. and Wylie, R. D.,** Total synthesis β-bulnesene and 1-epi-bulnesene by intramolecular photoaddition, *Helv. Chim. Acta*, 63, 1198, 1980.
202. **Oppolzer, W. and Burford, S. C.,** Synthesis of tricyclo [6.2.1.0.^{1,5}] undecadiones via intramolecular photoaddition of 5-(1-cyclopentenylmethyl)-3-alkoxy-2-cyclopentenones, *Helv. Chim. Acta*, 63, 788, 1980.
203. **Kueh, J. S. H., Mellor, M., and Pattenden, G.,** Photo-cyclisation of dicyclopent-1-enyl methanes to tricyclo [6.3.0.0^{2,6}] undecanes. A synthesis of the hirsutane carbon skeleton, *J. Chem. Soc. Chem. Commun.*, 5, 1978.
204. **Marini-Bettòlo, G., Sahoo, S. P., Poulton, G. A., Tsai, T. Y. R., and Wiesner, K.,** On the stereochemistry of photoaddition between α,β-unsaturated ketones and olefins — II, *Tetrahedron*, 36, 719, 1980.
205. **Wiesner, K.,** On the sterochemistry of photoaddition between α,β-unsaturated ketones and olefins, *Tetrahedron*, 31, 1655, 1975.
206. **Tsai, T. Y. R., Tsai, C. S. J., Sy, W. W., Shanbhag, M. N., Liu, W. C., Lee, S. F., and Wiesner, K.,** A stereospecific total synthesis of chasmanine, *Heterocycles*, 7, 217, 1977.
207. **Wiesner, K.,** The total synthesis of racemic talatisamine, *Pure Appl. Chem.*, 41, 93, 1975.
208. **Wiesner, K.,** Systematic development of strategy in the synthesis of polycyclic polysubstituted natural products: The aconite alkaloids, *Chem. Soc. Rev.*, 6, 413, 1977.
209. **Semmelhack, M. F., Tomoda, S., and Hurst, K. M.,** Synthesis of (±)-illudol, *J. Am. Chem. Soc.*, 102, 7567, 1980.
210. **Semmelhack, M. F., Tomoda, S., Nagaoka, H., Boettger, S. D., and Hurst, K. M.,** Synthesis of racemic fomannosin and illudol using a biosynthetically patterned common intermediate, *J. Am. Chem. Soc.*, 104, 747, 1982.
211. **March, J.,** *Advanced Organic Chemistry*, 2nd ed., McGraw-Hill, New York, 1977.
212. **Brady, W. T.,** Halogenated ketenes. Valuable intermediates in organic synthesis, *Synthesis*, 415, 1971.
213. **Brady, W. T.,** Synthetic applications involving halogenated ketenes, *Tetrahedron*, 37, 2949, 1981.
214. **Ghosez, L., Montaigne, R., Rossel, A., Vanlierde, H., and Mollet, P.,** Cycloadditions of dichlorketene to olefines, *Tetrahedron*, 27, 615, 1971.
215. **Krepski, L. R. and Hassner, A.,** An improved procedure for the addition of dichlorketene to unreactive olefins, *J. Org. Chem.*, 43, 2879, 1978.
216. **Krepski, L. R. and Hassner, A.,** Addition of dichloroketene to silyl enol ethers. Synthesis of functionalized cyclobutanones, *J. Org. Chem.*, 43, 3173, 1978.
217. **Bak, D. A. and Brady, W. T.,** Halogenated ketenes. 31. Cycloaddition of dichloroketene with hindered olefins, *J. Org. Chem.*, 44, 107, 1979.
218. **Greene, A. E.,** Iterative three-carbon annelations. Synthesis of (±)-hirsutene, *Tetrahed. Lett.*, 21, 3059, 1980.
219. **Houge, C., Frisque-Hesbain, A. M., Mockel, A., Ghosez, L., DeClerq, J. P., Germain, G., and Van Meerssche, M.,** Models for asymmetric [2 + 2] cycloaddition, *J. Am. Chem. Soc.*, 104, 2920, 1982.
220. **Trost, B. M.,** Transition metal templates for selectivity in organic synthesis, *Pure Appl. Chem.*, 53, 2357, 1981.
221. **Trost, B. M.,** Transition metal templates for selectivity in organic synthesis, *Aldrichim. Acta*, 14, 43, 1981.
222. **Trost, B. M. and Chan, D. M.,** Intramolecular carbocyclic [3 + 4] cycloaddition via organopalladium intermediates, *J. Am. Chem. Soc.*, 104, 3733, 1982.
223. **Knapp, S., O'Connor, U., and Mobilio, D.,** A [3 + 2] annulation procedure for methylenecyclopentanes, *Tetrahed. Lett.*, 21, 4557, 1980.
224. **Huisgen, R., Grashey, R., and Sauer, J.,** *The Chemistry of Alkenes*, Wiley-Interscience, New York, 1964.
225. **Titov, Yu. A.,** Orientation in diene synthesis and its dependence on structure, *Russian Chem. Rev. Engl. Transl.*, 267, 1962.
226. **Sauer, J.,** Diels-Alder Reaktionen: Zum Reaktionsmechanismus, *Angew. Chem.*, 79, 76, 1967.
227. **Eisenstein, O., Lefour, J.-M., and Anh, N. T.,** Simple prediction of regiospecificity in Diels-Alder reactions, *J. Chem. Soc. Chem. Commun.*, 969, 1971.
228. **Houk, K. N.,** Generalized frontier orbitals of alkenes and dienes. Regioselectivity in Diels-Alder reactions, *J. Am. Chem. Soc.*, 95, 409, 1973.
229. **Herndon, W. C.,** The theory of cycloaddition reactions, *Chem. Rev.*, 72, 157, 1972.
230. **Caramella, P., Rondan, N. G., Paddon-Row, M. M., and Houk, K. N.,** Origin of π-facial stereoselectivity in additions to π-bonds: Generality of the *anti*-periplanarity, *J. Am. Chem. Soc.*, 103, 2438, 1981.

231. **Franck, R. W., John, T. V., Olejniczak, K., and Blount, J. F.,** Stereochemical control by an allylic substituent of an acyclic dienophile in the Diels-Alder reaction, *J. Am. Chem. Soc.*, 104, 1106, 1982.
232. **Alston, P. V., Ottenbrite, R. M., and Shillady, D. D.,** Secondary orbital interactions determining regioselectivity in the Diels-Alder reaction, *J. Org. Chem.*, 38, 4075, 1973.
233. **Alston, P. V. and Shillady, D. D.,** A reexamination of the origin of regioselectivity in the dimerisation of acrolein. A frontier orbital approach, *J. Org. Chem.*, 39, 3402, 1974.
234. **Alston, P. V. and Ottenbrite, R. M.,** Secondary orbital interactions determining regioselectivity in the Lewis acid catalyzed Diels-Alder reactions II., *J. Org. Chem.*, 40, 1111, 1975.
235. **Anh, N. T., Canadell, E., and Einsenstein, O.,** La regle a'Alder generalise. Role privilegie du substituent donneur d'electrons, *Tetrahedron*, 34, 2283, 1978.
236. **Pancíř, J.,** Topological study of chemical reactivity. Diels-Alder reaction, *J. Am. Chem. Soc.*, 104, 7424, 1982.
237. **Williamson, K. L. and Li Hsu, Y.-F.,** The stereochemistry of the Diels-Alder reaction II. Lewis acid catalysis of *synanti* isomerism, *J. Am. Chem. Soc.*, 92, 7385, 1970.
238. **Burnier, J. S. and Jorgensen, W. L.,** Computer-assisted mechanistic evaluation of organic reactions. 7. Six electron cycloadditions, *J. Org. Chem.*, 48, 3923, 1983.
239. **Orsini, F., Pelizzoni, F., Pitea, D., Abbondanti, E., and Mugnoli, A.,** 9,10-syn-Podocarpene diterpenoids. An approach to the tricyclic skeleton by Diels-Alder cycloaddition. Related crystal structure determination and theoretical aspects, *J. Org. Chem.*, 48, 2866, 1983.
240. **Butz, L. W. and Rytina, A. W.,** The Diels-Alder reaction: Quinones and other cyclenones, *Org. Reactions*, 5, 136, 1949.
241. **Dickinson, R. A., Kubela, R., MacAlpine, G. A., Stojanac, Ž., and Valenta, Z.,** A stereospecific synthesis of ring A-aromatic steroids, *Can. J. Chem.*, 50, 2377, 1972.
241a. **Stojanac, Ž., Dickinson, R. A., Stojanac, N., Woznow, R. J., and Valenta, Z.,** Catalyzed orientation reversal in Diels-Alder reactions, *Can. J. Chem.*, 53, 617, 1975.
242. **Daniewski, A. R., White, P. S., and Valenta, Z.,** Total synthesis of 14β-hydroxy-4,9(11)-androstadiene-3,17-dione, *Can. J. Chem.*, 57, 1397, 1979.
243. **Hendrickson, J. B. and Singh, V.,** Catalysis and regioselectivity of quinone Diels-Alder reactions, *J. Chem. Soc. Chem. Commun.*, 837, 1983.
244. **Fringeelli, F., Minuti, L., Pizzo, F., Taticchi, A., Halls, T. D. J., and Wenkert, E.,** Diels-Alder reactions of cycloalkenones. 2. Preparation and structure of cyclohexadienone adducts, *J. Org. Chem.*, 48, 1810, 1983.
245a. **Viehe, H. G.,** *Acetylenes*, Marcel Dekker, New York, 1969.
245b. **Mach, K., Antropiusová, H., Petrusová, L., Tureček, F., Hanuš, V., Sedmera, P., and Schraml, J.,** Titanium-catalyzed Diels-Alder cycloaddition of conjugated diens to bis(trimethylsilyl)acetylene. 1,2-Bis(trimethylsilyl) cyclohexa-1,4-diene,1,2,-bis(trimethylsilyl)benzene, and their methyl derivatives, *J. Organometal. Chem.*, 289, 331, 1985.
246. **Watt, D. S. and Corey, E. J.,** A total synthesis of (±)-occidentalol, *Tetrahed. Lett.*, 4651, 1972.
247. **Ripoll, J. L.,** Applications recentes de la reaction de retro-Diels-Alder en synthese organique, *Tetrahedron*, 34, 19, 1978.
248. **Kametani, T. and Nemoto, H.,** Recent advances in the total synthesis of steroids *via* intramolecular addition reactions, *Tetrahedron*, 37, 3, 1981.
249. **Petrzilka, M. and Grayson, J. I.,** Preparation and Diels-Alder reactions of heterosubstituted 1,3-dienes, *Synthesis*, 753, 1981.
250. **Danishefsky, S.,** Siloxy dienes in total synthesis, *Acc. Chem. Res.*, 14, 400, 1981.
251. **Alston, P. V., Gordon, M. D., Ottenbrite, R. M., and Cohen, T.,** Secondary orbital interactions determining regioselectivity in the Diels-Alder reaction. 5. Thio-substituted 1,3-butadienes, *J. Org. Chem.*, 48, 5051, 1983.
252. **Overman, L. E. and Jessup, P. J.,** A short stereospecific total synthesis of *dl*-pumiliotoxin C, *Tetrahed. Lett.*, 1253, 1973.
253. **Overman, L. E. and Jessup, P. J.,** Synthetic applications of N-acyl-amino-1,3-dienes. An efficient stereospecific total synthesis of dl-pumiliotoxin C and a general entry to *cis*-decahydroquinoline alkaloids, *J. Am. Chem. Soc.*, 100, 5179, 1978.
254. **Overman, L. E. and Freerks, R. L.,** Short total synthesis of (+)-perhydrogephyrotoxin, *J. Org. Chem.*, 46, 2833, 1981.
255. **Danishefsky, S., Schuda, P. F., and Kato, K.,** Studies in the synthesis of vernolepin. A Diels-Alder approach to the angularly functionalized AB system, *J. Org. Chem.*, 41, 1081, 1976.
256. **Danishefsky, S., Schuda, P. F., and Kato, K.,** Diels-Alder reactions of o-benzoquinones. A route to derivatives of Δ^2-1-octalone, *J. Org. Chem.*, 41, 3468, 1976.
257. **Danishefsky, S., Kitahara, T., Schuda, P. F., and Etherredge, S. J.,** A remarkable epoxide opening. An expeditious synthesis of vernolepin and vernomenin, *J. Am. Chem. Soc.*, 98, 3028, 1976.

258. **Danishefsky, S., Kitahara, T., McKee, R., and Schuda, P. F.,** Reactions of silyl enol ethers and lactone enolates with dimethyl(methylene) ammonium iodide. The bis-α-methylenation of prevernolepin and prevernomenin, *J. Am. Chem. Soc.*, 98, 6715, 1976.
259. **Danishefsky, S., Schuda, P. F., Kitahara, T., and Etherredge, S. J.,** The total synthesis of *dl*-vernolepin and *dl*-vernomenin, *J. Am. Chem. Soc.*, 99, 6066, 1977.
260. **Danishefsky, S. and Khan, M.,** Regiospecificity in the Diels-Alder reactions of an enedione, *Tetrahed. Lett.*, 22, 489, 1981.
261. **Danishefsky, S., Morris, J., Mullen, G., and Gamill, R.,** Total synthesis of *dl*-tazettine and 6a-epipretazettine. A formal synthesis of *dl*-pretazettine. Some observations on the relationship of 6a-epipretazettine and tazettine, *J. Am. Chem. Soc.*, 104, 7591, 1982.
262. **Grieco, P. A., Yoshida, K., and Garner, P.,** Aqueous intramolecular Diels-Alder chemistry. Reactions of diene carboxylates with dienophiles in water at ambient temperature, *J. Org. Chem.*, 48, 3137, 1983.
263. **Garst, M. E.,** Epoxyannulation I.: New reaction path for butadienylsulfonium salts, *J. Org. Chem.*, 44, 1578, 1979.
264. **Crandall, J. K., Magaha, H. S., Widener, R. K., and Tharp, G. A.,** Intramolecular sulfur-ylide additions to ketones. A cyclopentane annulation, *Tetrahed. Lett.*, 21, 4807, 1980.
265. **Garst, M. E. and Johnson, A. T.,** Epoxyannulation II: Cyclization of ω-ketosulfonium salts, *Tetrahed. Lett.*, 21, 4811, 1980.
266. **Garst, M. E., McBride, B. J., and Johnson, A. T.,** Epoxyannulation. 4. Reactions of 1,5- , 1,6-, and 1,7-oxosulfonium-4H-imidazoles, *J. Org. Chem.*, 48, 8, 1983.
267. **Garst, M. E. and Arrhenius, P.,** Epoxyannulation. 5. Reactions of 1-butadienylsulfonium salts, *J. Org. Chem.*, 48, 16, 1983.
268. **Ireland, R. E., McGarvey, G. J., Anderson, R. C., Badoud, B., Fitzsimmons, B., and Thaisrivongs, S.,** A chiral synthesis of the left-side aldehyde for lasalocid A synthesis, *J. Am. Chem. Soc.*, 102, 6178, 1980.
269. **Hiranuma, H. and Miller, S. I.,** 1,4-Dimethoxy-1,3-butadiene as a diene donor in Diels-Alder cycloadditions, *J. Org. Chem.*, 47, 5083, 1982.
270. **Morrison, J. D. and Mosher, H. S.,** *Asymmetric Organic Reactions,* Prentice-Hall, New Jersey, 1971.
271. **ApSimon, J. and Sequin, R. P.,** Recent advances in asymmetric synthesis, *Tetrahedron,* 35, 2797, 1979.
272. **Trost, B. M., O'Krongly, D., and Belletire, J. L.,** A model for asymmetric induction in the Diels-Alder reaction, *J. Am. Chem. Soc.*, 102, 7595, 1980.
273. **Trost, B. M., Godleski, S. A., and Genêt, J. P.,** A total synthesis of racemic and optically active ibogamine. Utilization and mechanism of a new silver ion assisted palladium catalyzed cyclization, *J. Am. Chem. Soc.*, 100, 3930, 1978.
274. **Oppolzer, W., Kirth, M., Reichlin, D., Chapuis, C., Mohnhaupt, M., and Moffatt, F.,** Asymmetric induction in Diels-Alder reactions to acrylates derived from chiral sec-alcohols, *Helv. Chim. Acta,* 64, 2802, 1981.
275. **Oppolzer, W., Chapuis, C., and Kelly, M. J.,** Practical asymmetric Diels-Alder additions to camphor-10-sulfonic-acid-derived acrylates. Preliminary communication, *Helv. Chim. Acta,* 66, 2358, 1983.
276. **Choy, W., Reed, III, L. A., and Masamune, S.,** Asymmetric Diels-Alder reaction: Design of chiral dienophiles, *J. Org. Chem.*, 48, 1137, 1983.
277. **Jurczak, J., Bauer, T., Filipek, S., Tkacz, M., and Zygo, K.,** Asymmetric induction in the high-pressure cycloaddition of 2,3-o-isopropylidene-D-glyceraldehyde to 1-methoxybuta-1,3-diene, *J. Chem. Soc. Chem. Commun.*, 540, 1983.
278. **Oppolzer, W.,** Intaramoleculare [4 + 2] und [3 + 2]-cycloadditionen in der organischen synthese, *Angew. Chem.*, 16, 10, 1977.
279. **Oppolzer, W.,** Intramolecular cycloaddition reactions of *ortho*-quinodimethanes in organic synthesis, *Synthesis,* 793, 1978.
280. **Funk, R. L. and Vollhardt, K. P. C.,** Thermal, photochemical and transitionmetal mediated routes to steroids by intramolecular Diels-Alder reactions of o-xylenes (o-quinodimethanes), *Chem. Soc. Rev.*, 9, 41, 1980.
281. **Brieger, G. and Bennett, J. N.,** The intramolecular Diels-Alder reaction, *Chem. Rev.*, 80, 63, 1980.
282. **Wilson, S. R. and Mao, D. T.,** An intramolecular Diels-Alder route to eudesmane sesquiterpenes, *J. Am. Chem. Soc.*, 100, 6289, 1978.
283. **Burke, S., Powner, T. H., and Kageyama, M.,** A Diels-Alder approach to trans-fused, angularly methylated decalins, *Tetrahed. Lett.*, 24, 4529, 1983.
284. **Roush, W. R.,** Total synthesis of (±)-dendrobine, *J. Am. Chem. Soc.*, 100, 3599, 1978.
285. **Roush, W. R.,** Stereochemical aspects of the intramolecular Diels-Alder reactions of methyl (E,E,E)- and (Z,E,E)-6-alkoxy-11-methyldodeca-2,7,9-trienoate, *J. Org. Chem.*, 44, 4008, 1979.
286. **Roush, W. R.,** Total synthesis of (±)-dendrobine, *J. Am. Chem. Soc.*, 102, 1390, 1980.
287. **Roush, W. R., Ko, A. I., and Gillis, H. R.,** Stereochemical aspects of the intramolecular Diels-Alder reactions of methyl deca-2,7,9-trienoates. 1. Thermal cyclizations, *J. Org. Chem.*, 45, 4264, 1980.

288. **Roush, W. R. and Gilles, H. R.,** Stereochemical aspects of the intramolecular Diels-Alder reactions methyl deca-2,7,9-trienoates. 2. Lewis acid catalysis, *J. Org. Chem.*, 45, 4267, 1980.
289. **Roush, W. R. and Gillis, H. R.,** Improved synthesis of the perhydroindenone precursor of dendrobine, *J. Org. Chem.*, 45, 4283, 1980.
290. **Roush, W. R. and Peseckis, M. S.,** Intramolecular Diels-Alder reactions: The angularly methylated *trans*-perhydroindane ring system, *J. Am. Chem. Soc.*, 103, 6696, 1981.
291. **Roush, W. R., Gillis, H. R., and Ko, A. I.,** Stereochemical aspects of the intramolecular Diels-Alder reactions of deca-2,7,9-trienoate esters. 3. Thermal, Lewis acid catalyzed and asymmetric cyclizations, *J. Am. Chem. Soc.*, 104, 2269, 1982.
292. **Roush, W. R. and Gillis, H. R.,** Further stereochemical aspects of intramolecular Diels-Alder reactions in the undeca-2,8,10-trienoate ester series, *J. Org. Chem.*, 47, 4825, 1982.
293. **Oppolzer, W. and Flaskamp, E.,** An enantioselective synthesis and the absolute configuration of natural pumiliotoxin C, *Helv. Chim. Acta*, 60, 204, 1977.
294. **Oppolzer, W., Fehr, C., and Warneke, J.,** A new total synthesis of dl-pumiliotoxin C *via* an indanone, *Helv. Chim. Acta*, 60, 48, 1977.
295. **Oppolzer, W., Francotte, E., and Bättig, K.,** Total synthesis of (±)-lysergic acid by an intramolecular imino-Diels-Alder reaction, *Helv. Chim. Acta*, 64, 478, 1981.
296. **Oppolzer, W. and Snowden, R. L.,** A convergent synthesis of (±)-α- and β-himachalenes, *Helv. Chim. Acta*, 64, 2592, 1981.
297. **Oppolzer, W., Snowden, R. L., and Briner, P. H.,** Regioselective alkylation of the polyfunctional nucleophile 1-(methylthio)-3-triethylsilyloxypentadienyllithium, *Helv. Chim. Acta*, 64, 2022, 1981.
298. **Mach, K., Antropiusová, H., Sedmera, P., Hanuš, V., and Tureček, F.,** (6 + 2)Cycloadditions catalyzed by titanium complex, *Tetrahedron*, 40, 3295, 1984.
299. **Boeckman, R. K., Jr. and Ko, S. S.,** Stereocontrol in the intramolecular Diels-Alder reaction 1. An application to the total synthesis of (±)-marasmic acid, *J. Am. Chem. Soc.*, 102, 7146, 1980.
300. **Boeckman, R. K., Jr. and Ko, S. S.,** Stereochemical control in the intramolecular Diels-Alder reaction 2. Structural and electronic effects on reactivity and selectivity, *J. Am. Chem. Soc.*, 104, 1033, 1982.
301. **Boeckman, R. K., Jr. and Alessi, R. T.,** Stereocontrol in Diels-Alder reaction. 3. A potentially general method for the synthesis of *cis*-hydrindene by use of (Z)-diene units, *J. Am. Chem. Soc.*, 104, 3216, 1982.
302. **Boeckman, R. J., Jr. and Demko, D. M.,** Stereocontrol in Diels-Alder reaction. 4. A remarkable effect of overlap requirements in the connecting chain, *J. Am. Chem. Soc.*, 47, 1789, 1982.
303. **Boeckman, R. K., Jr. and Flann, C. J.,** Stereocontrol in the intramolecular Diels-Alder reaction VI. Use of mixed acetals as ester equivalents in the connecting chain, *Tetrahed. Lett.*, 24, 5035, 1983.
304. **Begley, M. J., Mellor, M., and Pattenden, G.,** A new approach to fused carbocycles. Intramolecular photocyclisations of 1,3-dione enol acetates, *J. Chem. Soc. Chem. Commun.*, 235, 1979.
305. **Jung, M. E. and Halweg, K. M.,** Intramolecular Lewis-acid promoted (2 + 2) cycloadditions: An efficient total synthesis of (±)-coronafacic acid via an internal Diels-Alder reaction, *Tetrahed. Lett.*, 22, 2735, 1981.
306. **Funk, R. L. and Zeller, W. E.,** Hypocholesterolemic agent compactin (ML-236 B). Total synthesis of the hexahydronaphtalene portion, *J. Org. Chem.*, 47, 180, 1982.
307. **White, J. D. and Sheldon, B. G.,** Intramolecular Diels-Alder reactions of sorbyl citraconate and mesaconate esters, *J. Org. Chem.*, 46, 2273, 1981.
308. **Taber, D. T. and Saleh, S. A.,** Control elements in the intramolecular Diels-Alder reactions. Synthesis of α-eudesmol, *Tetrahed. Lett.*, 23, 2361, 1982.
309. **Parker, K. A. and Iqbal, T.,** New approaches to the synthesis of vitamin D metabolites. 1. Stereocontrol in the intramolecular Diels-Alder reaction, *J. Org. Chem.*, 47, 337, 1982.
310. **Tietze, L. F., von Kiedriwski, G., and Berger, B.,** Stereo und regioselektive Synthese von enantiomerenreinem (+)-und (−)-hexahydrocannabinol durch intramoleculare cycloaddition, *Angew. Chem.*, 94, 222, 1982.
311. **Näf, F., Decorzant, R., and Thommen, W.,** A stereocontrolled entry to racemic eremophilane and valencane sesquiterpenes *via* an intramolecular Diels-Alder reaction, *Helv. Chim. Acta*, 65, 2212, 1982.
312. **Wilson, S. R. and Mao, D. T.,** An intramolecular Diels-Alder approach to diterpenes, *J. Org. Chem.*, 44, 3093, 1979.
313. **Schmidlin, T. and Tamm, C.,** Approaches to the total synthesis of cytochalasanes. A convergent synthesis of the octahydroindolone moiety related to proxiphomin, *Helv. Chim. Acta*, 61, 2096, 1978.
314. **Gschwend, H. W., Hillman, M. J., Kisis, B., and Rodebaugh, R. K.,** Intramolecular Diels-Alder reactions. Synthesis of 3a-phenylisoindolines as analgetic templates, *J. Org. Chem.*, 41, 104, 1976.
315. **Boeckman, R. K., Jr., Napier, J. J., Thomas, E. W., and Sato, R. I.,** Stereocontrol in the intramolecular Diels-Alder reaction. 5. Preparation of a tetracyclic intermediate for ikarugamycin, *J. Org. Chem.*, 48, 4153, 1983.
316. **Kurth, M. J., Burns, D. H., and O'Brien, M. J.,** Ikarugamycin: Total synthesis of the decahydro-as-indacene portion, *J. Org. Chem.*, 49, 73, 1984.

317. **Fallis, A. G.,** The intramolecular Diels-Alder reaction: Recent advances and synthesis applications, *Can. J. Chem.*, 62, 183, 1984.
318. **Wenkert, F. and Naemura, K.,** Synthesis of the himachalenes by an intramolecular Diels-Alder reaction route, *Synth. Commun.*, 3, 45, 1973.
319. **Joseph, T. C. and Dev, S.,** Studies in sesquiterpenes-XXXI. The absolute stereochemistry of himachalenes, *Tetrahedron,* 24, 3841, 1968.
320. **Oppolzer, W. and Keller, K.,** The total synthesis of dl-chelidonine, *J. Am. Chem. Soc.*, 93, 3836, 1971.
321. **Oppolzer, W., Bättig, K., and Petrzilka, M.,** The enantioselective synthesis of 3-methoxy-1,3,5(10)-estratriene-11,17-dione by an intramolecular cycloaddition reaction, *Helv. Chim. Acta,* 61, 1945, 1978.
322. **Nicolaou, K. C., Barnette, W. E., and Ma, P.,** A remarkably simple, highly stereoselective synthesis of steroids and other polycyclic systems. Total synthesis of estra-1,3,5(10)-trien-17-one via intramolecular capture of o-quinodimethanes generated by cheletropic elimination of SO_2, *J. Org. Chem.*, 45, 1463, 1980.
323. **Kametani, T., Matsumoto, H., Nemoto, H., and Fukumoto, K.,** Asymmetric total synthesis of estradiol by an intramolecular cycloaddition of benzocyclobutene derivative, *J. Am. Chem. Soc.*, 100, 6218, 1978.
324. **Grieco, P. A., Takigawa, T., and Schillinger, W. J.,** Bicyclo[2.2.1]heptanes as intermediates in the synthesis of steroids. Total synthesis of estrone, *J. Org. Chem.*, 45, 2247, 1980.
325. **Djuric, S., Sakar, T., and Magnus, P.,** Silicon in synthesis: An exceptionally short synthesis of dl-11α-hydroxyestrone methyl ether, *J. Am. Chem. Soc.*, 102, 6885, 1980.
326. **Ito, Y., Nakajo, E., Nakatsuka, M., and Saegusa, T.,** A new approach to gephyrotoxin, *Tetrahed. Lett.*, 24, 2881, 1983.
327. **Wu. T.-C., Mareda, J., Gupta, Y. N., and Houk, K. N.,** Selective intramolecular (6 + 4) cycloadditions of aminodienylfulvenes, *J. Am. Chem. Soc.*, 105, 6996, 1983.
328. **Liu, C.-Y., Mareda, J., and Houk, N. K.,** Intramolecular (8 + 2) cycloadditions of alkenylheptafulvenes, *J. Am. Chem. Soc.*, 105, 6714, 1983.
329. **Gupta, Y. N., Doa, M. J., and Houk, K. N.,** Intramolecular (6 + 4) cycloadditions: intramolecular control of periselectivity, *J. Am. Chem. Soc.*, 104, 7336, 1982.
330. **Bloomfield, J. J., Owsley, D. C., and Nelke, J. M.,** The acyloin condensation, *Org. Reactions,* 23, 259, 1976.
331. **Bloomfield, J. J. and Fennessey, P. V.,** Dieckmann and Thorpe reactions in dimethyl sulfoxide, *Tetrahed. Lett.*, 2275, 1964.
332. **Coates, P. M. and Chung, S. K.,** Stereochemistry in the solvolytic ring contraction of 2,2,4aα-trimethyl-1-decalyl methanesulfonate. A model reaction pertaining to triterpene biogenesis, *J. Org. Chem.*, 38, 3677, 1973.
333. **Hückel, W. and Friedrich, H.,** Zur Stereochemie bicyclischer Ringsysteme III. Die stereoisomerie des Hydrindans und seiner Derivate I, *Justus Liebigs Ann. Chem.*, 451, 132, 1927.
334. **Finley, K. T.,** The acyloin condensation as a cyclization method, *Chem. Rev.*, 64, 573, 1964.
335. **Tureček, F. and Vystrčil, A.,** The syntheses of 8-alkyl derivatives of bicyclo[4.3.0]-3-nonenes, *Collect. Czech. Chem. Commun.*, 41, 1571, 1976.
336. **Grieco, P. A. and Ohfune, Y.,** An improved route to a key hydroazulenone intermediate for helenanolide synthesis, *J. Org. Chem.*, 45, 2251, 1981.
337. **Ohfune, Y., Grieco, P. A., Wang, C.-L., and Majetich, G.,** Stereospecific total synthesis of d,l-helenalin. A general route to helenanolides and ambrosanolides, *J. Am. Chem. Soc.*, 100, 5946, 1978.
338. **Grieco, P. A., Ohfune, G., and Majetich, G.,** Pseudoguaianolides. Stereospecific total synthesis of (±)-ambrosin, (±)-damsin and (±)-psilostachyin C, *J. Am. Chem. Soc.*, 99, 7393, 1977.
339. **Ziegler, F. E. and Faug, J.-M.,** Regiochemical control in the conjugate addition of dithianylidene anions. Total syntheses of (±)-aromatin and (±)-confertin, *J. Org. Chem.*, 46, 825, 1981.
340. **Stevens, K. E. and Paquette, L. A.,** Stereocontrolled total synthesis of (+)-$\Delta^{9(12)}$-capnellene, *Tetrahed. Lett.*, 4393, 1981.
341. **McMurry, J. E., Fleming, M. P., Kees, K. L., and Krepski, L. R.,** Synthesis of cycloalkanes by intramolecular titanium-induced dicarbonyl coupling, *J. Org. Chem.*, 42, 2655, 1977.
342. **McMurry, J. E. and Miller, D. D.,** Synthesis of isocaryophyllene by titanium-induced keto ester cyclization, *Tetrahed. Lett.*, 24, 1885, 1983.
343. **Munslow, W. D. and Reusch, W.,** Novel synthesis of 5,5-dimethyl-1-octalin derivatives, *J. Org. Chem.*, 47, 5096, 1982.
344. **Semmelhack, M. F., Yamashita, A., Tomesch, J. C., and Hirotsu, K.,** Total synthesis of confertin via metal-promoted cyclization-lactonization, *J. Am. Chem. Soc.*, 100, 5565, 1978.
345. **Wender, P. A., Eissenstadt, M. A., and Filosa, M. P.,** A general methodology for pseudoguaiane synthesis. Total synthesis of (±)-damsinic acid and (±)-confertin, *J. Am. Chem. Soc.*, 101, 2196, 1979.
346. **Lansbury, P. and Serelis, A. K.,** A facile entry to pseudoguanianolides. Total synthesis of damsinic acid, *Tetrahed. Lett.*, 1909, 1978.
347. **Lansbury, P. T., Wovkulich, P. M., and Gallagher, P. E.,** A direct synthesis of hydroazulenes, *Tetrahed. Lett.*, 65, 1973.

348. **House, H. O., Sayer, T. S. B., and Yau, C. C.,** Chemistry of carbanions. 32. Formation of the perhydroazulene systems by intramolecular alkylation, *J. Org. Chem.*, 43, 2153, 1978.
349. **Johnson, F., Duquette, L. G., Parker, W. L., and Nasutavicus, W. A.,** Cyclization of δ- and γ-alkenenitriles by triethyloxonium fluoroborate, *J. Org. Chem.*, 39, 1434, 1974.
350. **Eliel, E. L. and Pillar, C.,** The conformation of a six-membered ring *cis*-1,2 fused to a five membered ring, *J. Am. Chem. Soc.*, 77, 3600, 1955.
351. **Owen, L. N. and Smith, P. N.,** Alicyclic glycols. VI. Derivatives of cyclopentane-1:2-diol, *J. Chem. Soc.*, 4026, 1952.
352. **Merour, J. Y., Roustan, J. L., Charrier, C., Collin, J., and Benaim, J.,** Formation of cyclic ketones starting from the iron tetracarbonyl dianion, *J. Organometal. Chem.*, 51, C24, 1973.
353. **Tsuji, J.,** *Organic Synthesis by Means of Transition Metal Complexes*, Springer-Verlag, Berlin, 1975.
354. **Redmore, D., Gutsche, C. D.,** *Adv. Alicyclic Chem.*, 3, 1, 1971.
355. **Achmad, S. A. and Cavill, G. W. K.,** Insect venoms, attractants and repellents VII. A stereospecific synthesis of iridodial, *Aust. J. Chem.*, 18, 1989, 1965.
356. **Kurek, A., Kohout, L., Fajkoš, J., and Šorm, F.,** On Steroids CXLVI. B-Homosteroids VII. Favorskii rearrangement of some B-homosteroid bromo ketones, *Collect. Czech. Chem. Commun.*, 38, 279, 1973.
357. **Irwin, A. J. and Jones, J. B.,** Stereospecific Thalium (III) nitrate mediated conversion of bicyclo[3.2.1]-2-octanone to exo-2-norbornanecarboxylic acid methyl ester, *J. Org. Chem.*, 42, 2176, 1977.
358. **Fieser, M. and Fieser, L. F.,** *Reagents for Organic Synthesis*, Vol. 4, J. Wiley & Sons, New York, 492, 1974.
359. **Casadevall, E. and Pouet, Y.,** Contraction de cycle a partir de la tosyloxy-8 bicyclo[4.2.0]octanone-7 *cis*, *Tetrahedron*, 31, 757, 1975.
360. **Miura, H., Fujimoto, Y., and Tatsuno, T.,** Modification of α-santonin. V. A novel lead(IV) acetate promoted Favorskii-type rearrangement, *Synthesis*, 898, 1979.
361. **Wiberg, K. B., Olli, L. K., Golembeski, N., and Adams, R. D.,** Tricyclo[4.2.0.0^{1,4}]octane, *J. Am. Chem. Soc.*, 102, 7467, 1980.
362. **Eliel, E. L.,** *Stereochemistry of Carbon Compounds*, McGraw-Hill, New York, 1962.
363. **Eliel, E. L., Allinger, N. L., Angyal, S. J., and Morrison, G. A.,** *Conformational Analysis*, Wiley-Interscience, New York, 1965.
364. **Paukstelis, J. V. and Macharia, B. W.,** Ring contraction of bicyclo[2.2.1]heptanes, *J. Org. Chem.*, 38, 646, 1973.
365. **Woodward, R. B., Gosteli, J., Ernest, I., Friary, R. J., Nestler, G., Raman, H., Sitrin, R., Suter, Ch., and Whitesell, J. K.,** A novel synthesis of prostaglandin $F_{2\alpha}$, *J. Am. Chem. Soc.*, 95, 6855, 1973.
366. **Hanson, J. R., Hitchcock, P. B., and Wadsworth, H. J.,** Base cytalysed epoxidation and reduction of 3-formyl-A-norandrost-3(5)-en-17-one, *J. Chem. Soc. Perkin Trans.*, 1, 3025, 1981.
367. **Hikino, H., Suzuki, N., and Takemoto, T.,** Synthesis of cyperolone, *Chem. Pharm. Bull.*, 14, 1441, 1966.
368. **Saken, T., Isoe, S., Hyeon, S. B., Katsumura, R., Maeda, T., Wolinsky, J., Dieckerson, D., Slabaugh, M., and Nelson, D.,** The exact nature of metabilactone and the terpenes of *Nepeta cataria*, *Tetrahed. Lett.*, 4097, 1965.
369. **Wolinsky, J. and Baker, W.,** The synthesis of 1-acetyl-4-isopropenyl-1-cyclopentene, *J. Am. Chem. Soc.*, 82, 636, 1960.
370. **Brown, J. B., Henbest, H. B., Jones, E. R. H.,** Studies on compounds related to auxin-a and auxin-b. Part III. The preparation of the cyclopentene analogue to auxin-b lactone, *J. Chem. Soc.*, 3634, 1950.
371. **Corey, E. J. and Nozoe, S.,** The total synthesis of helminthosporal, *J. Am. Chem. Soc.*, 87, 5728, 1965.
372. **Corey, E. J., Andersen, N. H., Carlson, R. M., Paust, J., Vedejs, E., Vlattas, I., and Winter, R. E. K.,** Total synthesis of prostaglandins. Synthesis of the pure dl-E_1, -$F_{1\alpha}$, -$F_{1\beta}$, -A_1 and B_1 hormones, *J. Am. Chem. Soc.*, 90, 3245, 1968.
373. **Kende, A. S., Bentley, T. J., Mader, R. A., and Ridge, D.,** A simple total synthesis of (±)-dendrobine, *J. Am. Chem. Soc.*, 96, 4332, 1974.
374. **Šorm, F. and Dyková, H.,** On B-norcholesterol, *Collect. Czech. Chem. Commun.*, 13, 407, 1948.
375. **Dauben, W. G. and Fonken, G. J.,** Reactions of B-norcholesterol, *J. Am. Chem. Soc.*, 78, 4736, 1956.
376. **Gutsche, C. D. and Redmore, D.,** *Carbocyclic Ring Expansion*, Academic Press, New York, 1968.
377. **Gehlhaus, J., Černý, V., and Šorm, F.,** Über Steroide CXXXVII. Über Ringerweiterungsreaktionen von 3α,5-cyclo-5α-cholestan-6-on mit diazomethan, *Collect. Czech. Chem. Commun.*, 37, 1331, 1972.
378. **Velgová, H., Synáčková, M., and Černý, V.,** Preparation of some 3α,5-cycloandrostane derivatives with expanded ring B, *Collect. Czech. Chem. Commun.*, 44, 260, 1979.
379. **Greene, A. E.,** Iterative three-carbon annelations. Synthesis of (±)-hirsutene, *Tetrahed. Lett.*, 21, 3059, 1980.
380. **Dave, V. and Warnhoff, E. W.,** Regiospecific homologation of unsymmetrical ketones, *J. Org. Chem.*, 48, 2590, 1983.

381. **Coates, R. M. and Sowerby, R. L.,** Stereoselective total synthesis of (±)-zizaene, *J. Am. Chem. Soc.*, 94, 5386, 1972.
382. **Knapp, S., Trope, A. F., Theodore, M. S., Hirata, N., and Barchi, J. J.,** Ring expansion of ketones to 1,2-keto thioketals. Control of bond migrations, *J. Org. Chem.*, 49, 608, 1984.
383. **Kohout, L., Fajkoš, J., and Šorm, F.,** On steroids CIV. 6,7-Substituted B-homosteroids; stereochemistry and proof of configuration, *Collect. Czech. Chem. Commun.*, 32, 1210, 1967.
384. **Taguchi, H., Yamamoto, H., and Nozaki, H.,** A practical synthesis of polyhalomethyllithium carbonyl adducts, *J. Am. Chem. Soc.*, 96, 3010, 1974.
385. **Taguchi, H., Yamamoto, H., and Nozaki, H.,** β-Oxido carbenoids as synthetic intermediates. A facile ring enlargement reaction, *J. Am. Chem. Soc.*, 96, 6510, 1974.
386. **Sisti, A. J., Rusch, G. M., and Sukhon, H. K.,** Spontaneous ring enlargement during the free-radical bromination of 2-benzyl-1,3,3-trimethyl- and 2-benzyl-3,3-dimethylbicyclo[2.2.1]heptanol-2, *J. Org. Chem.*, 36, 2030, 1971.
387. **Sisti, A. J. and Meyers, M.,** A new ring expansion reaction. V. The decomposition of the magnesium salt of various 1-(1-bromo-1-methyl)-1-cycloalkanols. Electrophilic addition to ispropylidenecycloalkanes, *J. Org. Chem.*, 38, 4431, 1973.
388. **Woodward, R. B.,** Recent advances in the chemistry of natural products, *Pure Appl. Chem.*, 17, 519, 1968.
389. **Roberts, M. R. and Schlessinger, R. H.,** Total synthesis of *dl*-helenalin, *J. Am. Chem. Soc.*, 101, 7626, 1979.
390. **Quallich, G. J. and Schlessinger, R. H.,** Total synthesis of *dl*-confertin and *dl*-damsin, *J. Am. Chem. Soc.*, 101, 7627, 1979.
391. **Hudlický, T., Kutchan, T. M., Wilson, S. R., and Mao, D. T.,** Total synthesis of (±)-hirsutene, *J. Am. Chem. Soc.*, 102, 6351, 1980.
392. **Piers, E. and Banville, J.,** Five-membered ring annelation *via* thermal rearrangement of β-cyclopropyl-α,β-unsaturated ketones: A new total synthesis of (+)-zizaene, *J. Chem. Soc. Chem. Commun.*, 1138, 1979.
393. **Hudlický, T. and Short, R. P.,** Terpenic acid by cyclopentane annulation of exocyclic dienes. Synthesis of triquinane portion of retigeranic acid, *J. Org. Chem.*, 47, 1522, 1982.
394. **Danheiser, R. L., Davila, C. M., Auchus, R. J., and Kadonaga, J. T.,** A stereoselective synthesis of cyclopentene derivatives from 1,3-dienes, *J. Am. Chem. Soc.*, 103, 2443, 1981.
395. **Mazur, Y. and Nussim, N.,** Synthesis of perhydroazulenes, *J. Am. Chem. Soc.*, 83, 3911, 1961.
396. **Heathcock, C. H. and Ratcliffe, R.,** A stereoselective total synthesis of the guaiazulenic sesquiterpenoids α-bulnesene and bulnesol, *J. Am. Chem. Soc.*, 93, 1746, 1971.
397. **Heathcock, C. H., DelMar, E. G., and Graham, S. L.,** Synthesis of sesquiterpene antitumor lactones. 9. The hydronaphtalene route to pseudoguaianes. Total synthesis of (±)-confertin, *J. Am. Chem. Soc.*, 104, 1907, 1982.
398. **Büchi, G., Hofheinz, W., and Paukstelis, J. V.,** The total synthesis of (−)-aromadendrene, *J. Am. Chem. Soc.*, 88, 4113, 1966.
399. **Kirk, D. N. and Hartshorn, M. P.,** *Steroid Reaction Mechanisms*, Elsevier, Amsterdam, 1968.
400. **Tatsuta, K., Akimoto, K., and Kinoshita, M.,** A new, stereocontrolled synthesis of *cis,anti,cis* -tricyclo [6.3.0.0^{2,6}]undecanes. Total synthesis of (±)-hirsutene, *J. Am. Chem. Soc.*, 101, 6116, 1979.
401. **Nozoe, S., Furukawa, J., Sankawa, J., and Shibata, S.,** Isolation, structure and synthesis of hirsutene, a precursor hydrocarbon of coriolin biosynthesis, *Tetrahed. Lett.*, 195, 1976.
402. **Suchý, M., Herout, V., and Šorm, F.,** On terpenes CLXV. Proof of structure of guaianolides artabsin and arborescin, *Collect. Czech. Chem. Commun.*, 29, 1829, 1964.
403. **Kretchmer, R. A. and Thompson, W. S.,** The total synthesis of (±)-damsin, *J. Am. Chem. Soc.*, 98, 3379, 1976.
404. **Meth-Cohn, O., Reason, A. J., and Roberts, S. M.,** Carbocyclic analogues of penicillin, *J. Chem. Soc. Chem. Commun.*, 90, 1982.
405. **Marshall, J. A. and Partridge, J. J.,** A stereoselective hydroazulene synthesis, *Tetrahed. Lett.*, 2545, 1966.
406. **Marshall, J. A. and Partridge, J. J.,** The total synthesis of racemic bulnesol, *J. Am. Chem. Soc.*, 90, 1090, 1968.
407. **Marshall, J. A. and Partridge, J. J.,** The total synthesis of (±)-bulnesol and related studies, *Tetrahedron*, 25, 2195, 1969.
408. **Froborg, J., Magnusson, G., and Thorén, S.,** Synthesis of the [4.3.1] decan-10-one system by cycloalkylation of specific cyclohexanone enolates with reactive 1,4-dichlorides, *J. Org. Chem.*, 39, 848, 1974.
409. **Froborg, J., Magnusson, G., Thorén, S.,** Synthesis of the bicyclo[4.3.1]decan-10-one system by cycloalkylation of specific cyclohexanone enolates with reactive 1,4-dichlorides, *J. Org. Chem.*, 39, 848, 1974.

410. **Ohfune, Y., Shirahama, H., and Matsumoto, T.,** Biogenetic-like synthesis of d,l-hirsutene, *Tetrahed. Lett.*, 2795, 1976.
411. **Liu, H. J. and Lee, S. P.,** Total synthesis of 5-epikessane and dehydrokessane, *Tetrahed. Lett.*, 3699, 1977.
412. **Wender, P. A. and Lechleiter, J. C.,** A photochemically mediated [4C + 2C] annelation. Synthesis of (±)-10-epijunenol, *J. Am. Chem. Soc.*, 100, 4321, 1978.
413. **Kueh, J. S. H., Mellor, M., and Pattenden, G.,** Photocyclisations of dicyclopent-1-enyl methanes to tricyclo[6.3.0.$0^{2,6}$] undecanes. A synthesis of the hirsutane carbon skeleton, *J. Chem. Soc. Chem. Commun.*, 5, 1978.
414. **Heathcock, C. A., Tice, C. M., and Germroth, T. C.,** Synthesis of sesquiterpene antitumor lactones. 10. Total synthesis of (±)-parthenin, *J. Am. Chem. Soc.*, 104, 6081, 1982.
415. **Ito, Y., Fujii, S., and Saegusta, T.,** Reaction of 1-silyloxybicyclo[n.1.0]alkanes with $FeCl_3$. A facile synthesis of 2-cycloalkenones via ring enlargement of cyclic ketones, *J. Org. Chem.*, 41, 2073, 1976.
416. **Baldein, J. E. and Broline, B. M.,** Relative configuration of the chiral 2,7- and 3,7-dimethyl-7-(methoxymethyl) cyclohepta-1,3,5-trienes, *J. Org. Chem.*, 47, 1385, 1982.
417. **Chenier, P. J.,** An improved synthesis of bicyclo[4.2.1] nonan-2-one, *J. Org. Chem.*, 42, 2643, 1977.
418. **Devreese, A. A., Demuynck, M., De Clercq, P. J., and Vandewalle, M.,** Guaianolides I. Perhydroazulenic lactones as intermediates for total synthesis, *Tetrahedron*, 39, 3039, 1983.
418a. **Devreese, A. A., Demuynck, M., De Clercq, P. J., and Vandewalle, M.,** Guaianolides II. Total synthesis of (±)-compressanolide and (±)-estafiatin, *Tetrahedron*, 39, 3049, 1983.
419. **Johnson, A. P. and Rahman, M.,** A versatile synthesis of angularly substituted *cis*-2-methoxy-$\Delta^{1,6}$-hexalins, *Tetrahed. Lett.*, 359, 1974.
420. **Adames, G., Grigg, R., and Grover, J. N.,** A general method for the preparation of *cis*-annelated rings, *Tetrahed. Lett.*, 363, 1974.
421. **Evans, D. A. and Golob, A. M.,** [3,3] Sigmatropic rearrangements of 1,5-diene alkoxides. The powerful accelerating effects of the alkoxide substituent, *J. Am. Chem. Soc.*, 97, 4765, 1975.
422. **Rhoads, S. J. and Raulins, N. R.,** The Claisen and Cope rearrangements, *Org. React.*, 22, 1, 1975.
423. **Ziegler, F. E.,** Stereo- and regiochemistry of the Claisen rearrangement. Applications to natural product synthesis, *Acc. Chem. Res.*, 10, 227, 1977.
424. **Bennett, G. B.,** The Claisen rearrangement in organic synthesis, *Synthesis*, 589, 1977.
425. **Jung, M. E. and Hudspeth, J. P.,** Total synthesis of (±)-coronafacic acid. Use of anionic oxy-Cope rearrangements on aromatic substrates in synthesis, *J. Am. Chem. Soc.*, 102, 2463, 1980.
426. **Datsur, K. P.,** A stereoselective approach to eremophilane sesquiterpenes. A synthesis of (±)-nootkatone and (±)-α-vetivone, *J. Am. Chem. Soc.*, 96, 2605, 1974.
427. **Ziegler, F. E. and Piwinski, J. J.,** Tandem Cope-Claisen rearrangement. Scope and stereochemistry, *J. Am. Chem. Soc.*, 104, 7181, 1982.
428. **Jung, M. E. and Hatfield, G. L.,** Synthesis of syn-7-benzyloxy-4-methylbicyclo[2.2.1]hept-5-en-2-one, an intermediate for the synthesis of steroids and tricothecanes; tandem anionic [1,3]-[3,3] sigmatropic rearrangement, *Tetrahed. Lett.*, 24, 2931, 1983.
429. **Furuichi, K. and Miwa, T.,** Total synthesis of (±)-forsythide aglucone dimethyl ester, *Tetrahed. Lett.*, 3689, 1974.
430. **Ziegler, F. E. and Lim, H.,** A synthesis of (±)-estrone methyl ether via the tandem Cope-Claisen rearrangement, *J. Org. Chem.*, 47, 5229, 1982.
431. **Inubishi, Y., Kikuchi, T., Ibuka, T., Tanaka, K., Saji, I., and Tokane, K.,** Total synthesis of the alkaloid (±)-dendrobine, *Chem. Pharm. Bull.*, 22, 349, 1974.
432. **Thompson, H. W. and Shah, N. V.,** Stereochemical control of reduction. 7. Reagent hinges: cis Reduction of β-octalones by internal delivery of chromium II, *J. Org. Chem.*, 48, 1325, 1983.
433. **Greene, A. E. and Deprès, J.-P.,** A versatile three-carbon annelation. Synthesis of cyclopentanone and cyclopentanone derivatives from olefins, *J. Am. Chem. Soc.*, 101, 4003, 1979.
434. **Cooke, E., Paradellis, T. C., and Edward, J. T.,** Stereochemical studies, Part 10. Synthesis of some epimeric δ-lactones, *Can. J. Chem.*, 60, 29, 1982.
435. **Bull, J. R., Jones, E. R. H., and Meakins, G. D.,** Nitrosteroids. Part II. A new route to nitro-steroids, *J. Chem. Soc.*, 2601, 1965.
436. **Kočovský, P.,** Synthesis of 3,4-, 4,5- and 5,6-unsaturated 19-substituted cholestane derivatives and related epoxides, *Collect. Czech. Chem. Commun.*, 45, 3008, 1980.
437. **Hora, J., Lábler, L., Kasal, A., Černý, V., Šorm, F., and Sláma, K.,** Molting deficiencies produced by some sterol derivatives in an insect (*Pyrrhocoris apterus* L.), *Steroids*, 8, 887, 1966.
438. **Fried, J. and Edwards, J. A.,** *Organic Reactions in Steroid Chemistry*, Van Nostrand, New York, 1972.
439. **Caine, D.,** Reduction and related reactions of α,β-unsaturated carbonyl compounds with metals in liquid ammonia, *Organic Reactions*, 23, 1, 1979.
440. **Marshall, J. A.,** Progress in hydroazulene synthesis, *Synthesis*, 517, 1972.

441. **De Clerq, P. and Vandewalle, M.,** Total synthesis of (±)-damsin, *J. Org. Chem.*, 42, 3447, 1977.
442. **Marshall, J. A., Bundy, G. L., and Fanta, W. I.,** Studies relating to the syntheses of (+)-valeranone, *J. Org. Chem.*, 33, 3913, 1968.
443. **Roush, W. R. and Pesackis, M.,** Intramolecular Diels-Alder reactions: the angular methylated trans-perhydroindan, *J. Am. Chem. Soc.*, 103, 6696, 1981.
444. **Jung, M. E. and Halweg, K. M.,** Simple stereoselective synthesis of *trans* 7a-methylhydrind-4-en-1-one, a key intermediate in steroid total synthesis, *Tetrahed. Lett.*, 22, 3929, 1981.
445. **Bal, S. A. and Helquist, P.,** Synthesis of steroid precursors. Intramolecular Diels-Alder approach to the *trans*-hydrindenone system, *Tetrahed. Lett.*, 22, 3933, 1981.
446. **Bajorek, J. J. S. and Sutherland, J. K.,** A *cis*-perhydroindanone synthesis utilising an intramolecular Diels-Alder reaction, *J. Chem. Soc. Perkin Trans.*, 1, 1559, 1975.
447. **Bernasconi, S., Gariboldi, P., Jommi, G., Montanari, S., and Sisti, M.,** Total synthesis of pinguisone, *J. Chem. Soc. Perkin Trans.*, 1, 2394, 1981.
448. **Bernasconi, S., Ferrari, M., Gariboldi, P., Jommi, G., and Sisti, M.,** Synthetic study of pinguisane terpenoids, *J. Chem. Soc. Perkin Trans.*, 1, 1994, 1981.
449. **Nagata, W. and Yoshioka, M.,** Hydrocyanation of conjugated carbonyl compounds, *Org. Reactions*, 25, 255, 1977.
450. **Hosomi, A. and Sakurai, H.,** Conjugate addition of allysilanes to enones. A new method of stereoselective introduction of the angular allyl group in fused cyclic α,β-enones, *J. Am. Chem. Soc.*, 99, 1673, 1977.
451. **Dawson, D. J. and Ireland, R. E.,** A method for the stereoselective introduction of angular methyl group, *Tetrahed. Lett.*, 1899, 1968.
452. **Still, W. C. and Schneider, M. J.,** A convergent route to α-substituted acrylic esters and application to the total synthesis of (±)-frullanolide, *J. Am. Chem. Soc.*, 99, 948, 1977.
453. **Kočovŝký, P. and Černý, V.,** Structural requirements in Westphalen rearrangement, *Collect. Czech. Chem. Commun.*, 42, 2415, 1977.
454. **Kočovský, P., Černý, V., and Tureček, F.,** Westphalen rearrangement. Mechanism of formation of 5α-acetoxy derivatives, *Collect. Czech. Chem. Commun.*, 44, 234, 1979.
455. **Mihina, J. S.,** Dehydration of steroid 5,6-halohydrins, *J. Org. Chem.*, 27, 2807, 1962.
456. **Kočovský, P., Drašar, P., Pouzar, V., and Havel, M.,** Synthesis of some Westphalen-type cardenolide analogs, *Collect. Czech. Chem. Commun.*, 47, 108, 1982.
457. **Pouzar, V., Drašar, P., Kočovský, P., and Havel, M.,** Synthesis of some isocardenolide analogs and 17β-maleimidoandrostane of Westphalen structural type, *Collect. Czech. Chem. Commun.*, 47, 96, 1982.
458. **Kočovský, P.,** Backbone rearrangements of steroids and related systems, *Chem. Listy*, 73, 583, 1979 (in Czech).
459. **Kočovský, P.,** Synthesis of 14-deoxy-14α-strophanthidin, *Collect. Czech. Chem. Commun.*, 45, 2998, 1980.
460. **Kočovský, P., and Černý, V.,** Synthesis of 6β-19-dimethoxy-3α,5-cyclo-5α-pregnan-20-one, *Collect. Czech. Chem. Commun.*, 44, 2275, 1979.

461. **Lee, T. V. and Richardson, K. A.,** Novel annulation reactions: The direct synthesis of cyclopentanes using a Lewis acid activated bisfunctional annulating reagent, *Tetrahed. Lett.*, 26, 3629, 1985.
462. **Magnus, P. and Quagliato D.,** Silicon in synthesis. 21. Reagents fore thiophenyl-functionalized cyclopentenone annulations and the total synthesis of (±)-hirsutene, *J. Org. Chem.*, 50, 1621, 1985.
463. **Danheiser, R. L., Bronson, J. J., and Okano, K.,** Carbanion accelerated vinylcyclopropane rearrangement. Application in a general, stereocontrolled annulation approach to cyclopentene derivatives, *J. Am. Chem. Soc.*, 107, 4579, 1985.
464. **Kozikowski, A. P.,** The isoxazoline route to the molecules of nature, *Accounts Chem. Res.*, 17, 410, 1984.
465. **Funk, R. L. and Bolton, G. L.,** The nitrone route to linearly fused tricyclopentanoids. Another synthesis of hirsutene, *J. Org. Chem.*, 49, 5021, 1984.
466. **Curran, D. P. and Jacobson, P. B.,** Combination Claisen-nitrile oxide annulation. A strategy for ring construction with rigid stereocontrol dictated by an allylic hydroxyl group, *Tetrahed. Lett.*, 26, 2031, 1985.
467. **Denmark, S. E., Dappen, M. S., and Sternberg, J. A.,** Intromolecular [4 + 2] cycloadditions of nitrosoalkenes with olefins, *J. Org. Chem.*, 49, 4741, 1984.
468. **Pariza, R. J. and Fuchs, P. L.,** Studies related to the Robinson transposition reaction, *J. Org. Chem.*, 50, 4252, 1985.
469. **Huffman, J. W., Potnis, S. M., and Satish, A. V.,** A silyl enol ether variation of the Robinson annulation, *J. Org. Chem.*, 50, 4266, 1985.
470. **Richter, F. and Otto, H.-H.,** Synthesis of substituted decalones by Diels-Alder reaction or by sequential Michael reaction—which one is more selective, *Tetrahed. Lett.*, 26, 4351, 1985.
471. **Danishefsky, S., Chacklamannil, S., Harrison, P., Silvestri, M., and Cole, P.,** Synthetic studies toward aflavinine: a synthesis of 3-desmethylaflavinine via a (2 + 2 + 2) annulation, *J. Am. Chem. Soc.*, 107, 2474, 1985.

472. **Posner, G. H. and Lu, S.-B.,** An extremely efficient method for one-pot three-component 2 + 2 + 2 construction of functionalized cyclohexenes, *J. Am. Chem. Soc.*, 107, 1424, 1985.
473. **Meyer, W. L., Brannon, M. J., da C. Burgos, C., Goodwin, T. E., and Howard, R. W.,** Annulation of α-formyl α,β-unsaturated ketones by a Michael addition-cyclization sequence. A versatile synthesis of alicyclic six-membered rings, *J. Org. Chem.*, 50, 438, 1985.
474. **Begley, M. J., Jackson, C. B., and Pattenden, G.,** Investigation of transannular cyclisation of verticillanes to the taxane ring system, *Tetrahed. Lett.*, 26, 3396, 1985.
475. **McMurry, J. E. and Erion, M. D.,** Stereoselective total synthesis of the complement inhibitor K-76, *J. Am. Chem. Soc.*, 107, 2712, 1985.
476. **Curran, D. P. and Rakiewicz, D. M.,** Tandem radical approach to linear condensed cyclopentanoids. Total synthesis of (±)-hirsutene, *J. Am. Chem. Soc.*, 107, 1448, 1985.
477. **Curran D. P., and Chen, M.-H.,** Radical initiated polyolefinic cyclizations in condensed cyclopentanoid synthesis. Total synthesis of (±)-$\Delta^{9(12)}$-capnellene, *Tetrahed. Lett.*, 26, 4991, 1985.
478. **Snider, B. B. Mohan, R., and Kates, S. A.,** Manganese(III)-based oxidative free radical cyclization. Synthesis of (±)-podocarpic acid, *J. Org. Chem.*, 50, 3659, 1985.
479. **Gise, B.,** The addition of radicals to alkenes, *Angew. Chem. Int. Ed. Engl.*, 24, 553, 1985.
480. **Andersen, N. H., Hadley, S. W., Kelly, J. D., and Bacon, E. R.,** Intramolecular olefinic aldehyde Prins reactions for the construction of five-membered rings, *J. Org. Chem.*, 50, 4144, 1985.
481. **Khand, I. U., Knox, G. R., Pauson, P. L., Watts, W. E., and Foreman, M. I.,** Organocobalt complexes. Part II. Reaction of acetylenehexacarbonyldicobalt complexes, $(R^1c_2R^2)Co(CO)_6$, with norbornene and its derivatives, *J. Chem. Soc. Perkin Trans.*, 1, 977, 1973.
482. **Schore, N. E. and Groudace, M. C.,** Preparation of bicyclo[3.3.0]oct-1-en-3-one and bycyclo[4.3.0]non-(9)-en-8-one via intramolecular cyclization of α,ω-enynes, *J. Org. Chem.*, 46, 5436, 1981.
483. **Knudsen, M. J. and Schore, N. E.,** Synthesis of the angularly fused triquinane skeleton via intramolecular organometallic cyclization, *J. Org. Chem.*, 49, 5025, 1984.
484. **Billington, D. C. and Willison, D.,** A simple organocobalt mediated synthesis of substituted 3-oxabicyclo[3.3.0] oct-6-en-7-ones, *Tetrahed. Lett.*, 25, 4041, 1984.
485. **Trost, B. M. and Lautens, M.,** Cyclization via isomerization: a palladium(2+)-catalyzed carbocyclization of 1,6-enynes to 1,3- and 1,4-dienes, *J. Am. Chem. Soc.*, 107, 1781, 1985.
486. **Trost, B. M. And Chung, J. Y. L.,** An unusual substituent effect of a palladium mediated cyclization: A total synthesis of (±)-streptolide, *J. Am. Chem. Soc.*, 107, 4586, 1985.
487. **Oppolzer, W.,** Asymmetric Diels-Alder and ene reactions in organic synthesis, *Angew. Chem. Int. Ed. Engl.*, 23, 876, 1984.
488. **Tobe, Y., Yamashita, Y., and Odaira, Y.,** Stereocontrolled total synthesis of (±)-isocomene and (±)-β-isocomene *via* ring enlargement, *J. Chem. Soc. Chem. Commun.*, 898, 1985.
489. **Corey, E. J., Desai, M. C., and Engler, T. A.,** Total synthesis of (±)-retigeranic acid, *J. Am. Chem. Soc.*, 107, 4339, 1985.
490. **Snider, B. B., Hui, R. A. H. F., and Kulkarni, Y. S.,** Intramolecular [2 + 2] cycloadditions of ketenes, *J. Am. Chem. Soc.*, 107, 2194, 1985.
491. **Markó, I., Ronsmans, B., Hesbain-Frisque, A.-M., Dumas, S., and Ghosez, L.,** Intramolecular [2 + 2] cycloadditions of ketenes and kekeniminium salts to olefins, *J. Am. Chem. Soc.*, 107, 2192, 1985.
492. **Kulkarni, Y. S., Burbaum, B. W., and Snider, B. B.,** Intramolecular [2 + 2] cycloaddition reactions of vinylketenes. Stereo- and regiospecific preparation of vinylketenes from α,β-unsaturated acid chlorides, *Tetrahed. Lett.*, 26, 5619, 1985.
493. **Greene, A. E. and Charbonnier, F.,** Asymmetric induction in the cycloaddition reaction of dichloroketene with chiral enol ethers. A versatile approach to optically active cyclopentenone derivatives, *Tetrahed. Lett.*, 26, 5525, 1985.
494. **Lin, T.-T. and Houk, K. N.,** Intramolecular Diels-Alder reactions of 1,3,8-nonatriene and 1,3,9-decatriene, *Tetrahed. Lett.*, 26, 2269, 1985.
495. **Wu, T.-C. and Houk, K. N.,** Intramolecular Diels-Alder reactions of dienamines with acrylates: trends in stereoselectivity upon substitution, *Tetrahed. Lett.*, 26, 2293, 1985.
496. **Brown, F. K. and Houk, K. N.,** The influence of substituents induced asynchronicty on the stereochemistries of intramolecular Diels-Alder reactions, *Tetrahed. Lett.*, 26, 2297, 1985.
497. **Ghosh, S. and Saha, S.,** Photo-induced Diels-Alder reactions. A novel route to transfused benzobicyclo[5.3.0] decanes and [5.4.0] undecanes, *Tetrahed. Lett.*, 26, 5325, 1985.
498. **Martin, S. F., Grzejszcak, S., Rüeger, H., and Williamson, S. A.,** Total synthesis of (±)-reserpine, *J. Am. Chem. Soc.*, 107, 4072, 1985.
499. **DeLucchi, O. and Modena, G.,** Acetylene equivalents in cycloaddition reactions, *Tetrahedron*, 40, 2585, 1984.
500. **Ciganek, E.,** The intramolecular Diels-Alder reaction, *Org. React.*, 32, 1, 1984.
501. **Taber, D. F.,** Intramolecular Diels-Alder and Alder ene reactions, in *Reactivity and Structure Concepts in Organic Chemistry*, Vol. 18, Springer-Verlag, Berlin, 1984.

502. **Poll, T., Sobczak, A., Hartmann, H., and Helmchen, G.,** Diastereoface discrimination metal coordination in asymmetric synthesis: D-pantolactone as practical chiral auxiliary for Lewis acid Diels-Alder reactions, *Tetrahed. Lett.*, 26, 3095, 1985.
503. **Williams, P. D. and LeGoff, E.,** Oxidation of furanes II. Use of furanes as masked dienophiles in the intramolecular Diels-Alder reaction, *Tetrahed. Lett.*, 26, 1367, 1985.
504. **Kurth, M. J., O'Brien, M. J., Hope, H., and Yanuck, M.,** Intramolecular Diels-Alder reaction of 1-nitrodeca-1,6,9-trienes, *J. Org. Chem.*, 50, 2626, 1985.
505. **Roush, W. R., Gillis, H. R., and Essenfeld, A. P.,** Hydrofluoric acid catalyzed intramolecular Diels-Alder reactions, *J. Org. Chem.*, 49, 4674, 1984.
506. **Evans, D. A., Chapman, K. T., and Bisaha, J.,** Diastereofacial selectivity in intramolecular Diels-Alder reactions of chiral triene-N-acyloxazolidones, *Tetrahed. Lett.*, 25, 4071, 1984.
507. **Oppolzer, W. and Dupuis, D.,** Asymmetric intramolecular Diels-Alder reactions of N-acyl-camphor-sultan trienes, *Tetrahed. Lett.*, 26, 5437, 1985.
508. **Roush, W. R. and Kageyama, M.,** Enantioselective synthesis of the bottom-half of chlorothricolide, *Tetrahed. Lett.*, 26, 4327, 1985.
509. **Nicolaou, K. C. and Li, W. S.,** An intramolecular Diels-Alder strategy to forskolin, *J. Chem. Soc. Chem. Commun.*, 421, 1985.
510. **Jenkins, P. R., Menear, K. A., Barraclough, P., and Nobbs, M. S.,** An intramolecular Diels-Alder approach to forskolin, *J. Chem. Soc. Chem. Commun.*, 1423, 1984.
511. **Weinreb, S. M.,** Alkaloid total synthesis by intramolecular imino Diels-Alder cycloadditions, *Accounts Chem. Res.*, 18, 16, 1985.
512. **Lasne, M.-C. and Ripoll, J.-L.,** New synthetic developments of the $[4_\pi + 2_\pi]$ cycloreversion, *Synthesis*, 121, 1985.
513. **Bach, R. D. and Klix, R. C.,** On the geometric requirements for concerted 1,2-carbonyl migration in α,β-epoxy ketones, *Tetrahed. Lett.*, 26, 985, 1985.
514. **Clark, R. G. and Thiensathit, S.,** A palladium-catalyzed rearrangement of 1-vinyl-1-cyclobutanols, *Tetrahed. Lett.*, 26, 2503, 1985.
515. **Hudlický, T., Kutchan, T. M., and Naqvi, S. M.,** The vinylcyclopropane-cyclopentene rearrangement, *Org. React.*, 33, 247, 1985.
516. **Smith III, A. B., Wexler, B. A., Tu, C.-Y., and Konopelski, J. P.,** Stereoelectronic effects in the cationic rearrangements of [4.3.2]propellanes, *J. Am. Chem. Soc.*, 107, 1308, 1985.
517. **Smith III, A. B. and Konopelski, J. P.,** Total synthesis of (+)-quadrone: assignment of absolute stereochemistry, *J. Org. Chem.*, 49, 4094, 1984.
518. **Watkins, J. C. and Rosenblum, M.,** 3 + 2 Cyclopentane annulation reactions using organoiron reagents. Hydrazulene synthesis, *Tetrahed. Lett.*, 25, 2097, 1984.

Chapter 4

SPIROCYCLIC COMPOUNDS

I. GENERAL CONSIDERATIONS

The stereoselective synthesis of spirocyclic compounds has been largely centered around the problem of forming the pivotal quaternary carbon center in a stereospecific manner.[1] The formation of quaternary carbon centers encounters specific problems[1] and the required stereoselectivity adds to the complexity of the task. In the previous chapter we have classified the synthetic methods as mono- and ditopic, depending on the number of bonds formed in the cyclization process. Although a ditopic formation of a spirocyclic system can be conceived, e.g. as an interaction of a cyclic carbene with a ditopic reagent (Figure 1), common synthetic methods are as a rule based on stepwise formation of bonds linked to the quaternary carbon center. The construction of spirocyclic systems (Figure 2) is usually achieved by intramolecular alkylation (A,B),[2] cycloaddition reactions (C),[3] ring closure in a preformed, geminally substituted derivative (D), or by rearrangements.[4]

II. ALKYLATION METHODS

A. α,α′-Dialkylation

Treating a cyclic ketone with an α,ω-dihalogen derivative and a base is a classical method of forming spirocyclic compounds (Figure 3)[5] (for a review see Reference 2).

In a modified procedure[6] more reactive enamines can be used instead of ketones. Malonate esters can also be employed as one-carbon synthons in forming spirocyclic systems. This approach is illustrated with the synthesis of a chain of spirocycles composed of cyclobutane rings[7] (Figure 4).

If the dihalogeno derivative possesses electrophilic centers of different reactivity, the alkylation can be conducted in a highly stereoselective manner as exemplified by the Stork synthesis of β-vetivone (3), (Figure 5).[8] Intramolecular alkylation in the intermediate 1 takes place from the less-hindered face of the enolate ring, yielding spiro ketone 2 with the correct relative configuration at the chirality centers.

A modified procedure[9] (Figure 6) employs a ''nonbasic'' generation of the requisite enolate via decarboxylation of keto ester 4. The intramolecular alkylation is controlled by the adjacent methyl group and leads preferentially (9:1) to ketone 5. This intermediate was further converted to (±)-β-vetispirene (6) and (±)-β-vetivone (3).[9]

Formation of the spiro center can be also accomplished by a double Michael addition as shown in Figure 7.[10]

B. α-Alkylation

Intramolecular alkylation in an α-substituted ketone can be realized in several ways.[2] For instance, enol lactones react with ynamines under catalysis of Lewis acids to form spirocyclic ketones (Figure 8).[11] The alkyl group controls the approach of the ketenimine electrophile to the enolate double bond in the intermediate 7, thus imparting stereoselectivity to the overall reaction.[11] The method was used in the synthesis of (±)-acoradiene III (8).[12]

Alkylation of enamine 9 with the bifunctional electrophile 10 (Figure 9)[19] led to a single diastereoisomer of spirocyclic ketone 11. The stereoselectivity of the second ring-forming step is due to steric control exerted by the methyl group. The ketone 11 served as a key intermediate in the synthesis of eremophilane-like sesquiterpenes.[14]

An interesting synthetic approach to spirocyclic systems comprises an iron-carbonyl catalyzed alkylation-condensation of 2,4-dibromo-3-pentanone with enamines (Figure 10).[15]

FIGURE 1.

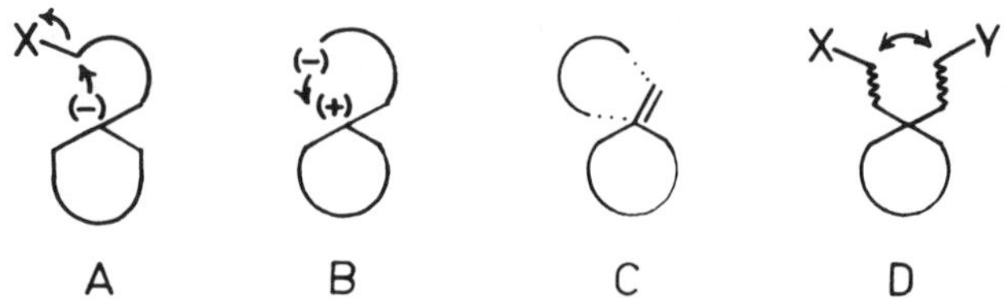

FIGURE 2.

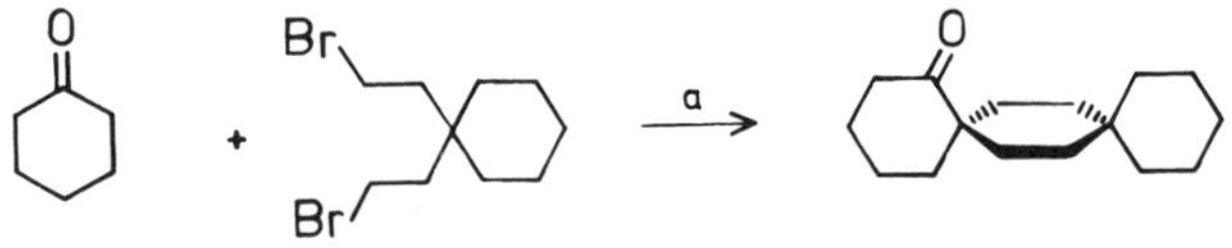

FIGURE 3. (a) t-BuOK, toluene.

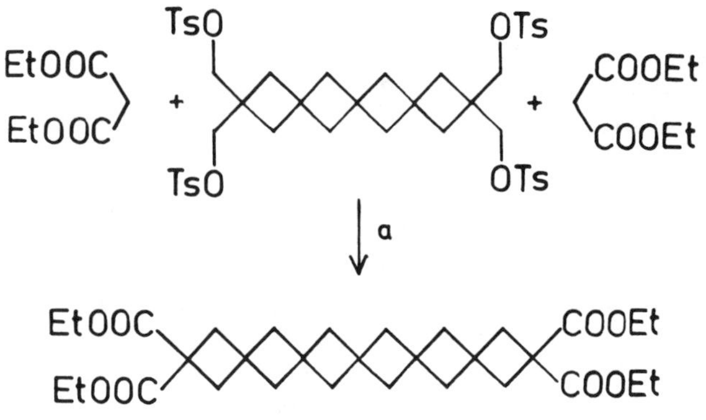

FIGURE 4. (a) Na, xylene.

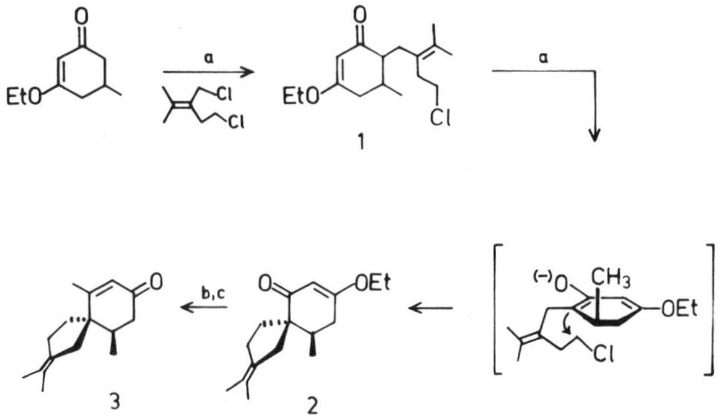

FIGURE 5. (a) $(i\text{-}Pr)_2NLi$; (b) CH_3Li, Et_2O; (c) 1*M*-HCl.

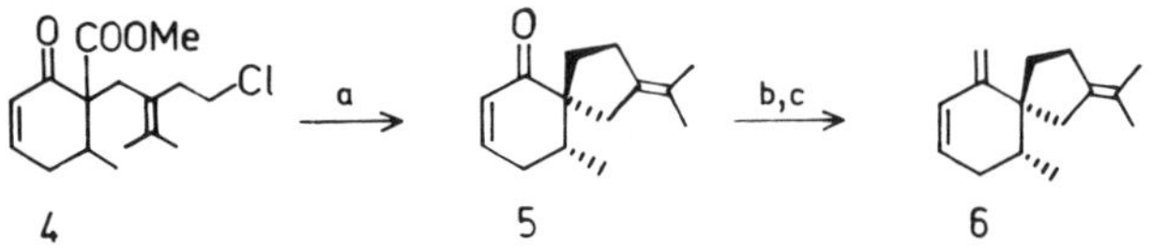

FIGURE 6. (a) LiCl, HMPA; (b) CH_3Li, Et_2O; (c) TsOH.

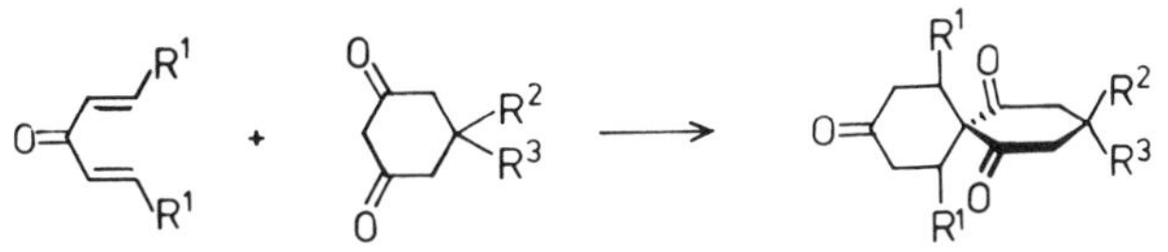

FIGURE 7.

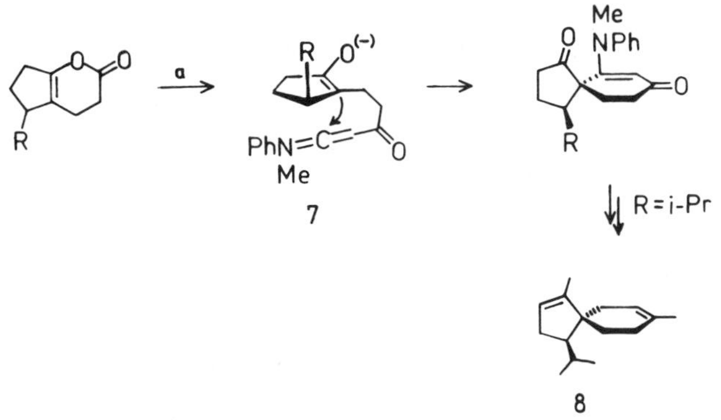

FIGURE 8. (a) Ph – N(CH_3) – C≡CH, $MgBr_2$.

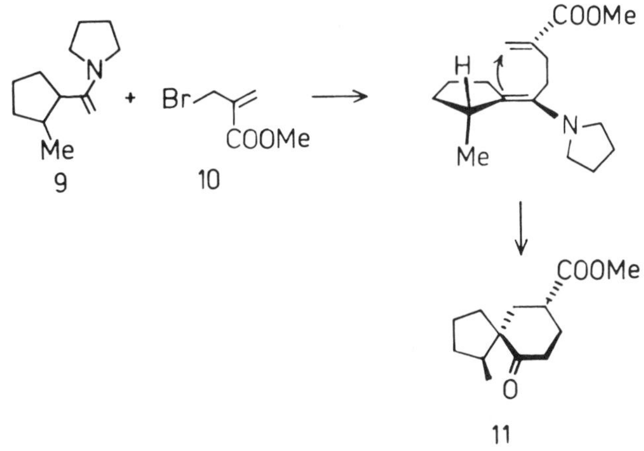

FIGURE 9.

Intermediate carbanions to be alkylated can be generated by the conjugate addition of dialkylcuprates to α,β-enones (Figure 11)[16] or fulvenes (Figure 12).[17] In both cases the subsequent aldon condensation (ring-forming process) is highly stereoselective, a feature which was utilized in the synthesis of (±)-β-vetivone.[17]

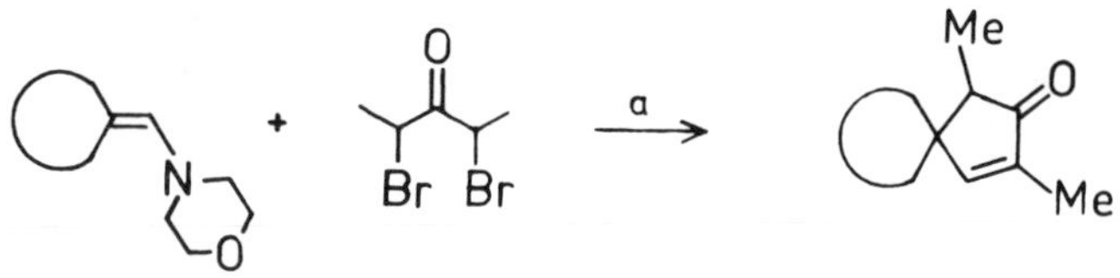

FIGURE 10. (a) $Fe_2(CO)_9$.

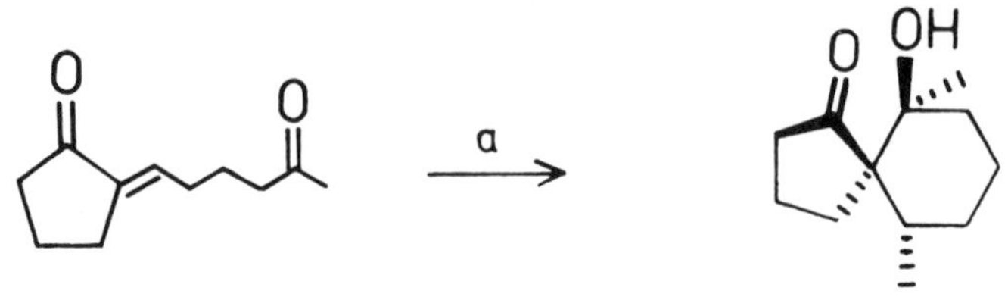

FIGURE 11. (a) Me_2CuLi, Et_2O.

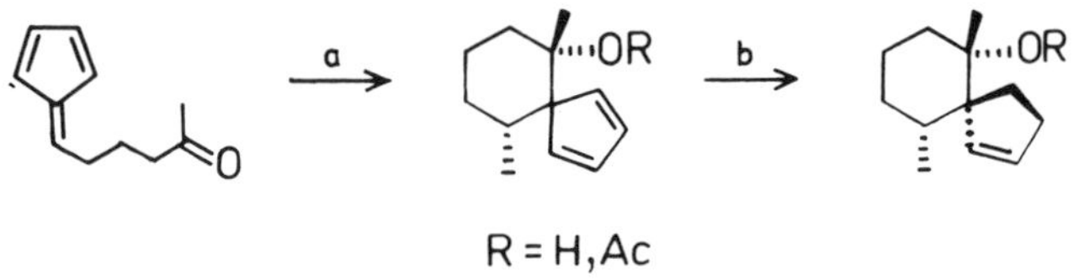

R = H,Ac

FIGURE 12. (a) Me_2CuLi; Et_2O; (b) N_2H_2.

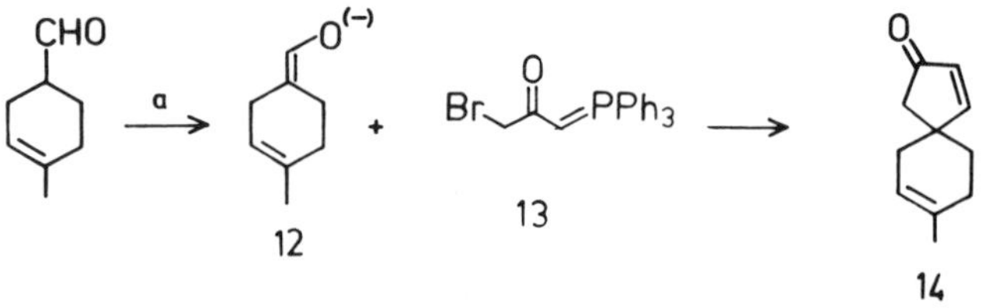

FIGURE 13. (a) KH, THF.

The Wittig reaction of aldehyde enolate 12 with the bifunctional reagent 13, followed by ring closure, has been reported to afford spirocyclic ketone 14, albeit in a low yield (Figure 13).[18] The three-carbon synthon in the Wittig reaction can be realized as the cyclopropane derivative 15 (Figure 14).[19] The vicinal methyl group again serves as the control element, directing the alkylation to occur at the reverse side of the cyclohexenone ring. This methodology was used by Dauben et al.[19] in their syntheses of spirovetivane sesquiterpenes β-vetinone, hinesol, β-vetispirene, and α-vetispirene.

The alkylating, electrophilic center can be represented by an allylpalladium complex, as shown in Figure 15.[20] β,γ-Unsaturated ketones can be cyclized to spirocyclic derivatives under the action of a strong base (Figure 16).[21] For other methods see References 22 to 26.

C. *Ipso* Alkylations and Oxidative Coupling of Phenols

Another synthetic approach to spirocyclic compounds relies upon the intramolecular *ipso* alkylation of 4-substituted phenols (Figure 17).[27,28] The problem of stereoselectivity is reduced to establishing correct relative configuration of the substituents on the five-membered ring. Although the cyclization yields a mixture of isomers, the more stable *trans*-isomer can

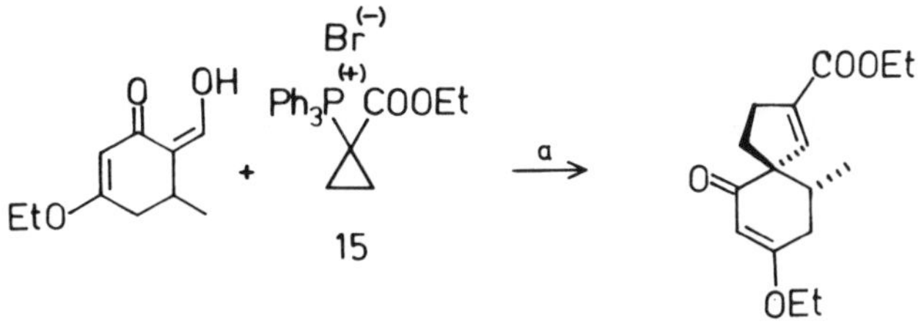

FIGURE 14. (a) NaH.

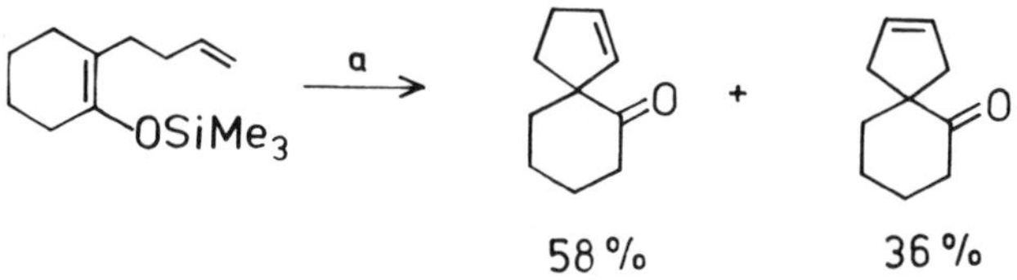

FIGURE 15. (a) $(AcO)_2Pd$, CH_3CN.

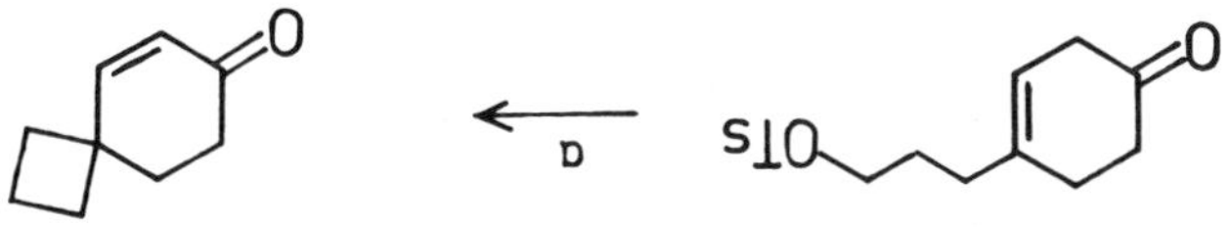

FIGURE 16. (a) NaH, t-BuOH, toluene.

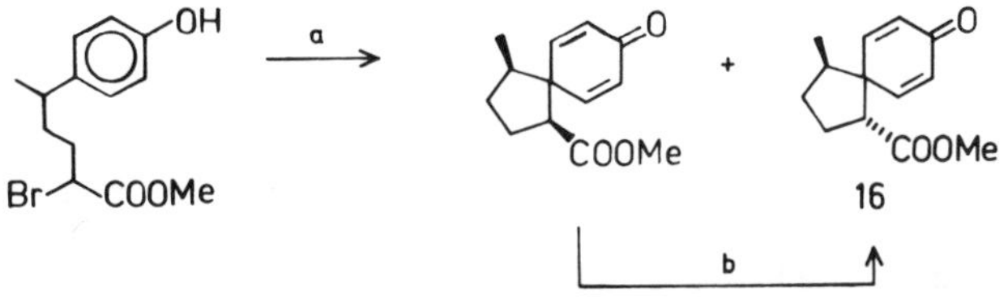

FIGURE 17. (a) t-BuOK, t-BuOH; (b) CH_3ONa.

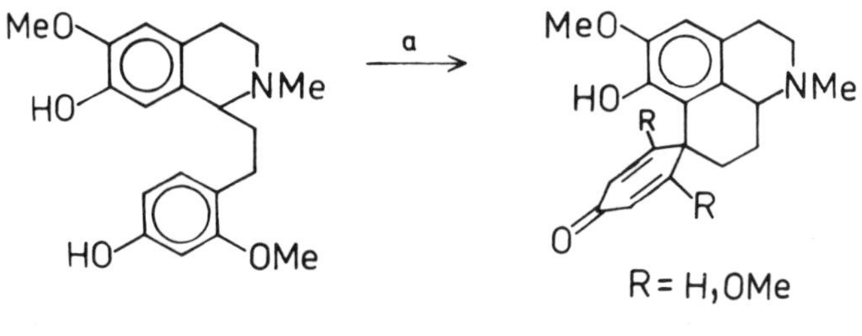

FIGURE 18. (a) $FeCl_3$, H_2O.

be obtained by equilibration. The spirodienone 16 served as an intermediate in the synthesis of (±)-α-cedrene.[27,28]

Formation of a spiro center in the *ipso* position of a substituted phenol can be achieved alternatively by oxidative coupling (Figure 18)[29-32] (for a review see Reference 33), or by photochemical debromination (Figure 19).[34]

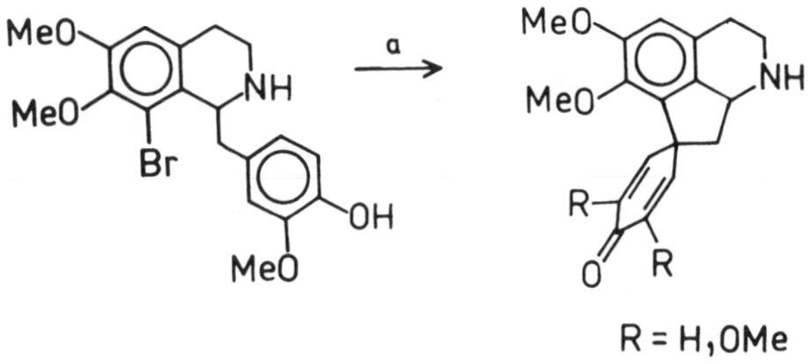

FIGURE 19. (a) hν, NaOH, CH_3OH, H_2O.

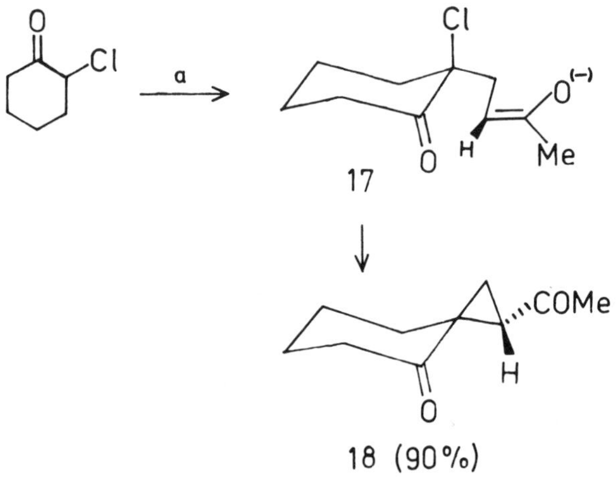

FIGURE 20. (a) NaH, C_6H_6, HMPA, then $CH_3COCH=CH_2$.

III. INVERSE ALKYLATIONS AND RELATED METHODS

In this Section we will deal with ring closures in which the spiro atom plays the role of an electrophilic center. This is exemplified by the stereoselective formation of the spiro [2.5]octane system 18 from α-chlorocyclohexanone and 3-buten-2-one (Figure 20).[35] The acyl group in the intermediate enolate 17 is oriented away from the cyclohexanone keto group to avoid repulsive interaction; this results in the preferential formation of the *anti*-isomer 18.[35]

An interesting method, developed by Semmelhack and Yamashita,[36] is based on intramolecular alkylation of a carbanion α to the nitrile group by the electrophilic *ipso* carbon atom of the arenechromium complex 19 (Figure 21). The spirocyclic enones 20 and 21 were further employed in the synthesis of acorenone and acorenone B.

Cyclohexenones bearing a functionalized substituent at C-3 can be converted to spirocycles via an intramolecular conjugate addition. This approach was employed in the synthesis of acorone (Figure 22).[37] In the case of a less reactive nucleophile, e.g. stannane 22 (Figure 23), the cyclization may be induced by the Lewis-acid catalysis.[38]

Other methods include the intramolecular nucleophilic displacement in allylic acetates,[39] cyclodehydration of unsaturated alcohols,[40] reactions of cyclic anhydrides with bifunctional Grignard reagents,[41,42] and light-induced cyclizations of iminium salts (Figure 24) which was developed in a model study aimed at the synthesis of harringtonine alkaloids.[43]

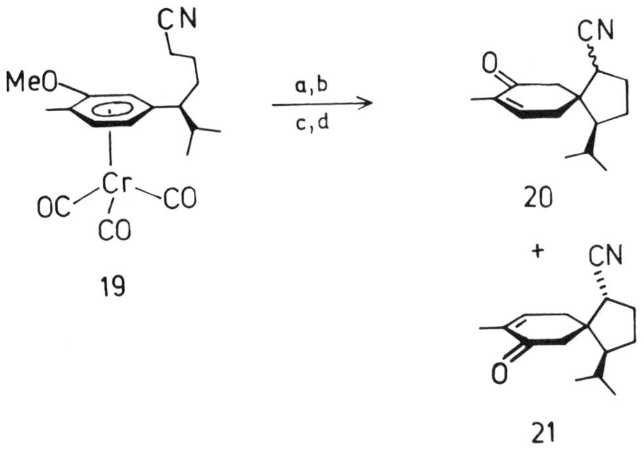

FIGURE 21. (a) $(i\text{-}Pr)_2NLi$; (b) CF_3SO_3H; (c) NH_4OH; (d) H_3O^+.

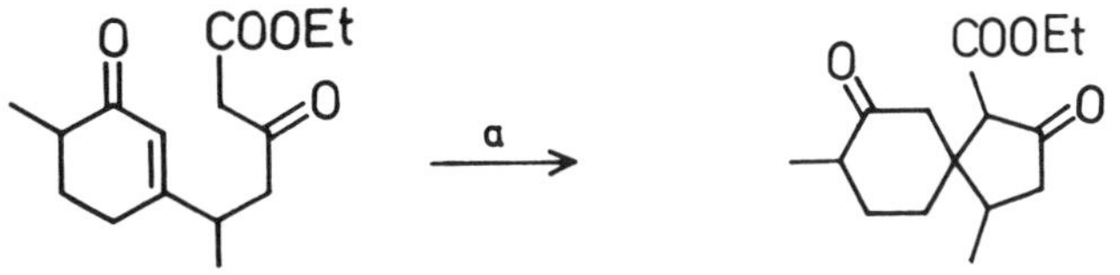

FIGURE 22. (a) CH_3ONa, CH_3OH.

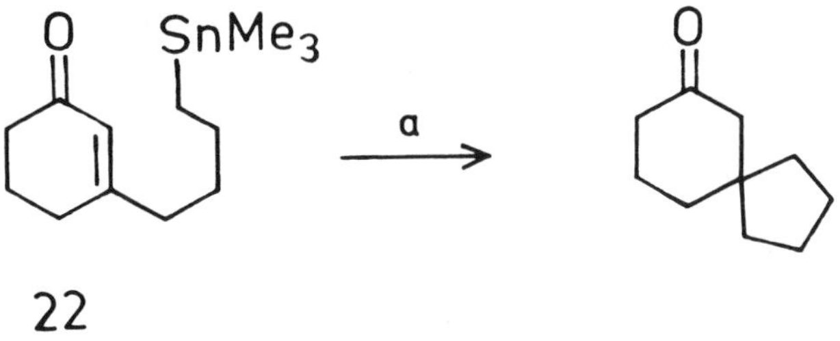

FIGURE 23. (a) $TiCl_4$.

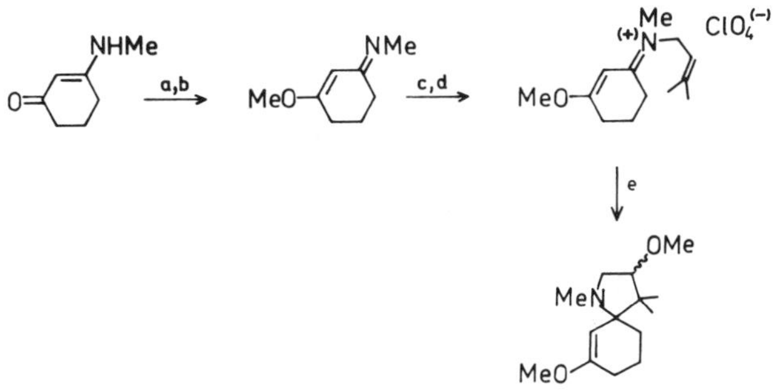

FIGURE 24. (a) CH_3I; (b) 10%-NaOH; (c) Prenylbromide; (d) Dowex in ClO_4^- form; (e) $h\nu$, CH_3OH.

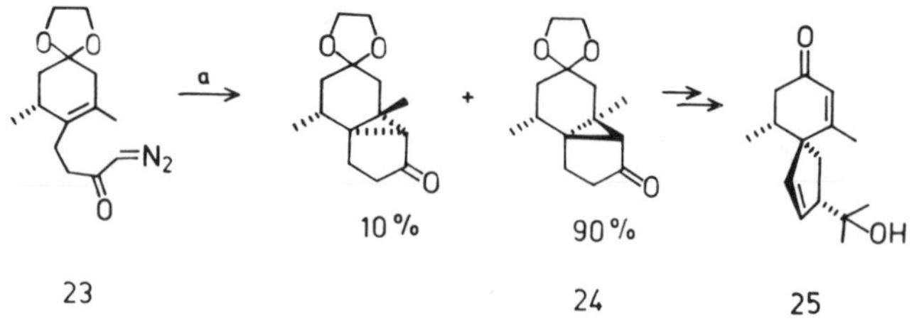

FIGURE 25. (a) Cu, C_6H_6, reflux.

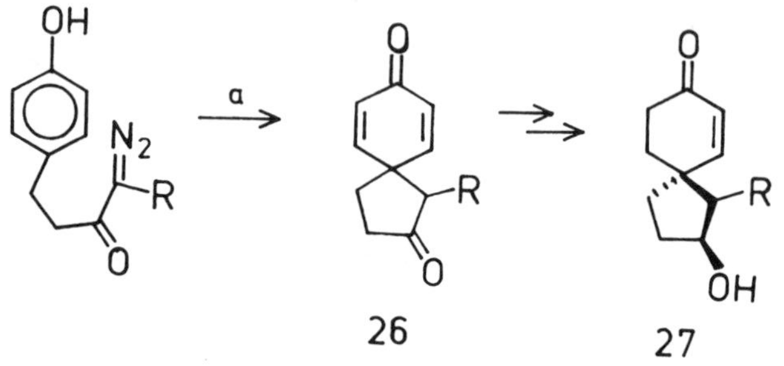

FIGURE 26. (a) CuI, C_6H_6, reflux.

IV. CYCLOADDITIONS

The steric course of cycloadditions is often quite sensitive to the spatial arrangement of substituents in the vicinity of the double bond in the $_{\pi}2_s$ component. This feature can be utilized for a stereoselective formation of the quaternary spiro center,[3] provided the existing ring contains a suitable control element.

A. [2 + 1] Cycloadditions

The intramolecular [2 + 1] addition of a diazoketone to a cyclic olefin represents one of the fundamental strategies in the synthesis of spirocyclic sesquiterpenes. Figure 25 shows the reaction sequence used by Deslongchamps et al. in the synthesis of agarospirol.[44] The relative configuration at the quaternary carbon spiro atom in 25 is determined by the approach of the carbene reagent to the double bond in 23, which is in turn controlled by the methyl group attached to the sp^3 center. The tricyclic skeleton in the intermediate 24 is transformed into the desired spirocycle by cleaving the perimeter bond of the cyclopropane ring.

A similar strategy, based on an initial [2 + 1] cycloaddition, has been used in syntheses of (±)-chamigrene,[45] acorone,[46] and acorenone.[45,46]

Phenols bearing a diazoalkane group can be alkylated in the *ipso* position (Figure 26) affording spiro dienones 26.[47] Since the cyclohexadienone moiety has a plane of symmetry, the problem of stereoselection occurs at the stage of partial reduction and introduction of substituents in 26 to form enone 27. The solution to this problem[48] opened a synthetic route to spirovetivane, alaskane, and acorane sesquiterpenes.[49]

B. [2 + 2] Cycloadditions

[2 + 2] Dimerization of cycloalkenes possessing an exocyclic double bond generally leads to spirocyclic compounds:[3] such an example is the formation of the cyclobutanedione derivative (Figure 27).[50]

Another strategy rests on a photoinduced intramolecular [2 + 2] cycloaddition followed by cleavage of the cyclobutane ring in the next step (Figure 28).[51] This approach has been successfully employed in the synthesis of acorenone (Figure 29).[52]

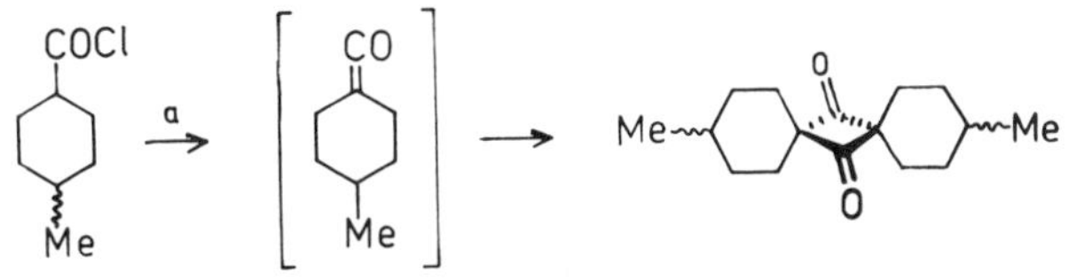

COCl a CO Me Me Me Me O O

FIGURE 27. (a) Et_3N, toluene, reflux.

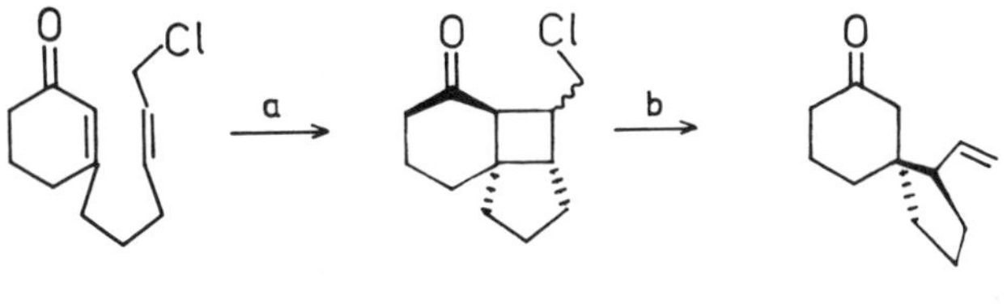

O Cl a O Cl b O

FIGURE 28. (a) $h\nu$; (b) Li, NH_3, THF.

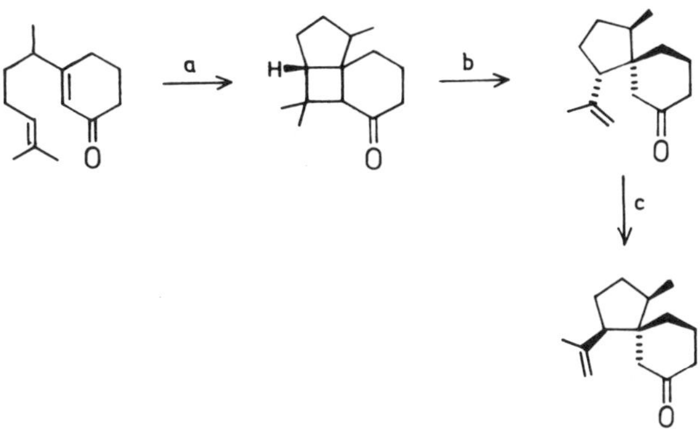

O a H O b O c O

FIGURE 29. (a) $h\nu$, $-50°C$; (b) $h\nu$, $0°C$; (c) $AgNO_3$, CH_3OH.

Intramolecular [2 + 2] addition of the triple bond to the enone system in 28 (Figure 30)[53] afforded the tricyclic cyclobutene derivative 29 which was cleaved to spirocycles 30 and 31. The minor isomer 30 was then converted to ketone 32, an intermediate in the synthesis of perhydrohistrionicotoxin.[53]

C. [4 + 2] Cycloadditions

The construction of a spirocyclic system via [4 + 2] cycloaddition requires the presence of an activated exocyclic double bond in the $O_{\pi}2_s$ component.[3] The double bond can also be present in a latent form, e.g. as a β-halogeno, acetoxy, hydroxy, or dialkylaminogroup (Figure 31),[54] from which it is liberated *in situ* and immediately consumed in the [4 + 2] cycloaddition (for a review see Reference 3). Even in simple cases, the reaction can be highly stereoselective, as illustrated by the formation of a single isomer from α-methylenecyclohexanone and cyclopentadiene (Figure 32).[55]

The [4 + 2] cycloaddition route has been employed in the synthesis of (±)-chamigrene (Figure 33).[56] Although this synthesis does not confront problems of stereoselectivity, the regioselectivity of the Diels-Alder reaction is highly desirable in order to secure the correct position of the enol oxygen atom.

Catalyzed addition of isoprene to α-methylenecyclopentanone 33 led preferentially to two spirocyclic ketones, 34 and 35, which were further converted to α- and β-alaskene, respectively (Figure 34).[57,58]

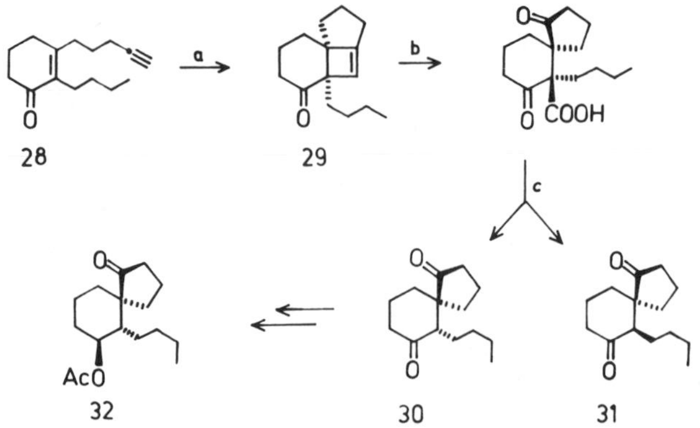

28 29 30 31 32

FIGURE 30. (a) hν, CH_3OH, AcONa; (b) RuO_4, $NaIO_4$; (c) reflux in toluene.

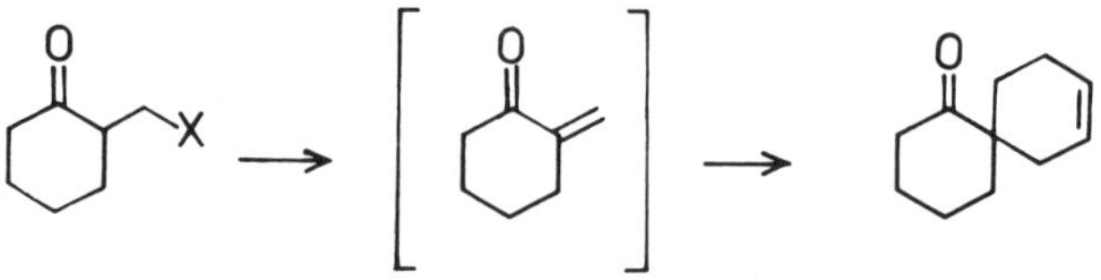

FIGURE 31. X = Cl, OH, OAc, NMe_2.

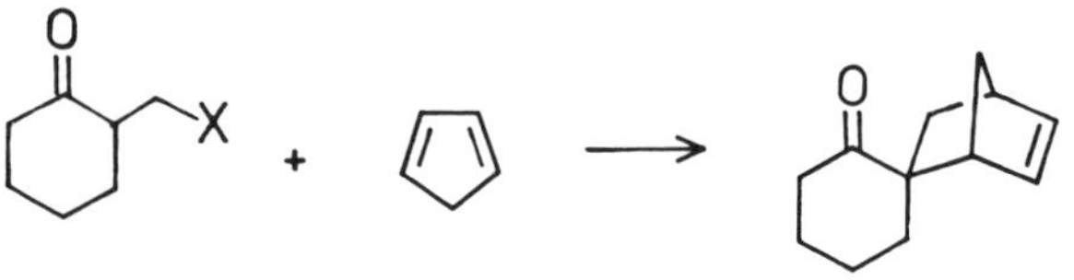

FIGURE 32. X = OAc.

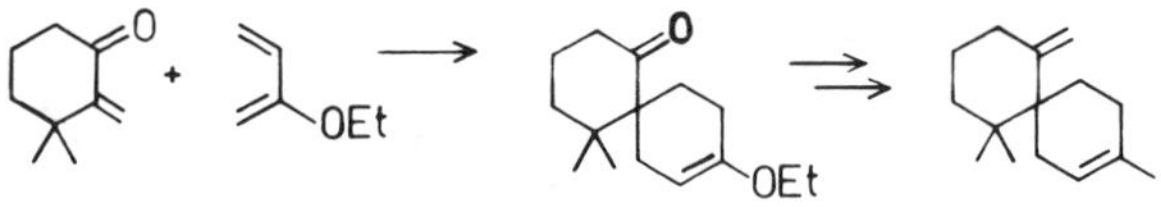

FIGURE 33.

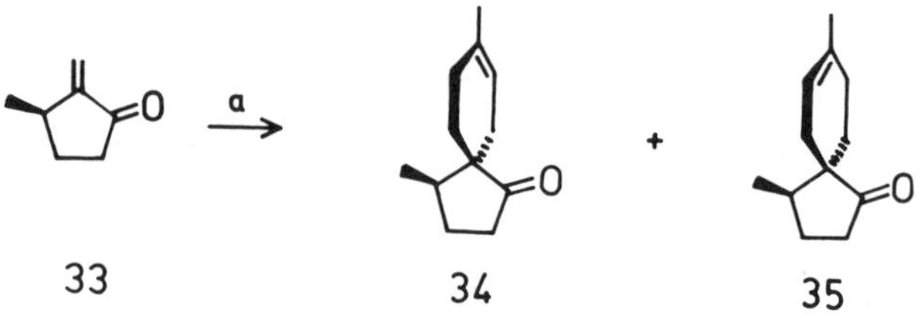

33 34 35

FIGURE 34. (a) Isoprene, $SnCl_4$.

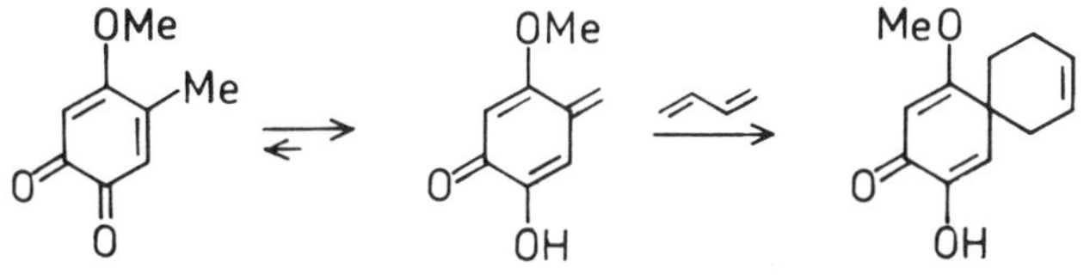

36 37

FIGURE 35.

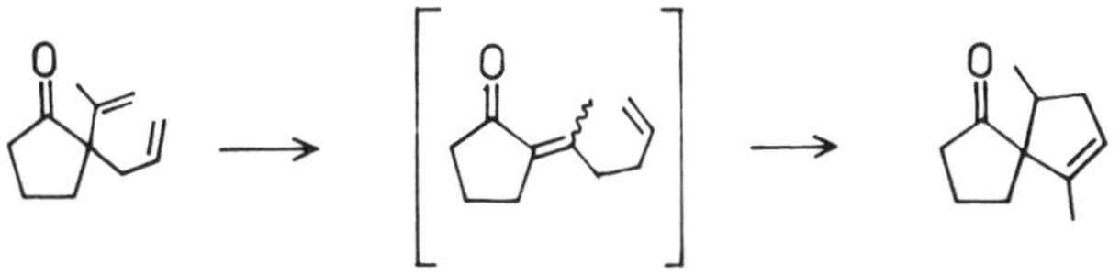

FIGURE 36.

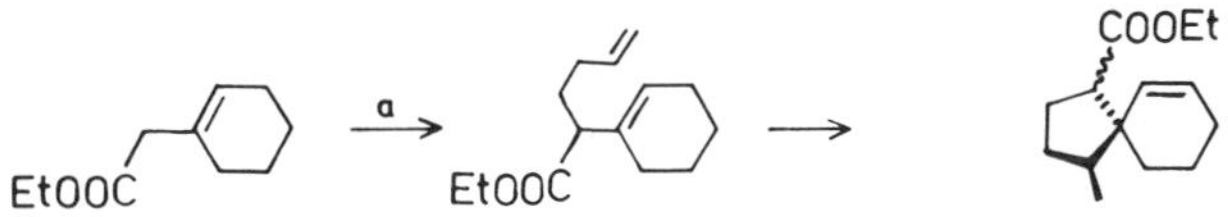

FIGURE 37. (a) Lithium cyclohexylisopropylamide, 1-bromo-3-butene, DME, HMPA.

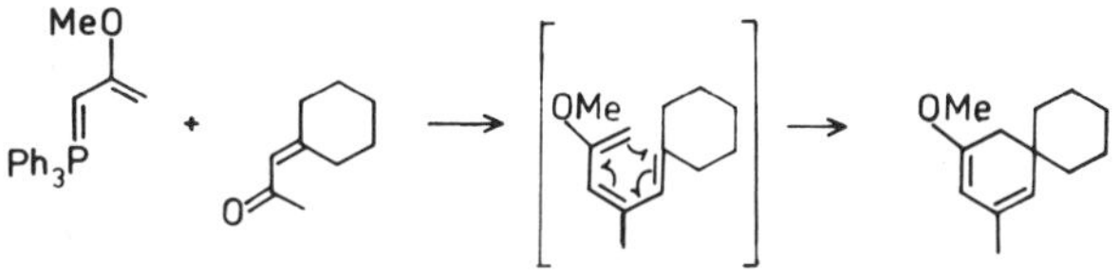

38

FIGURE 38.

Substituted o-quinones like 36 can also be used as $O_{\pi}2_{s}$ components owing to the existence of the quinone methide enol form 37 in equilibrium with the o-quinone (Figure 35).[59]

D. Conia, Ene, and Related Cyclizations

Cope rearrangements in tandem with the ene reaction have been used in the syntheses of spirocyclic ketones (Figure 36).[60-64] A simpler version, the Conia cyclization of α-(ω-alkenyl)cycloalkanones, has also been reported.[65,66] The ene reaction of 1-substituted cyclohexenes (Figure 37) was utilized by Oppolzer et al. in their syntheses of (±)-β-acoradiene and (±)-β-acorenol.[67]

Also noteworthy is the electrocyclic closure of the cyclohexadiene ring in the intermediate 38 prepared by the Wittig reaction (Figure 38).[68]

V. CYCLIZATIONS OUTSIDE THE SPIROATOM

This synthetic strategy is based on forming the quaternary carbon center first and then connecting the geminal side chains to create the spirocycle. The problem of stereoselectivity is thus reduced to obtaining the correct configuration at the quaternary center.[1] Of many applications of this basic strategy[69-79] we present here the construction of the spiro system needed for the synthesis of perhydrohistrionicotoxin (Figure 39).[80,81]

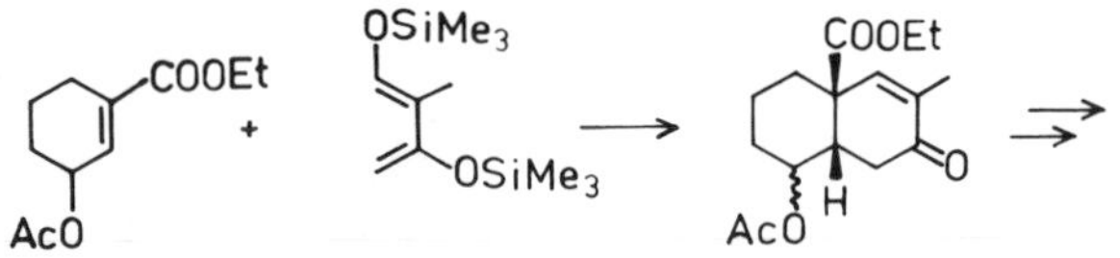

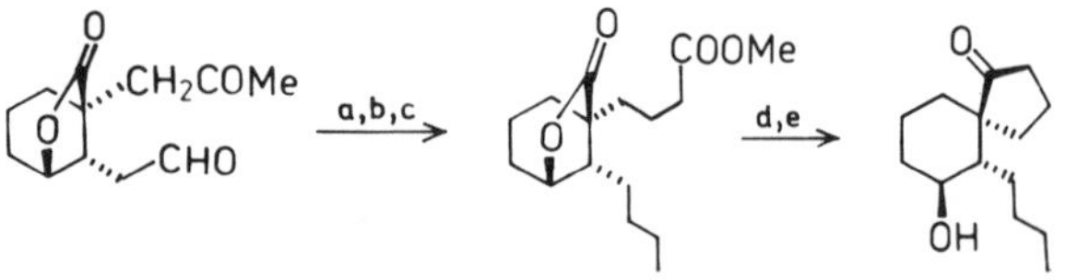

FIGURE 39. (a) $Ph_3P=CH-CH_3$; (b) H_2, Pt; (c) (1) carboxylation, (2) $NaBH_4$, CH_3OH, (3) CH_3SO_2Cl, Et_3N, (4) DBU, Et_3N, (5) H_2, Pt; (d) KH, THF; (e) DABCO, xylene, reflux.

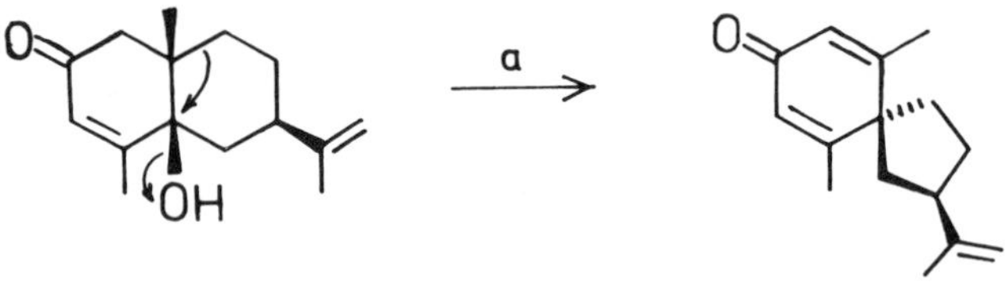

FIGURE 40. (a) $POCl_3$, C_5H_5N.

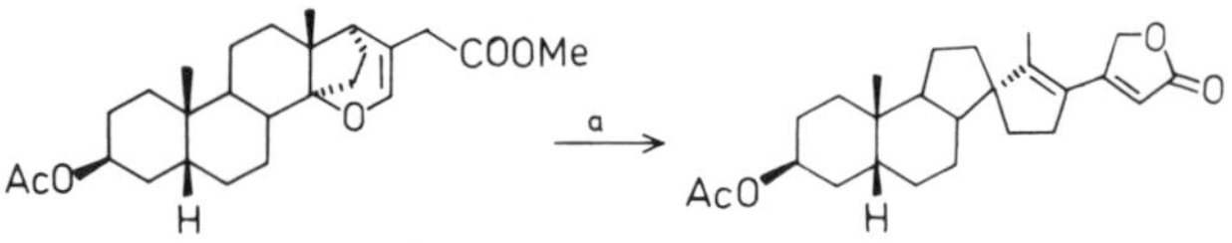

FIGURE 41. (a) p-TsOH, C_6H_6.

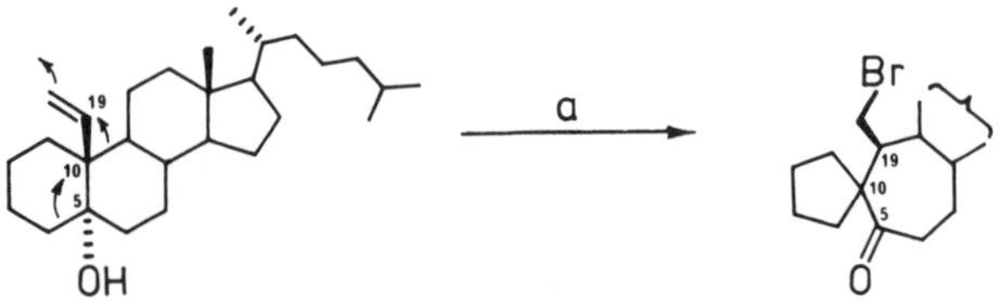

FIGURE 42. (a) $CH_3CONH{\cdot}Br$, $HClO_4$, H_2O, dioxane.

VI. REARRANGEMENTS

Simple rearrangements of ortho-condensed systems to spirocycles, e.g. the pinacol[82] or Demyanov[83] rearrangement, have been reviewed in detail.[4] Acid-catalyzed rearrangements of γ-hydroxy-α,β-enones[84] represent an important synthetic route to spirocyclic natural products, as exemplified by the synthesis of β-vetispirene (Figure 40).[85,86] Similarly, the acid-catalyzed rearrangement of steroids bearing an oxygen function at C-14 provides a stereoselective approach to bufadienolide analogs[87] (Figure 41).[88]

A tandem of Wagner-Meerwein rearrangements, initiated by the addition of a bromonium ion to the 10β-vinyl group, has been reported to convert 5α-hydroxy-10β-vinyl-19-norcholestane (39) to the spiroketone 40 (Figure 42).[89,90]

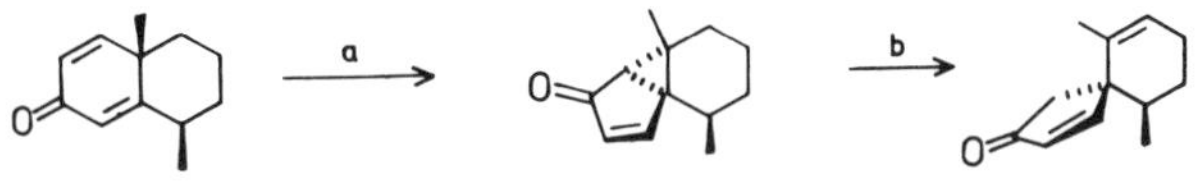

FIGURE 43. (a) hν; (b) H_2SO_4, AcOH, Ac_2O.

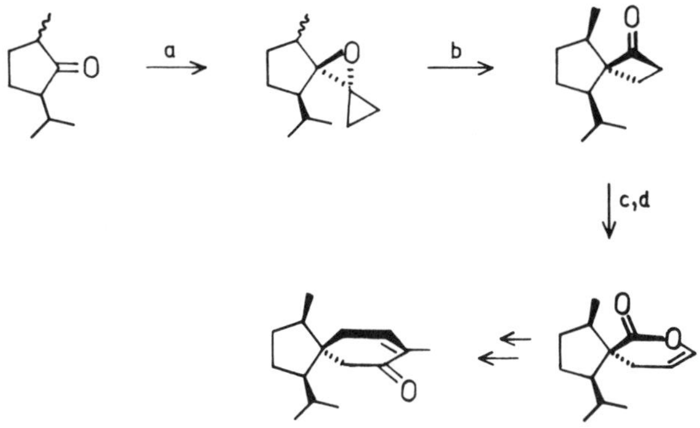

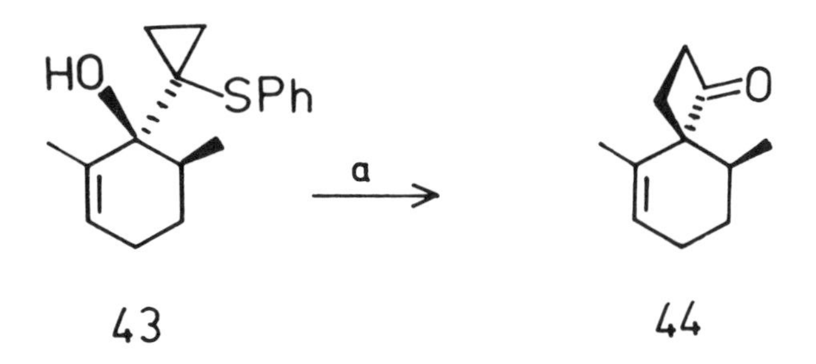

FIGURE 44. (a) $S^+Ph_2BF_4^-$, KOH, Me_2SO; (b) $LiBF_4$, C_6H_6; (c) HCO_2Et, NaH, C_6H_6, CH_3OH; (d) TsOH, C_6H_6, H_2O.

FIGURE 45. (a) $SnCl_4$.

The cyclopropane ring in the tricyclic enone 41 can be cleaved by acids to afford the spirocyclic enone 42.[91] The rearrangement, originally discovered with lumisantonin derivatives[91] was later used in the synthesis and structure revision of β-vetivone (Figure 43).[92,93]

Trost's spiroannulation procedure, based on the oxaspiropentane rearrangement (Figure 44),[94,95] represents another useful method for the stereoselective construction of spirocyclic systems. It was employed in the synthesis of acorenone B.[96] A related method relies upon the stereospecific rearrangement of cyclopropyl carbinols (43) (Figure 45).[97,98] The relative configuration at the spiro atom in 44 is determined by the arrangement of substituents at the quaternary carbon in the reactant 43. For a related methodology see Reference 99.

An interesting application of the Grob fragmentation of γ-diolmonomesylates is shown in Figure 46.[100] In this case the total synthesis proved the structure of (±)-hinesol.[100]

Photorearrangements of cross-conjugated dienones[4,101,102] represent a powerful strategy leading to spirocyclic systems. A selected example (Figure 47) shows the key step in the synthesis of (±)-α-vetispirene.[103]

We close this Chapter with an example of photoinduced enamide cyclization leading to the spirocyclic lactam 46 (Figure 48).[104,105]

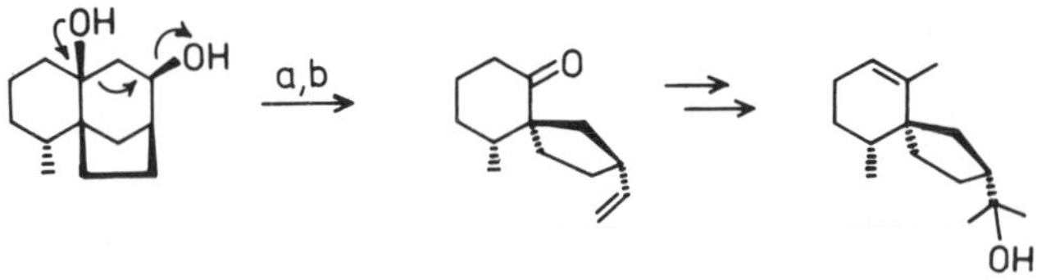

FIGURE 46. (a) CH_3SO_2Cl, C_5H_5N; (b) t-BuOK.

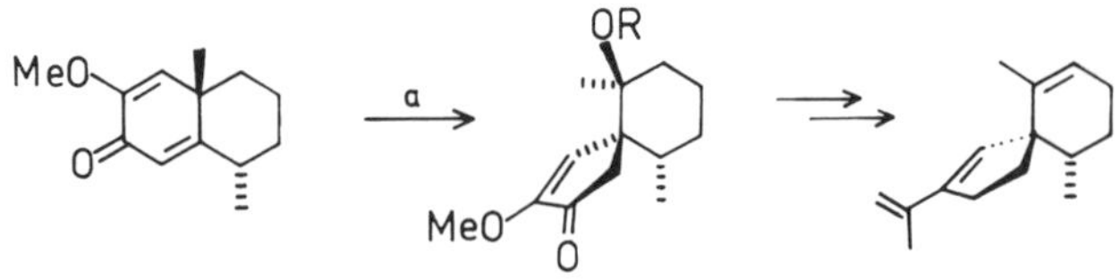

FIGURE 47. (a) hν, ROH.

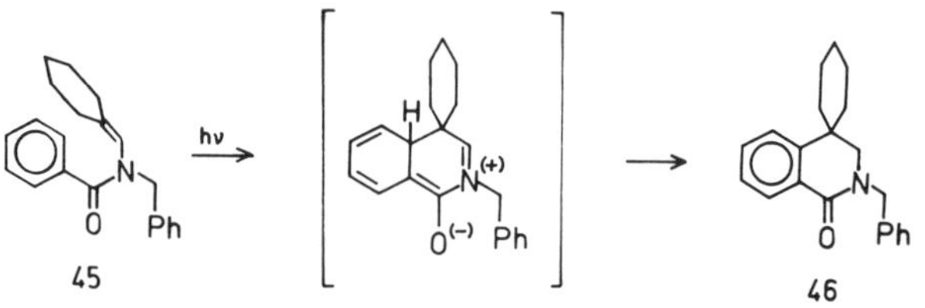

FIGURE 48.

VII. NOTES ADDED IN PROOF

Radical cyclization has been used to create the spiro-carbon.[106] An expedient approach to functionalized spiro [4.5] decanes has been developed using the intramolecular Diels-Alder addition as the key step followed by ozonolysis.[107]

REFERENCES

1. **Martin, S. F.,** Methodology for the construction of quaternary carbon centers, *Tetrahedron,* 36, 419, 1980.
2. **Krapcho, A. P.,** Synthesis of carbocyclic spiro compounds *via* intramolecular alkylation routes, *Synthesis,* 383, 1974.
3. **Krapcho, A. P.,** Synthesis of carbocyclic spiro compounds *via* cycloaddition routes, *Synthesis,* 77, 1976.
4. **Krapcho, A. P.,** Synthesis of carbocyclic spiro compounds *via* rearrangement routes, *Synthesis,* 425, 1978.
5. **DeJongh, H. A. P. and Wynberg, H.,** Spiranes III. Synthesis of polyspiro compounds consisting of cyclohexane rings, *Tetrahedron,* 20, 2533, 1964.
6. **Krieger, H., Routsalainen, H., and Montin, J.,** Notiz zur Umsetzung von 1-Pyrrolidino-cyclopenten mit 1,4-Dijod-butan, *Chem. Ber.,* 99, 3715, 1966.
7. **Buchta, E. and Merk, W.,** Spirocyclische Verbindungen. VIII. Symmetrische, nur aus Cyclobutan-Ringen bestehende Oligo- und Polyspirane, *Justus Liebigs Ann. Chem.,* 694, 1, 1966.
8. **Stork, G., Danheiser, R. L., and Ganem, B.,** Spiroannulation of enol ethers of cyclic 1,3-diketones. A simple stereospecific synthesis of β-vetivone, *J. Am. Chem. Soc.,* 95, 3414, 1973.
9. **Eilerman, R. G. and Willis, B. J.,** A new spiroannulation procedure. Intramolecular decarboxylation alkylation of β-ketoesters, *J. Chem. Soc. Chem. Commun.,* 30, 1981.
10. **DeJongh, H. A. P. and Wynberg, H.,** Spiranes. Synthesis of carbocycle polyspirans with six-membered rings, *Rec. Trav. Chim.,* 82, 202, 1963.

11. **Ficini, J., Revial, G., and Genet, J. P.,** Acylation of ynamines by enol lactones: a new method of stereoselective spiroannulation, *Tetrahedron Lett.*, 22, 629, 1981.
12. **Ficini, J., Revial, G., and Genet, J. P.,** A synthesis of (*d,l*) acoradiene III, *Tetrahed. Lett.*, 22, 633, 1981.
13. **Dunham, D. J. and Lawton, G. R.,** The synthesis of spiro systems by the α,α′-annelation process, *J. Am. Chem. Soc.*, 93, 2074, 1971.
14. **Dunham, D. J. and Lawton, R. G.,** Spiro intermediates in sesquiterpene rearrangements and synthesis, *J. Am. Chem. Soc.*, 93, 2075, 1971.
15. **Noyori, R., Yokoyama, K., Makino, S., and Hayakawa, Y.,** Reaction of α,α′-dibromo ketones and enamines with the aid of iron carbonyls. A novel cyclopentenone synthesis, *J. Am. Chem. Soc.*, 94, 1772, 1972.
16. **Näf, F., Decorzant, R., and Thommen, W.,** Regiospecific intramolecular aldol condensation induced by conjugate addition of lithium dimethylcuprate to ξ-oxo-α,β-enone, *Helv. Chim. Acta*, 58, 1808, 1975.
17. **Büchi, G., Berthet, D., Decorzant, R., Grieder, A., and Hauser, A.,** Spirovetivanes from fulvenes, *J. Org. Chem.*, 41, 3208, 1976.
18. **Altenbach, H.-J.,** (Bromacetylmethylen)triphenylphosphoranein neues Cyclopentanellierungsagens, *Angew. Chem.*, 91, 1005, 1979.
19. **Dauben, W. G. and Hart, D. J.,** A general method of preparing functionalized spirocycles. Synthesis of spirovetivane sesquiterpenes, *J. Am. Chem. Soc.*, 99, 7307, 1977.
20. **Kende, A. S., Roth, B., and Sanfilippo, P. J.,** Facile palladium (II)-mediated synthesis of bridged and spirocyclic bicycloalkenones, *J. Am. Chem. Soc.*, 104, 1784, 1982.
21. **Fassnacht, J. H. and Nelson, N. A.,** Preparation of spiran ring systems by intramolecular alkylation, *J. Org. Chem.*, 27, 1885, 1962.
22. **Davreux, J. P. and Bruylants, A.,** Compounds with multiple spiranic structure derived from 2,4-diisopropylidenecyclobutanone, *Bull. Univ. Sci. Acad. Roy. Belg.*, 54, 823, 1968; *Chem. Abstr.* 71, 112857, 1969.
23. **Johnson, C. R., Katekar, G. F., Huxol, R. F., and Janiga, E. R.,** Nucleophilic alkylidene transfer reagents. Synthesis of spiro compounds, *J. Am. Chem. Soc.*, 93, 3771, 1971.
24. **Kanno, S., Kato, T., and Kitahara, Y.,** Biogenetic-type synthesis of (±)-α-chamigrene, *Chem. Commun.*, 1257, 1967.
25. **Becker, H.-D., Bremholt, T., and Adler, E.,** Oxidative formulation and photochemical isomerization of spiro-epoxy-2,4-cyclohexadienones, *Tetrahed. Lett.*, 4205, 1972.
26. **Becker, H.-D. and Bremholt, T.,** Oxidative rearrangement of o-hydroxydiarylcarbinols, *Tetrahed. Lett.*, 197, 1973.
27. **Corey, E. J., Girotra, N. N., and Matthew, C. T.,** Total synthesis of *dl*-cedrene and *dl*-cedrol, *J. Am. Chem. Soc.*, 91, 1557, 1969.
28. **Crandall, T. G. and Lawton, R. G.,** A biogenetic-like synthesis of cedrene, *J. Am. Chem. Soc.*, 91, 2127, 1969.
29. **Kametani, T. and Satoh, F.,** Dienone synthesis by phenolic oxidation of dihydroxy-1-phenethyl-1,2,3,4-tetrahydroisoquinoline and sodium borohydride reduction, *Chem. Pharm. Bull.*, 17, 814, 1969.
30. **Afzal, M. N., Allbutt, A. D., Jordaan, A., and Kirby, G. W.,** Preparation and solvolysis of dispirotetraene-diones, *Chem. Commun.*, 996, 1969.
31. **Battersby, A. R. and Brown, T. H.,** Orientalinone, dihydroorientalinone, and salutaridine from *Papaver orientale* -related tracer experiments, *Proc. Chem. Soc.*, 85, 1964.
32. **Kametani, T., Satoh, F., Yagi, H., and Fukumoto, K.,** The syntheses of homoproaporphines by phenolic oxidative coupling. II. Separation of two isomeric dienones of homoproaporphines, *J. Org. Chem.*, 33, 690, 1968.
33. **Scott, A. I.,** Oxidative coupling of phenolic compounds, *Q. Rev.*, 19, 1, 1965.
34. **Kametani, T., Sugahara, T., Sugi, H., Shibuya, S., and Fukumoto, K.,** Total synthesis of (±)-pronuciferine, (±)-O-methylorientalinone, and (±)-O-methylkreysigione, *Tetrahedron*, 27, 5993, 1971.
35. **Causse-Zoller, M. and Fraise-Jullien, R.,** Extension d'une reaction de synthèse de composés cyclopropaniques polyfonctionels, *Bull. Soc. Chim. France*, 430, 1966.
36. **Semmelhack, M. F. and Yamashita, A.,** Arene-metal complexes in organic synthesis. Synthesis of acrenone and acorenone B, *J. Am. Chem. Soc.*, 102, 5924, 1980.
37. **Pinder, A. R., Rice, S. J., and Rice, R. M.,** Synthesis of the spiro[4.5]decane system. An approach to the acorane sesquiterpene group, *J. Org. Chem.*, 37, 2202, 1972.
38. **Macdonald, T. L. and Mahalingam, S.,** Alkyltin(IV)-mediated carbocyclisation, *J. Am. Chem. Soc.*, 102, 2113, 1980.
39. **Godleski, S. A., Meinhart, J. D., Miller, D. J., and VanWallendael, S.,** Palladium-assisted synthesis of 1-azaspirocycles, *Tetrahed. Lett.*, 22, 2247, 1981.

40. **Nojima, M., Nagai, T., and Tokura, N.,** Spiro compound formation II. The formation of spiro compounds and its mechanistic aspects by the acetolysis of 1-(Δ^4-pentenyl)cyclohexanol and 1-(Δ^4-pentenyl)cyclopentanol, *J. Org. Chem.*, 33, 1970, 1968.
41. **Canonne, P., Lemay, G., and Bélanger, D.,** Reaction of di(bromomagnesio)alkanes with unsymmetrically substituted cyclic anhydrides, *Tetrahed. Lett.*, 21, 4167, 1980.
42. **Cannonne, P., Bélanger, D., Lemay, G., and Foscolos, G. B.,** One-step spiroannulation. Synthesis of spiro γ- and δ-lactones, *J. Org. Chem.*, 46, 3091, 1981.
43. **Tiner-Harding, T., Ullrich, J. W., Chin, F. T., Chen, S. F., and Mariano, P. S.,** Electron-transfer initiated iminium salt photo spiro cyclization methodologies. Model studies for harringtonine alkaloid synthesis, *J. Org. Chem.*, 47, 3360, 1982.
44. **Mongrain, M., Lafontaine, J., Bélanger, A., and Deslongchamps, P.,** Stereoselective synthesis of (±)-epihinesol(agarospirol), *Can. J. Chem.*, 48, 3273, 1970.
45. **Ruppert, J. F. and White, J. D.,** Tricyclo[5.3.0.0^{1,6}]decan-5-one and tricyclo[5.4.0.0^{1,6}]undecan-5-one. Synthesis and selective transformation to spiro and fused bicyclic systems, *J. Am. Chem. Soc.*, 103, 1808, 1981.
46. **Ruppert, J. F., Avery, M. A., and White, J. D.,** Synthesis of (−)-acorenone B, *J. Chem. Soc. Chem. Commun.*, 978, 1976.
47. **Iwata, C., Yamada, M., Shinoo, Y., Kobayashi, K., and Okada, H.,** Intramolecular cyclisation of phenolic α-diazoketones. Novel synthesis of the spiro[4.5]decane carbon framework, *J. Chem. Soc. Chem. Commun.*, 888, 1977.
48. **Iwata, C., Myiashita, K., Ida, Y., and Yamada, M.,** Effects of neighboring hydroxy groups in metal-ammonia reductions of α,β-unsaturated carbonyl compounds, *J. Chem. Soc. Chem. Commun.*, 461, 1981.
49. **Iwata, C., Yamada, M., Shinoo, Y., Kobayashi, K., and Okada, H.,** Studies on the syntheses of spirodieneone compounds. VII. Novel synthesis of the spiro[4.5]decane carbon framework, *Chem. Pharm. Bull.*, 28, 1932, 1980.
50. **Farina, M. and DiSilvestro, G.,** Some remarks on high-symmetry molecules. Molecular structure of *cis* and *trans* 3,11-dimethylspiro[5.1.5.1]tetradecan-7,14-dione, *Tetrahed. Lett.*, 183, 1975.
51. **Oppolzer, W., Gorrichon, L., and Bird, T. G. C.,** A stereoselective approach to the spiro[4.5]decane system via intramolecular photocycloaddition and reductive fragmentation. Preliminary communication, *Helv. Chim. Acta,* 64, 186, 1981.
52. **Duc-Do-Khac-Manh, Ecoto, J., Fétizon, M., Colin, H., and Diez-Masa, J. C.,** A new approach to spiro sesquiterpenes of the acorane family, *J. Chem. Soc. Chem. Commun.*, 953, 1981.
53. **Koft, E. R. and Smith, III, A. B.,** Intramolecular (2+2) photochemical cycloadditions. 3. Perhydrohistrionicotoxin studies. Synthesis of spiro[4.5]decanones via intramolecular (2+2) photocycloaddition, *J. Org. Chem.*, 49, 832, 1984.
54. **Mousseron, M., Jacquier, R., and Christol, H.,** Réarrangements acidocatalysés. Nouvelles synthèses d'α-spirocétones, *Bull. Soc. Chim. France,* 346, 1957.
55. **Elagina, N. V., Martinkova, N. S., Kazanskii, B. A.,** Synthesis of 1,4-endomethylenspiro[5.5]undecane, *J. Gen. Chem. USSR,* 29, 3669, 1959.
56. **Tanaka, A., Uda, H., and Yoshikoshi, A.,** The total synthesis of (±)-chamigrene, *Chem. Commun.*, 188, 1967.
57. **Marx, J. N. and Norman, L. R.,** Synthesis of (−)-acorone and related spirocyclic sesquiterpenens, *J. Org. Chem.*, 40, 1602, 1975.
58. **McCurry, P. M., Jr. and Singh, R. K.,** Regio- and stereo-specificity in the Diels-Alder reaction of a Lewis acid-complexed 4-methylenecyclohex-2-enone, *J. Chem. Soc. Chem. Commun.*, 59, 1976.
59. **Mazza, S., Danishefsky, S., and McCurry, P.,** Diels-Alder reactions of o-benzoquinones, *J. Org. Chem.*, 39, 3610, 1974.
60. **Conia, J. M. and Le Perchec, P.,** La cyclisation thermique des cétones α,β,ξ-diéthyléniques, *Tetrahed. Lett.*, 3305, 1965.
61. **Conia, J. M. and LePerchec, P.,** Thermolyse et photolyse de cétones non saturées. Les transposition de Cope, puis cyclisation, des allyl-isopulegones, *Bull. Soc. Chim. France,* 278, 1966.
62. **Conia, J. M. and LePerchec, P.,** Thermolyse et photolyse de cétones non saturées. Les transpositions thermiques: transposition de Cope, puis cyclisation, d'alkyl-vinyl-cyclanones, *Bull. Soc. Chim. France,* 281, 1966.
63. **Conia, J. M. and LePerchec, P.,** Thermolyse et photolyse de cétones non saturées. Conditions structurales et méchanisme de la cyclisation thermique des cétones α,β-ϵ,ξ-diéthyleniques, *Bull. Soc. Chim. France,* 287, 1966.
64. **Conia, J. M., Drouet, J. P., and Gore, J.,** Thermolyse et photolyse de cétones non saturées. Sur l'obtention et al stereochimie de quelques cétones de la famille de l'acorane formées dans la thermocyclisation de la (+)(isopropenyl-1-pentene-4-yl)-2-méthyl 5-cyclohexanone, *Tetrahedron,* 27, 2481, 1971.
65. **Rouessac, F., Beslin, P., and Conia, J. M.,** La cyclisation thermique des cétones cycliques ϵ,ξ-ethyléniques. Voie d'accès aux cétones polycycliques, *Tetrahed. Lett.*, 3319, 1965.

66. **Conia, J. M. and Leyendeker, F.,** Thermolyse et photolyse de cétones non saturées. Le comportement thermique de cétones éthyléniques en fonciton du nombre de carbones separant les deux centres insaturés, *Bull. Soc. Chim. France,* 830, 1967.
67. **Oppolzer, W., Mahalanabis, K. K., and Battig, K.,** A flexible stereoselective synthesis of the spirosesquiterpenes (±)-β-acorenol, (±)-β-acoradiene, (±)-β-acorenone B, and (±)-acorenone via an intramolecular ene-reaction, *Helv. Chim. Acta,* 60, 2388, 1977.
68. **Martin, S. F. and Desai, S. R.,** 2-Ethoxyallylidene triphenylphosphorane. A new reagent for cyclohexenone annulation, *J. Org. Chem.,* 42, 1664, 1977.
69. **Martin, S. F. and Chou, T.,** Stereoselective total synthesis of racemic acorone, *J. Org. Chem.,* 43, 1027, 1978.
70. **Martin, S. F.,** Carbonyl homologation with α-substitution. A new approach to spiroannulation, *J. Org. Chem.,* 41, 3337, 1976.
71. **Corey, E. J. and Boger, D. L.,** New annulation processes for fused and spiro rings based on the chemistry of benzothiazoles, *Tetrahed. Lett.,* 13, 1978.
72. **Canonne, P., Foscolos, G. B., and Bélanger, D.,** One-step annelation. A convenient method for the preparation of diols, spirolactones and spiro ethers from lactones, *J. Org. Chem.,* 45, 1828, 1980.
73. **Kano, K., Hayashi, K., and Mitsuhashi, H.,** Syntheses of steroids having a bakkenolide type spirolactone ring. I. Synthesis of 4′-methylene dihydrospiro(5α-cholestane-3,3′(2′H)furan)-2′-one, *Chem. Pharm. Bull.,* 30, 1198, 1982.
74. **Burke, S. D., Murtiashaw, C. W., Dike, M. S., Strickland, S. M., and Saunders, J. O.,** Vinylsilane-mediated spiroannulation. Synthesis of spiro[4.5]decadienones, *J. Org. Chem.,* 46, 2400, 1981.
75. **Pons, M. and Simons, S. S.,** Facile, high-yield synthesis of spiro-C-17-steroidal oxetan-3′-ones, *J. Org. Chem.,* 46, 3262, 1981.
76. **Nakamura, E., Fukuzaki, K., and Kuwajima, I.,** Cyclopentenones by internal acylation of vinylsilanes. Rapid construction of trichothecane-type carbon frameworks, *J. Chem. Soc. Chem. Commun.,* 449, 1983.
77. **Matsumoto, T., Shirahama, H., Ichihara, A., Shin, H., Kagawa, S., Sakan, F., Matsumoto, S., and Nishida, S.,** Total synthesis of *dl*-illudin M, *J. Am. Chem. Soc.,* 90, 3280, 1968.
78. **Leriverend, P. and Conia, J. M.,** Étude des petits cycles. La réduction de α,α-bis(bromomethyl)cyclanones par le zinc. Voie d'accès aux cyclopropylcétones spiraniques et aux cétones étyléniques d'agrandissement de cycle. (1re partie), *Bull. Soc. Chim. France,* 116, 1966.
79. **Leriverend, P. and Conia, J. M.,** Étude des petits cycles. La réduction des α,α-bis(bromométhyl)cyclanones par le zinc. Voie d'accès aux cyclopropylcetones spiraniques et aux cétones éthyléniques d'agrandissement de cycle (2e partie), *Bull. Soc. Chim. France,* 121, 1966.
80. **Ibuka, T., Minekata, H., Mitsui, Y., Tabushi, E., Taga, T., and Inunushi, Y.,** Efficient stereoselective synthesis of *rel*-(6S,7S,8S)-7-butyl-8-hydroxy-1-azaspiro[5.5]undecan-2-one, a key intermediate for perhydrohistrionicotoxin, and its *rel*-(6R) isomer, *Chem. Lett.,* 1409, 1981.
81. **Ibuka, T., Mitsui, Y., Hayashi, K., Minakata, H., and Inubushi, Y.,** A new stereoselective synthetic route to perhydrohistrionicotoxin, *Tetrahed. Lett.,* 22, 4425, 1981.
82. **Mundy, B. P. and Otzonberger, R. D.,** Studies on the pinacol rearrangement, *J. Orig. Chem.,* 38, 2109, 1973.
83. **Hückel, W. and Blohm, M.,** Zur Stereochemie bicyclischer Ringsysteme. VII. Die Stereochemie des Dekahydro-Naphtalins und Seiner Derivate. IV. 9-Substitutierte Dekahydro-naphtaline, *Justus Liebigs Ann. Chem.,* 502, 114, 1933.
84. **Burkinshaw, G. F., Davis, B. R., Hutchinson, E. G., Woodgate, P. D., and Hodges, R.,** The synthesis and acid-catalysed rearrangements of 4-hydroxycyclohexa-2,5-dienones, *J. Chem. Soc. C.,* 3002, 1971.
85. **Andersen, N. H., Falcone, M. S., and Syrdal, D. D.,** Structure of vetivenenes and vetispirenes, *Tetrahed. Lett.,* 1759, 1970.
86. **Hikino, H., Aota, K., Kuwano, D., and Takemoto, T.,** Structure of α-rotunol and β-rotunol, *Tetrahed. Lett.,* 2741, 1969.
87. **Shoppee, C. W., Hughes, N. W., Lack, R. E., and Newmann, B. C.,** The structure of digacetigenin, *Tetrahed. Lett.,* 3171, 1967.
88. **Pettit, G. R., Kasturi, T. R., Knight, J. C., and Occolowitz, J.,** Bufadienolides. 8. 12(13→14)*abeo* skeletal rearrangements, *J. Org. Chem.,* 35, 1404, 1970.
89. **Kočovský, P. and Tureček, F.,** Neighboring group participation and rearrangement in hypobromous acid addition, *Tetrahed. Lett.,* 22, 2699, 1981.
90. **Kočovský, P. and Tureček, F.,** An unusual rearrangement in hypobromous acid addition to 10β-vinylcholestane derivatives, *Collect. Czech. Chem. Commun.,* 46, 2892, 1981.
91. **Kropp, P. J.,** The acid-catalyzed cleavage of cyclopropyl ketones related to lumisantonin, *J. Am. Chem. Soc.,* 87, 3914, 1965.
92. **Marshall, J. A. and Johnson, P. C.,** The total synthesis of (±)-β-vetivone, *J. Chem. Soc. Chem. Commun.,* 391, 1968.

93. **Marshall, J. A. and Johnson, P. C.,** The structure and synthesis of β-vetivone, *J. Org. Chem.*, 35, 192, 1970.
94. **Trost, B. M. and Bogdanowicz, M. J.,** New synthetic reactions. Geminal alkylation, *J. Am. Chem. Soc.*, 95, 2038, 1973.
95. **Trost, B. M. and Bogdanowicz, M. J.,** New synthetic reactions. A versatile cyclobutanone (spiroannelation) and γ-butyrolactone (lactone annelation) synthesis, *J. Am. Chem. Soc.*, 95, 5321, 1973.
96. **Trost, B. M., Hiroi, K., and Holy, N.,** A new stereocontrolled approach to spirosesquiterpenes. Synthesis of acorenone B, *J. Am. Chem. Soc.*, 97, 5873, 1975.
97. **Trost, B. M.,** New alkylation methods, *Acc. Chem. Res.*, 7, 85, 1974.
98. **Trost, B. M. and Keeley, D. E.,** On the stereoselectivity and regiospecificity of spiroannelations with 1-lithiocyclopropyl phenyl sulfide, *J. Am. Chem. Soc.*, 96, 1252, 1974.
99. **Trost, B. M. and Mao, M. K.-T.,** α-Substitution-spiroannulation of saturated ketones, *J. Am. Chem. Soc.*, 105, 6753, 1983.
100. **Marshall, J. A. and Brady, S. F.,** Stereochemical relationships in spirovetivane sesquiterpenes. The total synthesis of hinesol, *Tetrahed. Lett.*, 1387, 1969.
101. **Kropp, P. J. and Erman, W. F.,** Photorearrangements of cyclic crossconjugated dienones, *J. Am. Chem. Soc.*, 85, 2456, 1963.
102. **Domb, S. and Schaffner, K.,** Photochemical reactions. The photo rearrangement of 2,5-diene-1,7-dione, *Helv. Chim. Acta*, 53, 1765, 1970.
103. **Caine, D., Boucugnani, A. A., Chao, S. T., Dawson, J. B., and Ingwalson, P. F.,** Stereospecific synthesis of 6,c-10-dimethyl-(r-5-c)-spiro[4.5]dec-6-en-2-one and its conversion into (±)-α-vetispirene, *J. Org. Chem.*, 41, 1539, 1976.
104. **Gramain, J. C., Troin, Y., and Valle, D.,** A new photochemical spiroannelation method. Access to substituted piperidines, *J. Chem. Soc. Chem. Commun.*, 832, 1981.
105. **Lenz, G. R.,** The photochemistry of enamides, *Synthesis*, 489, 1978.
106. **Set, L., Cheshire, D. R., and Clive, D. L. J.,** Synthesis of spiro-compounds: use of diselenoacetals for generation of quaternary centers by alkylation and radical cyclization, *J. Chem. Soc. Chem. Commun.*, 1205, 1985.
107. **Nyström, J.-E., McCanna, T. B., Helquist, P., and Iyer, R. S.,** Short intramolecular Diels-Alder approach to functionalized spiro 4.5 decanes, *Tetrahed. Lett.*, 26, 5393, 1985.

Chapter 5

BRIDGED SYSTEMS

With bicyclic systems we can conceive of three annulation types: (1) *ortho*-condensed rings in which the peripheral bonds are linked with a cross-piece bond (A in Figure 1); (2) bridged rings in which the cross-piece is made up of *n* atoms ($n \geqslant 1$) (B in Figure 1); and (3) spirocyclic with a single pivot carbon atom connecting both rings (C in Figure 1).

While syntheses of the first and third type of compounds have been discussed in the previous chapters, the second type will be the objective of this chapter.

The situation becomes more complex with tricyclic systems (Figure 2). Depending on the number of quaternary carbon atoms we can derive four types of *ortho*-condensed systems (Figure 2, D to G). Note that only type D can be built up unequivocally from a bicyclic *ortho*-condensed system, while types E and G contain spirocyclic subsystems, and F may be regarded as a bridged bicyclic subsystem with one cross-piece bond added. Furthermore, we can consider ring systems that would arise by bridging *ortho*-condensed (H, I) or spirocyclic (J) bicyclic systems, or those in which the central ring is spanned by two bridges (K to N). The latter tricycles (K to N) can be further divided into cross-bridged (L) and parallel-bridged (K) systems, and those having one (M) or two (N) quaternary bridgehead carbon atoms. Let us say further that natural products could hardly be expected to contain skeletons of the N type. Depending on the length of the bridges in K, L, and M, these systems may also involve *ortho*-condensed subsystems.

The basic principles and strategies in the stereoselective synthesis of cyclic systems have been given in Chapter 2. Here we will focus our attention on particular syntheses of bi-, tri-, and polycyclic bridged systems or subunits composing a more complex molecule of a natural product. This will include cases in which the synthetic problem is to build up a bridged bicyclic fragment, regardless of whether the other part of the molecule in question contains other *ortho*-condensed or spirocyclic systems.

I. BICYCLIC BRIDGED SYSTEMS

We have already outlined the general principles of construction of bridged bicyclic systems in Chapter 2, Figure 13. As with the *ortho*-condensed skeletons we will distinguish cyclizations at or outside the annulation sites; the tactics will include monotopic and ditopic processes, speaking in terms of retrosynthetic analysis.

A. Cyclization at Bridgehead Positions

In order to close a ring at bridgehead sites we can use methods of mono- and ditopic cyclization. In the former case we would start from a monocyclic dipolar synthon (1 in Figure 3). The starting point of a ditopic annulation depends on whether the process is inter- or intramolecular. Intermolecular ditopic cyclization starts from a monocyclic precursor; on the other hand, an intramolecular ditopic process would form both rings in a single synthetic step, starting from an acyclic precursor (see Chapter 2, Figure 1).

1. Monotopic Cyclizations

The general strategy of forming a bicyclic bridged system by monotopic cyclization at the bridgehead position is shown in Figure 3. In this section we will discuss the choice of suitable synthetic equivalents of the dipolar synthon 1.

The cyclization shown in Figure 3 can be performed as intramolecular nucleophilic substitution. Such an example is shown in Figure 4. The electrophilic center in 2 is located on the terminal carbon atom of the side chain bearing the chlorine atom, while the nucleophilic

A B C

FIGURE 1.

D E F G

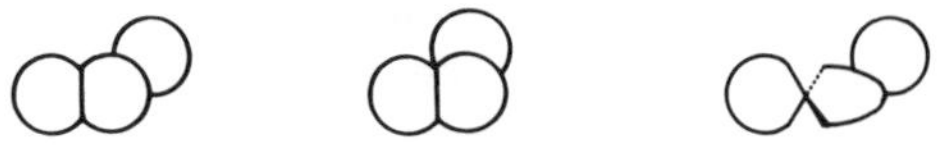

H I J

K L M N

FIGURE 2.

1

FIGURE 3.

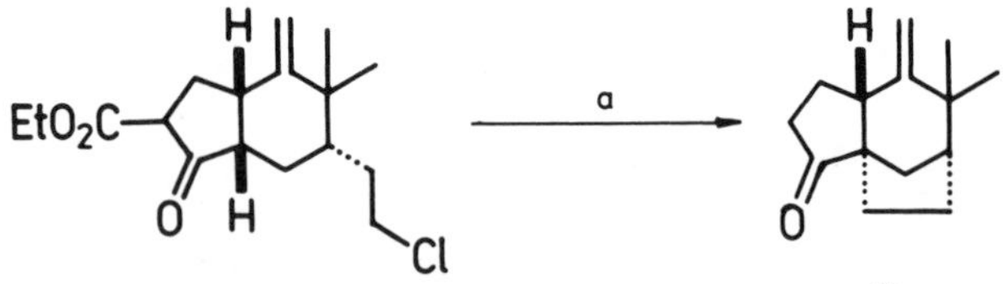

EtO2C H H O 2 Cl a H O 3

FIGURE 4. (a) NaOH, CH_3OH, H_2O, reflux.

center is at the α-position to the carbonyl group. Under basic conditions, compound 2 undergoes cyclization and hydrolysis-decarboxylation affording khusimonine 3.[1,2] For further examples see References 3 and 4. Intramolecular Michael addition has also been used in the synthesis of bridged systems.[5] The stereoelectronic control in cyclizations of this type has been discussed in Reference 6.

An alternative way to bridged systems makes use of intramolecular aldolization or related reactions (Figure 5).[7-12] The Wittig reaction has been employed in synthesis of strained bridged systems, even in cases where the molecule being formed violated Bredt's rule (Figure 6).[13-17]

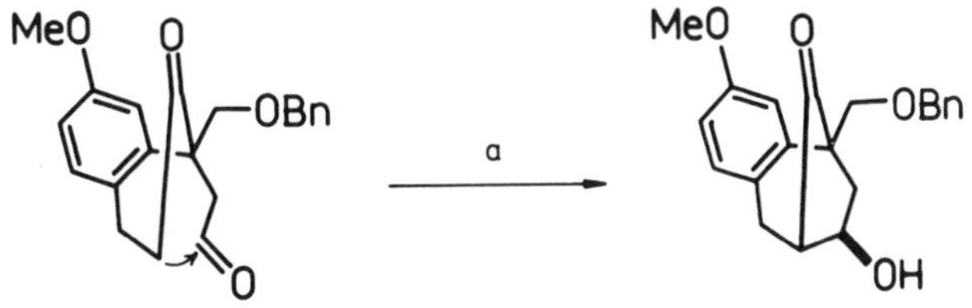

FIGURE 5. (a) NaOH, CH_3OH, H_2O, 55°C, 30 hr.

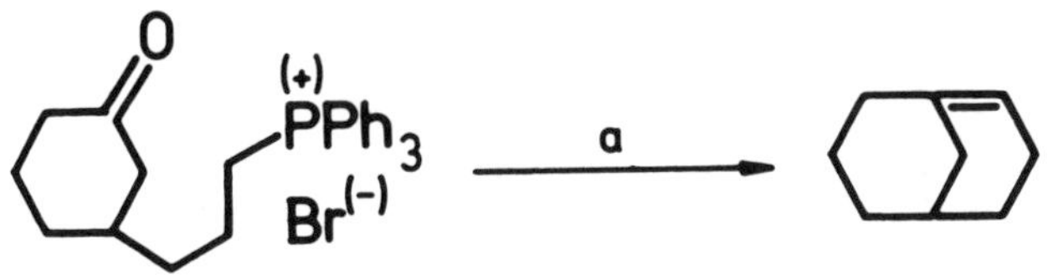

FIGURE 6. (a) NaH, t-$C_5H_{11}OH$.

FIGURE 7. (a) $KHCO_3$, dioxane, H_2O, 80°C, 4 hr.

As with *ortho*-condensed systems, we can use the addition of an electrophilic carbon center to a double bond to create a bridged system (Figure 7).[18] This cyclization can also be regarded as a solvolysis of the mesylate assisted by participation by the neighboring double bond. The steroid mesylate 4 is cleanly solvolyzed to form the [3.2.1] bridged system. By contrast, the Δ^5-isomer 5 is inert under the same or even more drastic conditions. Stereochemical analysis indicates that in 4 the distance between the electrophilic center at C-19a and C-3 is ca. 0.25 nm, with the angle of approach being estimated as ca. 10 to 15°. The cyclization in the Δ^5-isomer 5 is disfavored by geometrical factors. The distance between C-19a and C-6 is about 0.30 nm (from Dreiding models) and also the angle of approach is larger (ca. 30°). Hence, it follows that stereoelectronic effects should be carefully analyzed in designing a cyclization step in a synthesis.

Another example of the intramolecular electrophilic addition is presented in Figure 8. The bicyclic compound 6 arises by attack of the acylium ion to the double bond.[19,20] Reactive positions on an aromatic ring can be also attacked by an electrophile[21,22] as shown in Figure 9.[23,24]

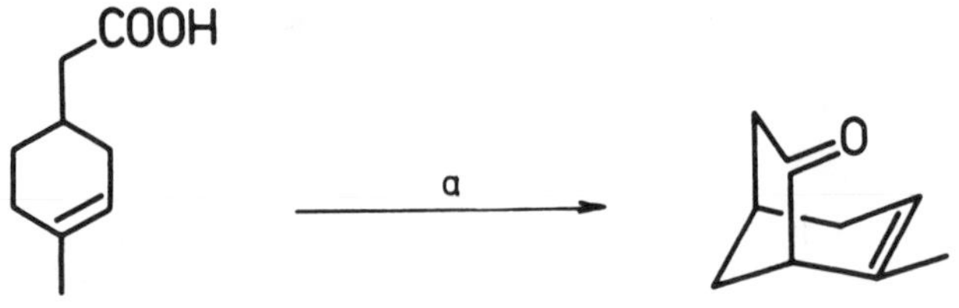

FIGURE 8. (a) $(CF_3CO)_2O$.

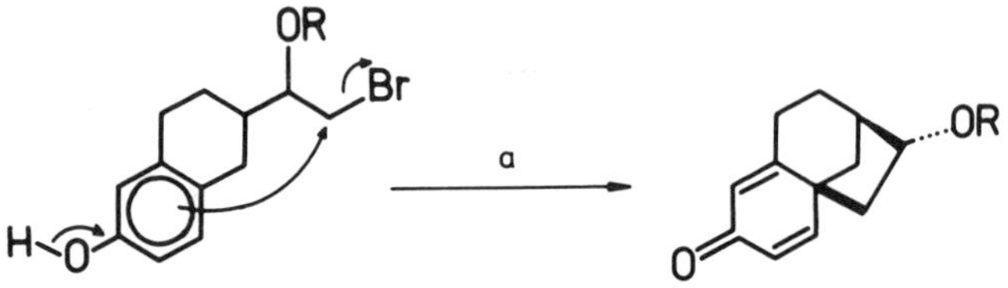

FIGURE 9. (a) t-BuOK, t-BuOH, reflux.

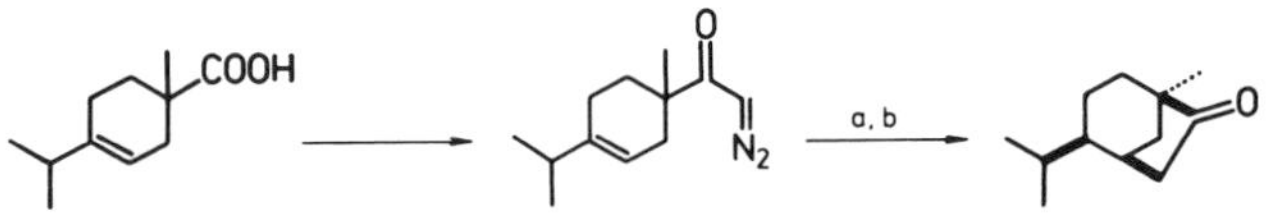

FIGURE 10. (a) $BF_3.Et_2O$; (b) H_2, Pd.

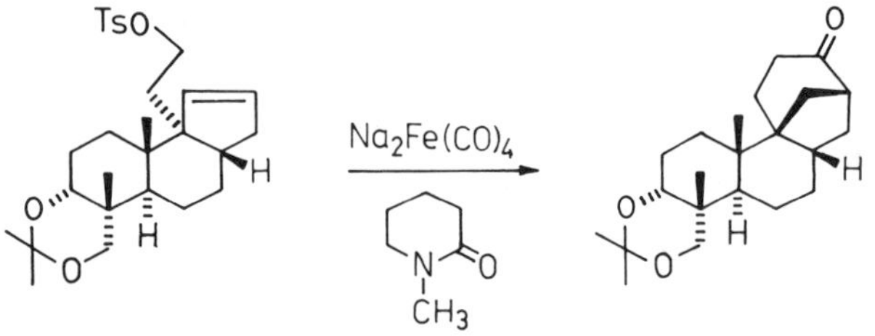

FIGURE 11.

Intramolecular cyclization of unsaturated diazoketones (itself a ditopic reaction) can be used to prepare a bridged system, provided the intermediate cyclopropane ring is regioselectively cleaved by hydrogenolysis (Figure 10).[25] Further examples of cyclization are given in References 26 to 29.

The general scheme of cyclization (Figure 3) can be modified by inserting a bidental synthon between the reactive centers of the dipolar synthon. Such a stitching is shown in Figure 11.[27-30] The carbonyl group provided by sodium tetracarbonylferrate[30] links the electrophilic carbon atom (from the tosylate) with the nucleophile (one of the carbon atoms of the original double bond).

2. Ditopic Cyclizations

The most common means of construction of bicyclic bridged systems is the Diels-Alder reaction (Figure 12). The orientation and substituent effects in the Diels-Alder reaction have been discussed in detail elsewhere[31,32] (see also the discussion in Chapter 3, Section I. B.4

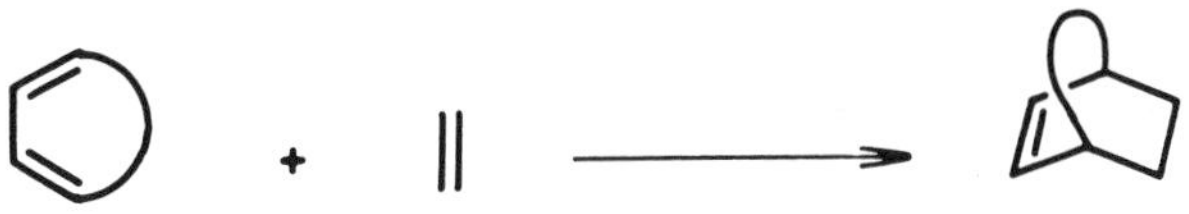

FIGURE 12.

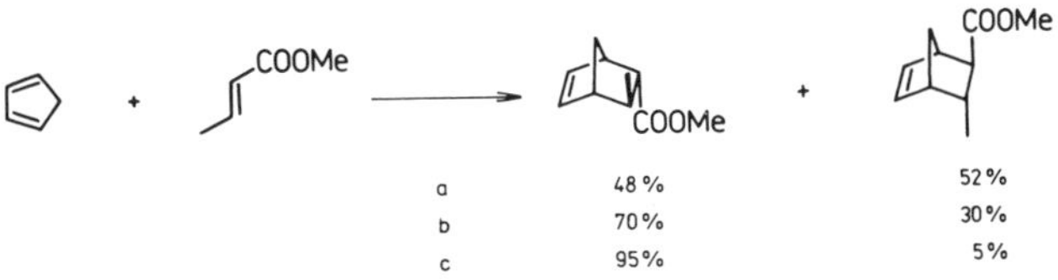

FIGURE 13. (a) Decaline, 30°C; (b) AcOH, 30°C; (c) $AlCl_3$, BF_3.

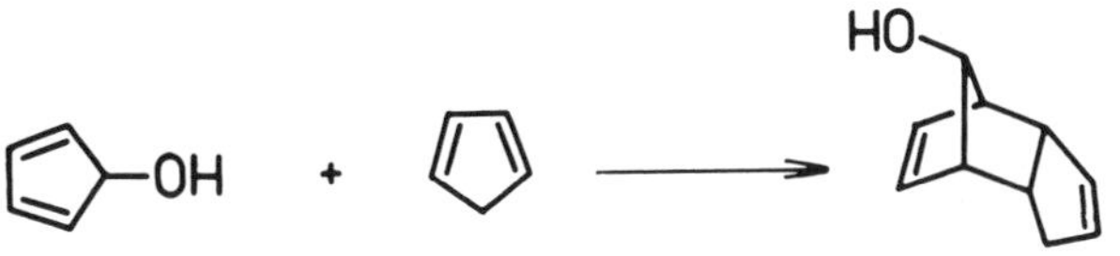

FIGURE 14.

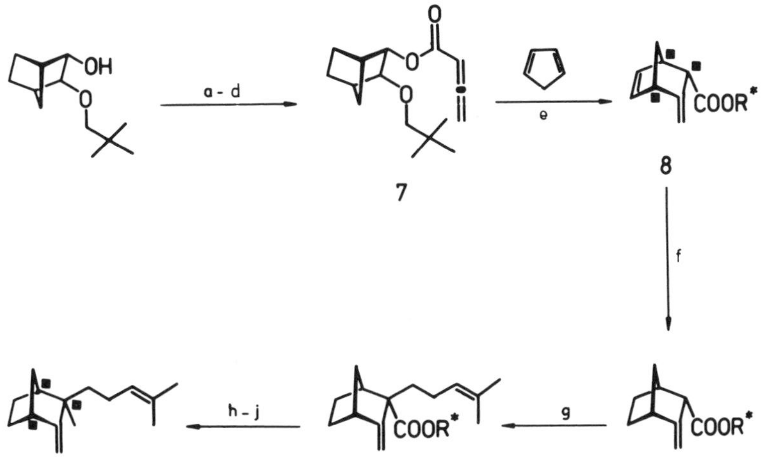

FIGURE 15. (a) $BrCH_2COBr$, AgCN, C_6H_6, reflux 20 min; (b) Ph_3P, C_6H_6, (c) reflux 15 hr; (d) CH_3COCl, Et_3N, CH_2Cl_2, 0 → 40°C; (e) $(i\text{-}PrO)_2TiCl_2$, CH_2Cl_2, −20°C, 6 hr; (f) H_2, $NaBH_4$, $(AcO)_2Ni$; (g) $(i\text{-}Pr)_2NLi$, then $Me_2CH=CHCH_2CH_2I$, THF, HMPA; (h) $LiAlH_4$; (i) PCC; (j) Wolf-Kizhner.

and Chapter 1 in Volume II). Here we will restrict ourselves to the illustration of the effects of catalysts on the steric course of the formation of substituted [2.2.1] systems (Figure 13).[33]

The Diels-Alder addition of cyclopentadienol to cyclopentadiene yields an *anti*-adduct, i.e. the addition takes place at the side opposite to the hydroxyl group (Figure 14).[34] Other examples of the Diels-Alder addition of various dienophiles to cyclopentadiene can be found in References 26, 31, and 35 to 41 (for review see Reference 42).

An elegant homochiral synthesis of (−)-β-santalene, based on an asymmetric Diels-Alder reaction of cyclopentadiene to a chiral allene system 7, has been reported by Oppolzer (Figure 15).[43] The optical yield of the Diels-Alder adduct 8 was 99% (!), whereas the *endo/exo* selectivity was 98:2. For different strategy resting on rearrangement in bridged system see Reference 44.

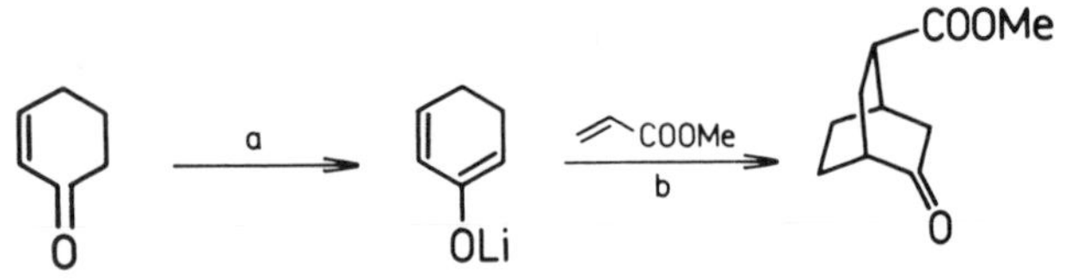

FIGURE 16. (a) $(i\text{-}Pr)_2NLi$, THF, C_6H_{14} (4:1), −23°C, 1 hr; (b) −23°C, 1 hr.

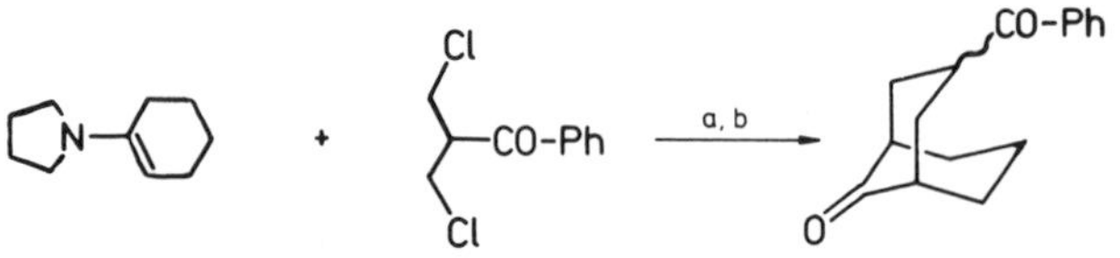

FIGURE 17. (a) Et_3N, CH_3CN, 1.5 hr; (b) H_2O, r.t., 12 hr.

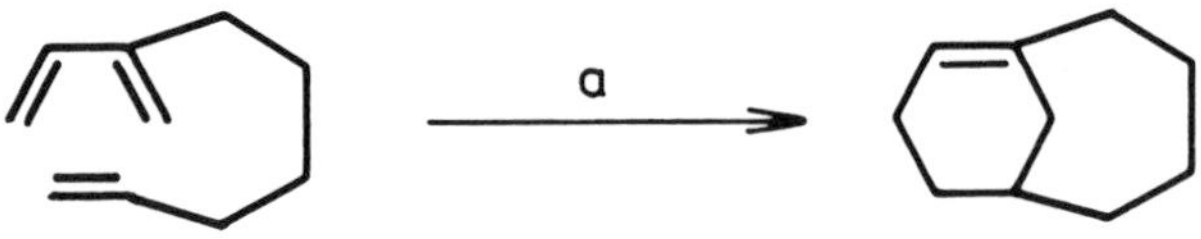

FIGURE 18. (a) 455°C, 5 sec.

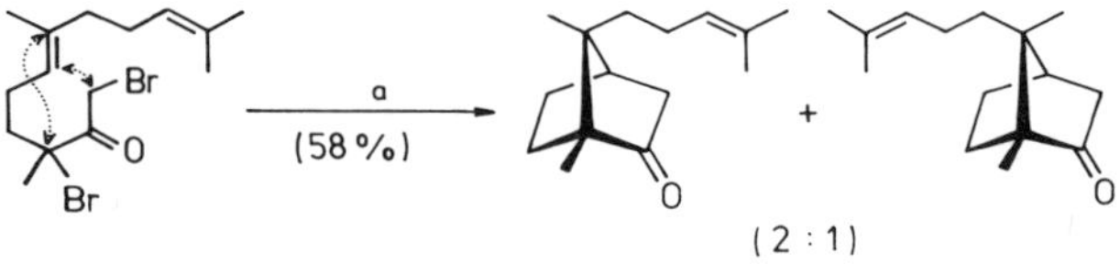

FIGURE 19. (a) $Fe(CO)_5$, 100°C (58%).

Michael addition of the enolate derived from cyclohexenone to methyl acrylate, followed by cyclization (Figure 16)[45-47] is analogous to Diels-Alder cyclization, yielding a bicyclo[2.2.2]octane system. Cyclizations of this type can also be accomplished by some related processes.[48-59] Bidental synthons, such as the dichloroketone in Figure 17, react with cyclic enamines to form bridged bicyclic systems.[52,53]

As already shown in Figure 12, the intermolecular Diels-Alder reaction may be used for constructing a bicyclic system if the synthesis starts from a monocyclic and an acyclic synthon. If the diene and the dienophile moieties are linked by a chain of appropriate length, it is possible to perform an intramolecular Diels-Alder cyclization yielding directly a bicyclic bridged system (Figure 18)[54,55] (for review see Reference 56).

The intramolecular version of [3 + 2] dipolar addition (Figure 19)[57] has also been used in a synthesis of a substituted bicyclo[2.2.1]heptanone. The dipolar $_{\pi}3_s$ component was prepared *in situ* under catalysis by iron pentacarbonyl.[57]

3. *Rearrangements in the Synthesis of Bridged Systems*

As we have mentioned, rearrangements of a Wagner-Meerwein type afford various possibilities for transformations of *ortho*-condensed systems to bridged ones and *vice versa*. Interconversions of bridged systems are known as well and often provide an elegant means of the synthesis of otherwise difficultly accessible skeletons. However, the albene case teaches us,[58] that one has to be extremely careful when using this strategy (see Chapter 3).

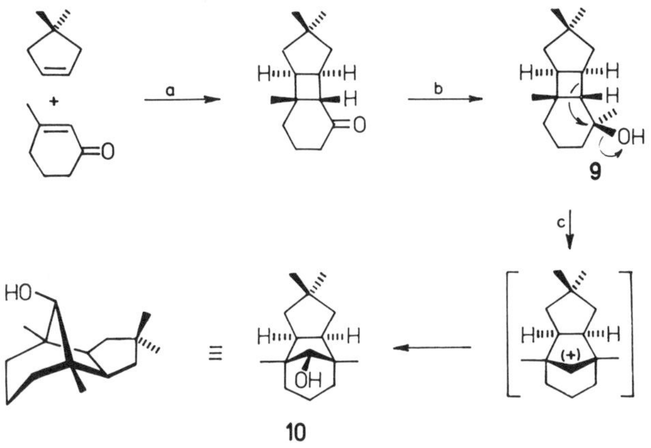

FIGURE 20. (a) hν; (b) CH_3Li; (c) H^+.

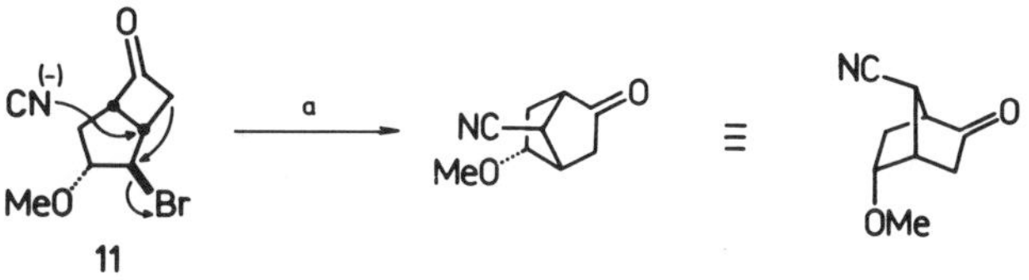

FIGURE 21. (a) KCN, cat. CH_3ONa, CH_3OH.

a. Rearrangements of ortho-Condensed Systems to Bridged Systems

In this subsection we will show two examples of rearrangements of a [x.y.0] → [x,y - 1,1] type. Corey and Nozoe[59,60] have implemented the stereospecific rearrangement of a [4.2.0] subsystem to a [3.2.1] subsystem (Figure 20) in their synthesis of α-caryophyllene alcohol. The rearrangement was induced by protonation of the tertiary alcohol 9. An analogous rearrangement ([4.2.0] to [3.2.1]), was used in the synthesis of hirsutene.[61]

ortho-Condensed [3.2.0] systems, easily accessible by photoaddition of ketenes to cyclopentadiene (see Chapter 3, Section I.B.2), can be transformed by rearrangement to norbornane derivatives. The rearrangement shown in Figure 21[62] is triggered by nucleophilic substitution of the exobromine atom. Since S_N2 *endo*-attack of cyanide ion at the carbon bearing bromine is hampered for steric reasons, the strained [3.2.0] system 11 rearranges and the nucleophile enters the accessible apical position. This scheme represents an alternative route to the [2.2.1] system (see Figure 13).

b. Rearrangements in Bridged Systems

In Chapter 3, Section III.C.2 we have shown some rearrangements of bridged skeletons leading to *ortho*-condensed systems. Bridged skeletons of a [x.y.z] type can also be rearranged to [x + 1,y,z − 1] systems. An example of such a procedure (12→ 13) is visualized in Figure 22.[63] Another example of the acid-catalyzed rearrangement is shown in Figure 23.[64] Further examples resting on this elegant strategy can be found in References 26 and 44.

In addition to rearrangements involving simultaneous enlargement and contraction of vicinal rings, there are also reactions that result in expansion of only one ring. The rearrangement of aminoalcohol 14 (Figure 24) was the key step in the synthesis of α-cedrene (15).[65] This topic has been treated in detail in Chapter 3, Section III.

4. Fragmentation Methods in the Synthesis of Bridged Systems

The fragmentation of the cross-piece bond in an *ortho*-condensed polycyclic skeleton has

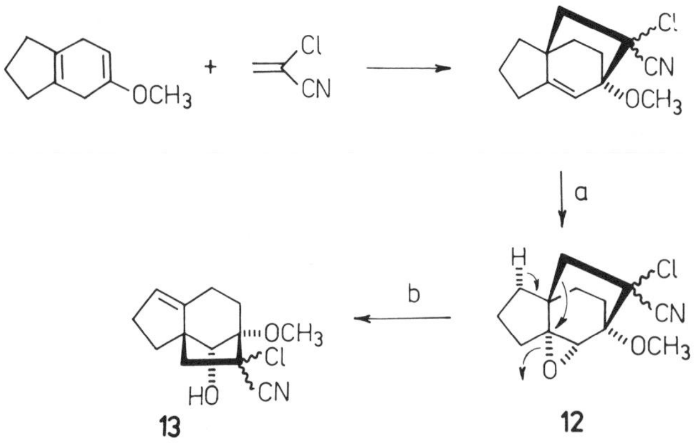

FIGURE 22. (a) m-Cl-$C_6H_4CO_3H$; (b) $BF_3.Et_2O$.

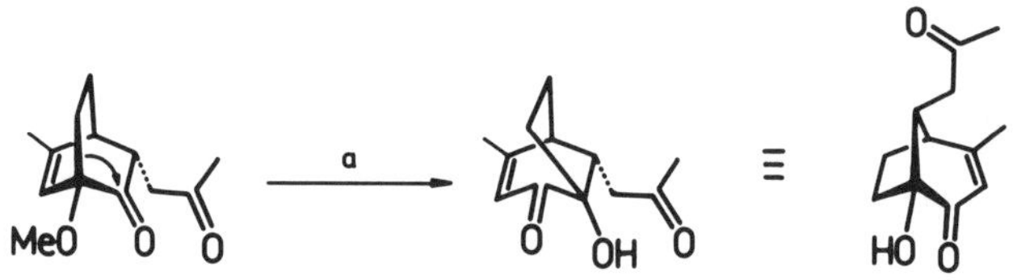

FIGURE 23. (a) HCO_2H.

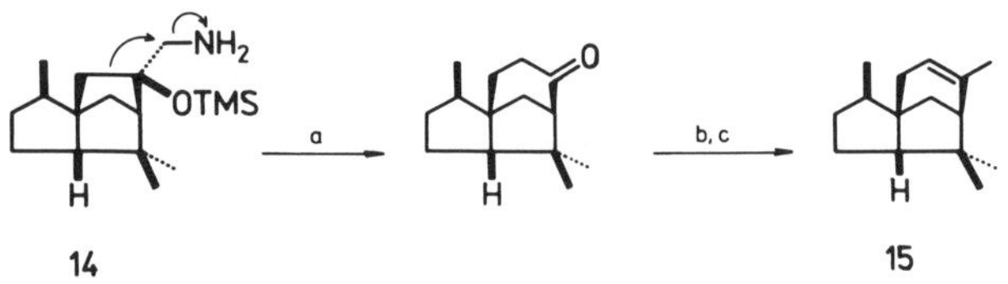

FIGURE 24. (a) HNO_2; (b) CH_3Li; (c) $SOCl_2$, C_5H_5N.

already been mentioned as a way to new skeletons, especially those containing medium or large rings (see Chapter 3, Section III.D). This step can of course be incorporated into the synthetic strategy aiming at a bridged system, as documented by the synthesis of taxane skeleton 21 by Trost and Hiemstra (Figure 25).[66] For other examples see References 67 and 68.

B. Cyclization Outside the Annulation Sites

The basic methods enabling cyclization of a suitable precursor (especially one with defined stereochemistry of substituents) outside the annulation sites have been discussed in Chapter 3, Section II. The situation can be generally described by Figure 26.

The choice of a suitable synthon to be cyclized depends on the availability of the respective synthetic equivalents of 22. For instance, Stork and Clarke, have employed Claisen condensation in 23 (Figure 27)[69] to clamp the six-membered ring in the molecule of cedrol (24).

Intramolecular nucleophilic substitution represents one of the pivotal steps in Danishefsky's synthesis of the quadrone skeleton (Figure 28).[70,71] Other examples of this sort can be found in Reference 26.

II. TRICYCLIC BRIDGED SYSTEMS

In principle, tricyclic bridged systems can be synthesized by methods of monotopic or

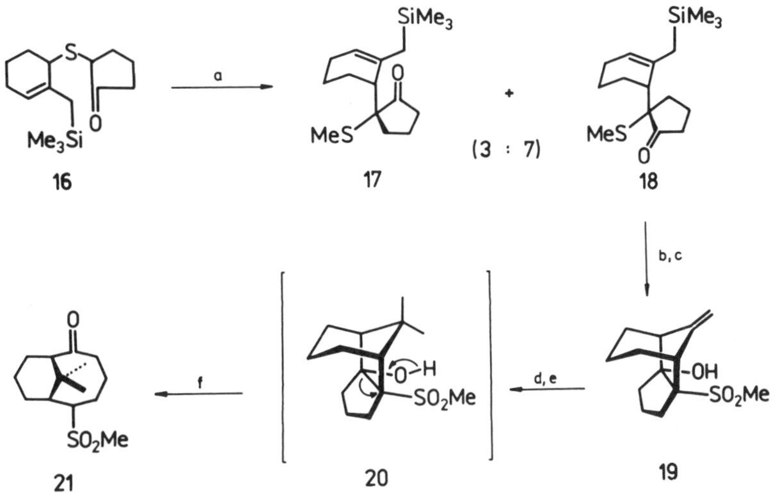

FIGURE 25. (a) CH_3I, KH, DME, reflux; (b) m−Cl−$C_6H_4CO_3H$; (c) $EtAlCl_2$, C_6H_6; (d) CH_2I_2, Zn−Cu; (e) H_2, Pt; (f) t-BuOK (cat.), Me_2SO.

FIGURE 26.

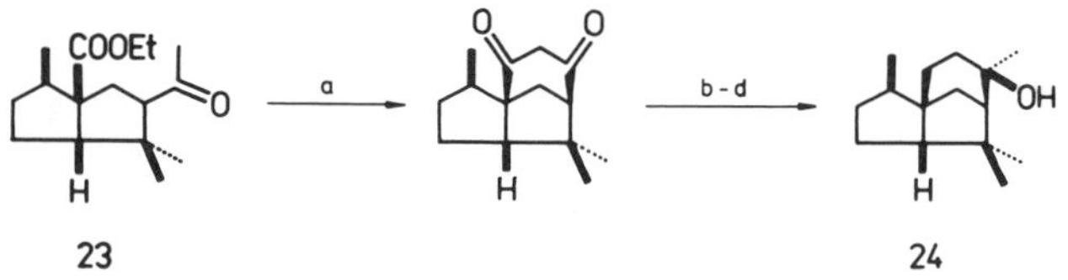

FIGURE 27. (a) t-BuOK, t-BuOH; (b) $LiAlH_4$; (c) $CrO_3 \cdot C_5H_5N$; (d) CH_3Li.

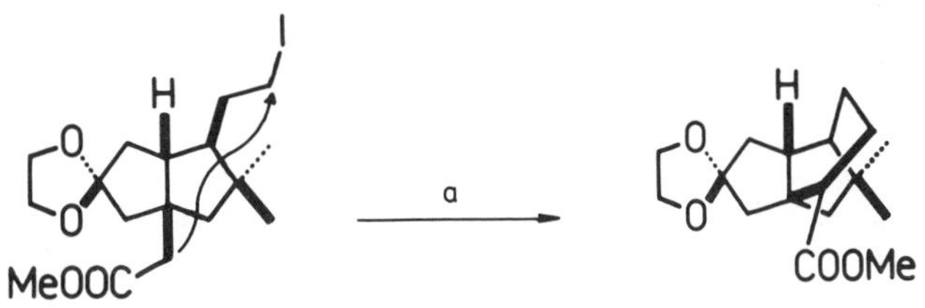

FIGURE 28. (a) $(Me_3Si)_2NLi$.

ditopic cyclization, the new bond being formed at or outside the annulation sites. Nevertheless, the more complex nature of tricyclic systems makes the retrosynthetic analysis more flexible as to the number of strategic bonds to be broken, and thus broadens the spectrum of possible synthetic approaches. Therefore, we will pay more attention to retrosynthetic analysis of these systems.

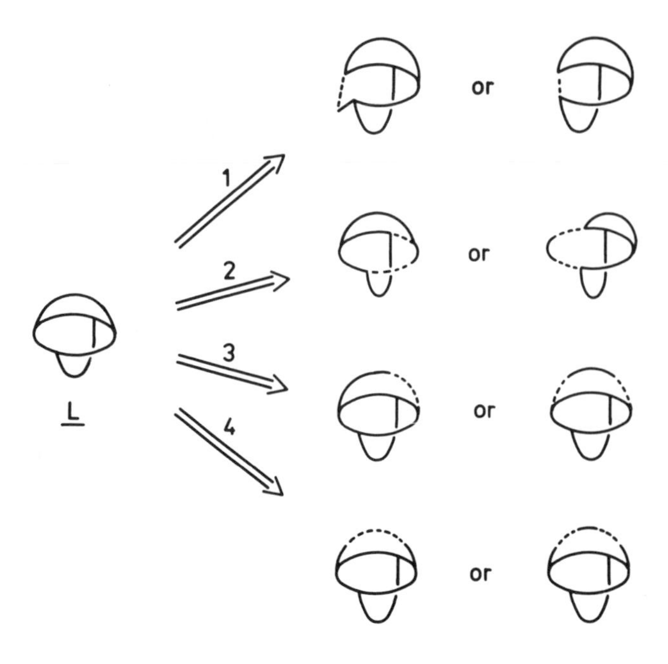

FIGURE 29.

When designing a synthesis of a tricyclic system (or subsystem), the point to begin with is to locate the central ring, i.e. that spanned by two bridges. Location of the central ring in complex systems may or may not be a trivial matter, and sometimes two or more central rings can be found, especially in systems with molecular symmetry. In the next step of the retrosynthetic analysis we disconnect one of the bonds in the central ring and look for an appropriate synthon and its synthetic equivalent, the cyclization of which would reform the central ring.

Another possible approach consists of disconnecting one of the rings that bridge the central ring. In addition to these basic possibilities we can of course consider rearrangements.

In this Chapter we shall focus our attention on syntheses of tricyclic systems of the L and M type (cf. Figure 2). The bridged tricycles belonging to the H to J types are composed of bicyclic subunits and this would be reflected in the retrosynthetic analysis. Tricycles under K and N are uncommon among natural products and thus synthetic approaches to them will be omitted.

The possible means of retrosynthetic analysis of a tricyclic bridged system L are shown in Figure 29: (1) one disconnects one of the bonds in the central ring and then plans to close the ring via monotopic cyclization; (2) consists of disconnecting two bonds in the central ring followed by reverse ring closure by a ditopic process (inter- or intramolecular); (3) it is supposed that the central ring has already been built and it is necessary to span it by the second bridge, via either monotopic or ditopic cyclization; and (4) comprises closing the ring outside the annulation sites, again via a monotopic or ditopic cyclization. The tricyclic system M (see Figure 2) would be analyzed in the same way.

Figure 30 depicts tricyclic bridged systems of L type in which the central ring is a cyclobutane, cyclopentane, cyclohexane, or a cycloheptane, i.e. the systems which are encountered most frequently in natural products. The structures that would involve bridging vicinal ring position may be regarded as being composed of *ortho*-condensed subsystems and are therefore omitted. Obviously, tricyclic structures derived from cyclobutane (4_1) and cyclopentane (5_1) can be each realized as a single isomer regardless of the length of the bridge. Cyclohexane offers three alternatives (6_1 to 6_3), while with cycloheptane there are

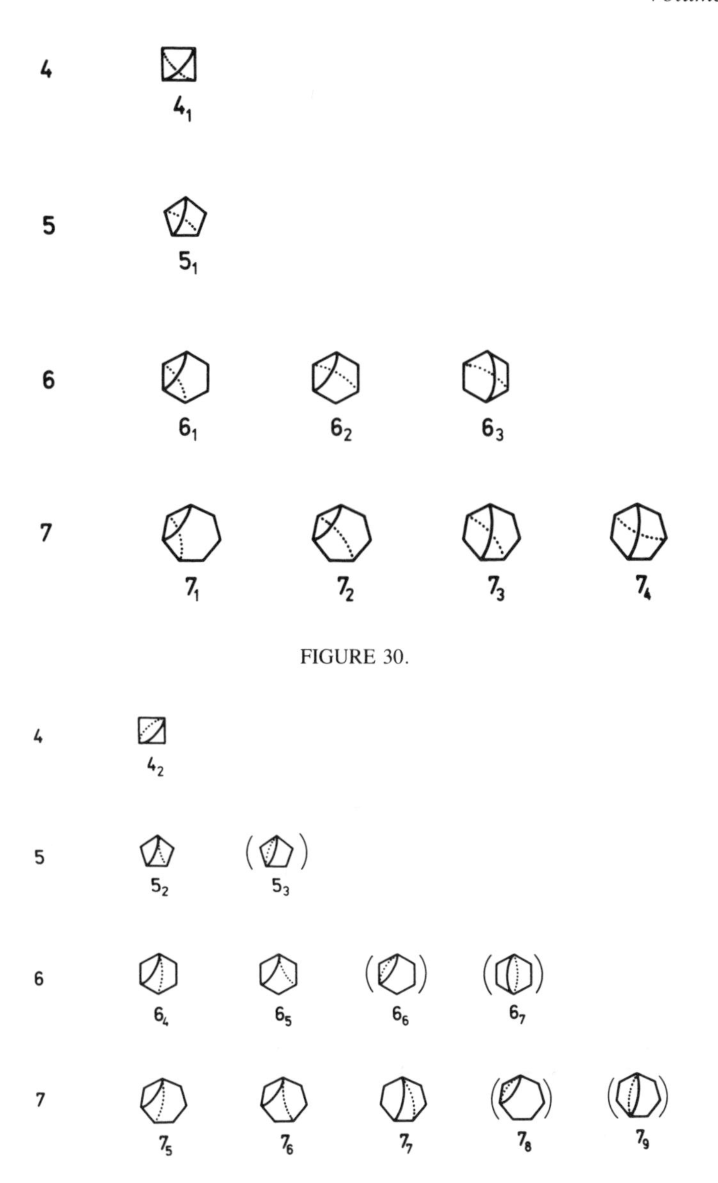

FIGURE 30.

FIGURE 31.

four isomers (7_1 to 7_4). Tricyclic systems of M and N type which contain one or two quaternary bridgehead positions due to double branching may exist in a number of variations (Figure 31). However, the types in parentheses can hardly be found in natural products and will not be treated in detail.

A. Syntheses Including Formation of the Central Ring

We have already briefly summarized the types of tricyclic skeletons which are found in nature and have also shown retrosynthetic analysis using mono- and ditopic processes to close the central ring. The outlined general scheme will be elaborated in the following sections.

1. Monotopic Cyclization

Let us first consider the principal possibilities of the retrosynthetic analysis on the basis of connectivity of the individual positions on a given skeleton.

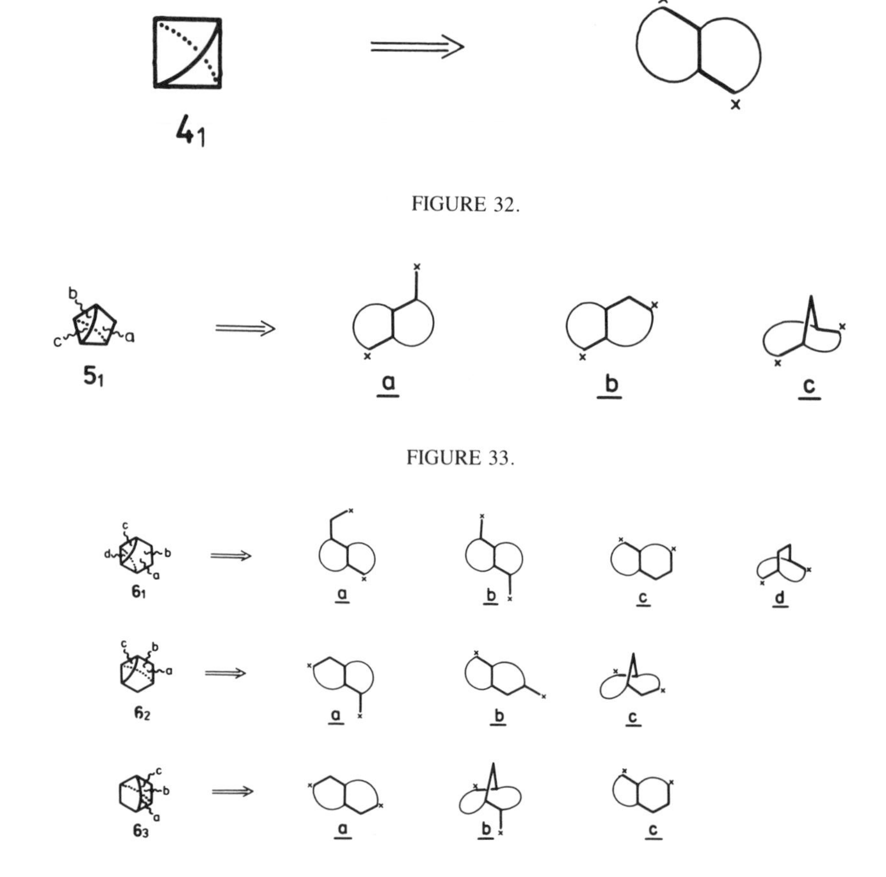

FIGURE 32.

FIGURE 33.

FIGURE 34.

The systems with a cyclobutane central ring (4_1) can be retroanalyzed, regardless of substituents, by disconnecting any bond of the central ring (Figure 32). In this case, disconnecting either bond leads to a single synthon, a bicyclic *ortho*-condensed derivative (Figure 32). The positions to be linked with a new bond are denoted by crosses while the bonds constituting the central ring are drawn in bold lines.

Systems possessing a cyclopentane central ring (5_1, Figure 33) afford three different structures which arise by disconnecting bonds a, b, or c. The first two structures (a, b) involve *ortho*-condensed synthons while the third structure (c) is bridged.

The retrosynthetic analysis of systems with six- and seven-membered central rings follows the same lines as that of 4_1 and 5_1 skeletons (Figures 34 and 35).

The situation is different with systems of the M type in which one of the bridgehead positions on the central ring is a quaternary carbon due to double bridging. The retrosynthetic analysis shows that the synthons obtained by disconnecting the pertinent bonds of the central rings cannot be *ortho*-condensed systems, but spirocycles or bridged systems. For illustration we show three examples of retrosynthetic analysis in systems with a cyclopentane (5_2) or a cyclohexane (6_4, 6_5) central ring (Figures 36 and 37 respectively).

It should be noted here that the synthesis of compounds possessing skeletons of the L type often follows the guidelines provided by retrosynthetic analysis, even though not all of the possibilities are synthetically feasible due to steric reasons. On the other hand, compounds with skeletons of the M type are usually synthesized according to a different strategy because of the presence of the quaternary bridgehead carbon atom (*vide infra*). The

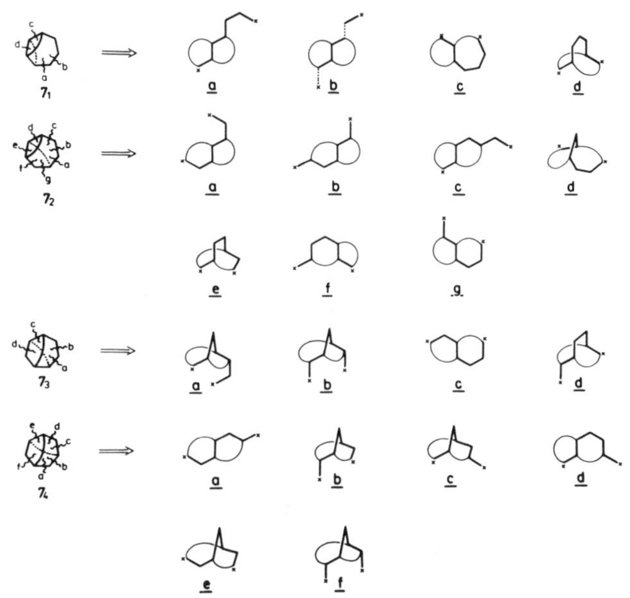

FIGURE 35.

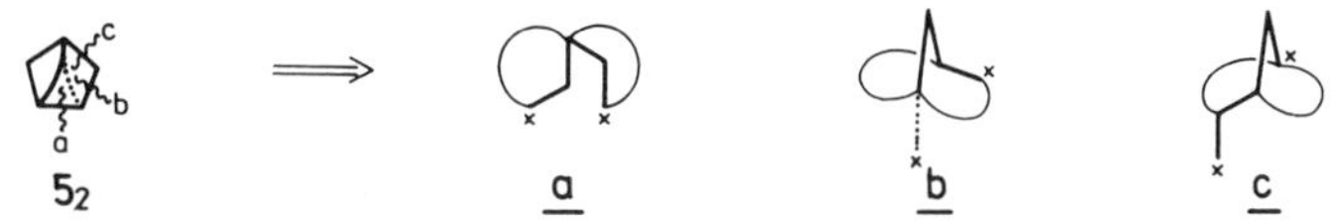

FIGURE 36.

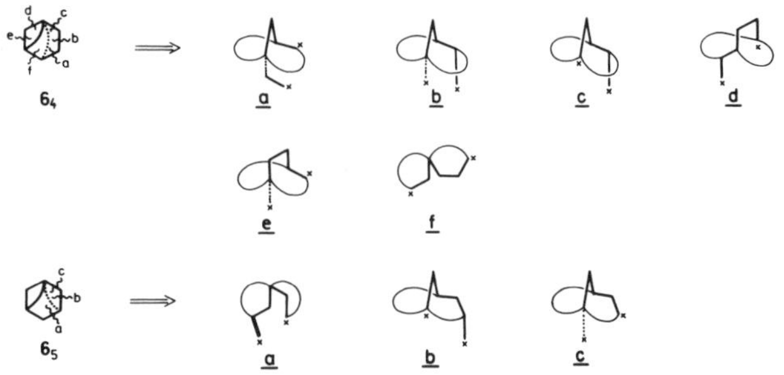

FIGURE 37.

following paragraphs will show selected examples of syntheses that follow the strategy outlined in Figures 32 to 35.

Copaene (29) and ylangene (30) are isomeric sesquiterpenes containing a doubly bridged central ring of the 4_1 type. According to Figure 32, a possible synthetic route would start from a bicyclic *ortho*-condensed derivative furnished with an electrophilic and a nucleophilic center at the connection sites. Since the bridges in the target compounds each contain three carbon atoms, the starting synthon should be a decalin system. The corresponding synthetic equivalent is represented by the enolate 26 (Figure 38)[72,73] generated from the tosyloxy ketone 25. The tosyloxy group is placed at one position of the future connection (electrophilic center), while the nucleophilic center is formed by generating the enolate. Linking these

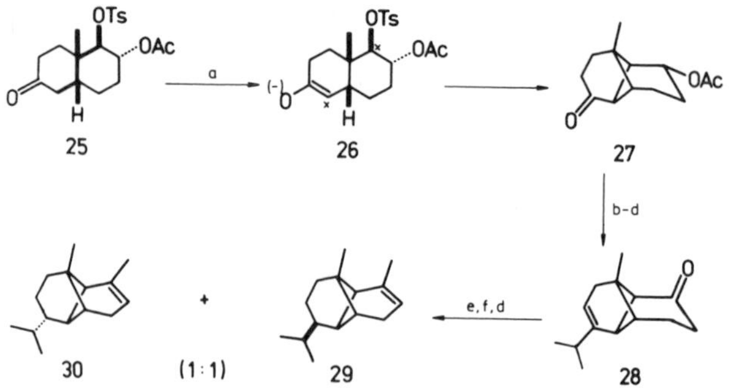

FIGURE 38. (a) Base; (b) i-PrLi; (c) CrO_3, H_2SO_4, Me_2CO, H_2O; (d) $POCl_3$, C_5H_5N; (e) CH_3Li; (f) H_2, Pd–C.

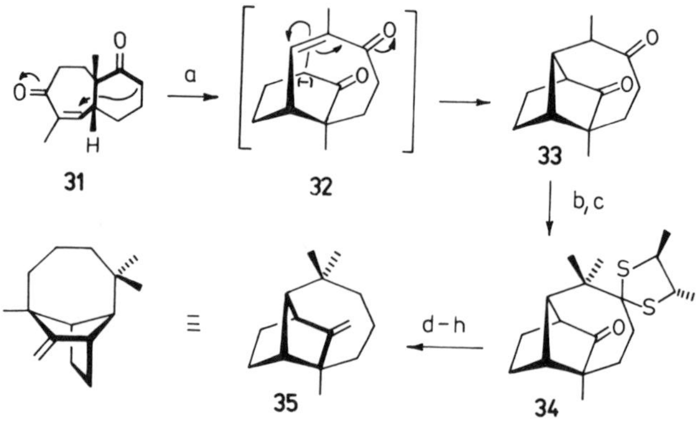

FIGURE 39. (a) Et_3N, ethylene glycol, 225°C; (b) CH_3I, Ph_3CNa; (c) L-(+)–$CH_3CH(SH)$–$CH(SH)$–CH_3 and resolution; (d) $LiAlH_4$; (e) Na, ethylene glycol, N_2H_4; (f) CrO_3; (g) CH_3Li; (h) $SOCl_2$, C_5H_5N.

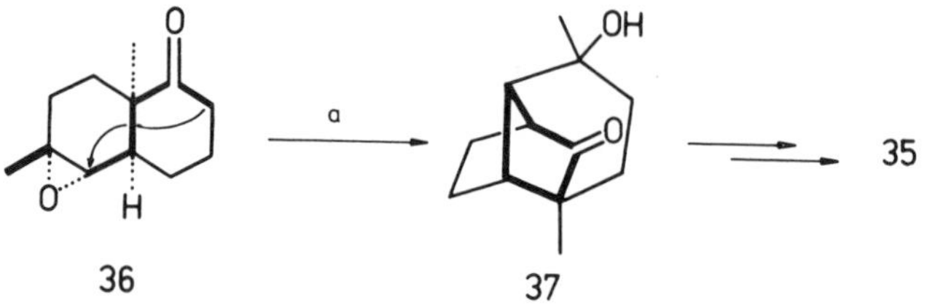

FIGURE 40. (a) $CH_3SOCH_2^-$ Na^+, Me_2SO.

two centers by a chemical bond creates the basic skeleton in 27 which is furnished with substituents and transformed to target sesquiterpenes 29 and 30.

Longifolene (35) contains a cyclopentane central ring, bridged by a two-carbon and a four-carbon bridge. The Corey synthesis was represented by the *ortho*-condensed [5.4.0] system 5_1b. The synthetic equivalent of this synthon was the endione 31 (Figure 39)[74,75] which was cyclized in basic medium to give compound 33 possessing the desired skeleton. In several steps, including resolution of the racemic product to enantiomers, the intermediate 33 was converted to natural (+)-longifolene.

Another synthesis of longifolene, based on the same retrosynthetic analysis (5_1b), was reported by McMurry (Figure 40).[76] An analogy can be found in the base-catalyzed cyclization of santonin to santoninic acid.[77] For another example of the 5_1b strategy employing, however, a different synthetic equivalent (see Reference 78).

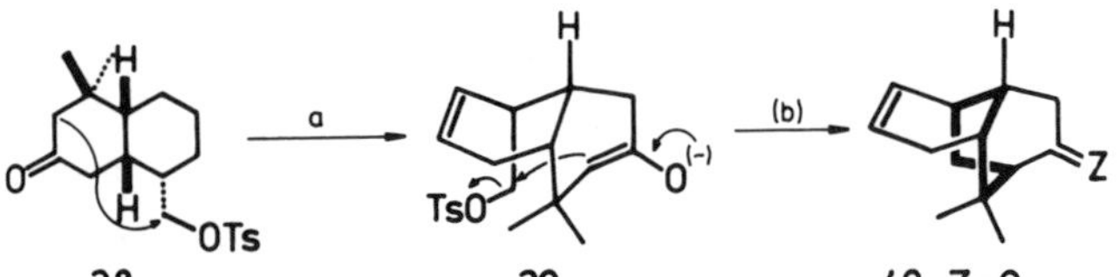

FIGURE 41. (a) NaH, Me_2SO; (b) N_2H_4, KOH.

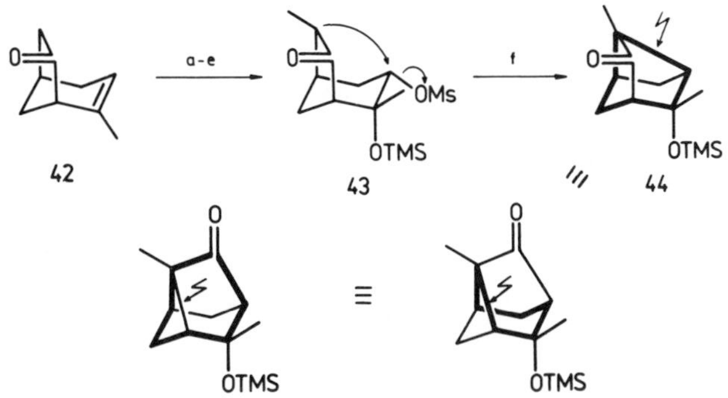

FIGURE 42. (a) OsO_4, N-methylmorpholine oxide, THF, H_2O; (b) MsCl, Et_3N; (c) Me_3SiCl, Et_3N; (d) $(i\text{-}Pr)_2NLi$; (e) CH_3I; (f) t-$C_5H_{11}OK$, t-$C_5H_{11}OH$.

A six-membered central ring of 6_2 type (see Figure 34) is contained in the molecule of deoxynorpatchoulenol (41). The synthesis[79] is based on the 6_2a retrosynthetic analysis. The synthetic equivalent of the corresponding synthon is the unsaturated tosyloxy ketone 38 (Figure 41). The enolate 39 prepared from 38 is cyclized to afford the intermediate ketone 40 which already contains the basic skeleton. The target hydrocarbon 41 was obtained from 40 by Huang-Minlon reduction.

Tricyclic derivative 44, an intermediate in the synthesis of α-*cis*-bergamotene, represents another compound containing the cyclohexane central ring of the 6_2 type. The synthesis (Figure 42),[17,80] starts from the bridged ketone 42, according to retrosynthetic analysis 6_2c. The cyclization was realized as intramolecular nucleophilic substitution of the mesyloxy group by the enolate of 43, giving rise to compound 44 with the basic bergamotene skeleton (the bond formed is denoted by an arrow). The formula 44a points to the six-membered central ring which is doubly bridged according to the general type 6_2. On the other hand, the same compound could be regarded as having a seven-membered central ring (44b), where two positions are linked with a cross-piece bond, forming thus an *ortho*-condensed subsystem.

Systems belonging to the 6_3 type, e.g. twistane,[81-83] have been synthesized according to scheme 6_3c.

A seven-membered central ring of 7_4 type can be found in the tricyclic derivative 46 (Figure 43).[84] The synthesis by Japanese authors is based on retrosynthetic analysis according to 7_4d. The formation of the central ring was achieved by intramolecular alkylation of the enolate 45 (generated from the corresponding ketone) with the tosyloxymethylene group, giving rise to ketone 46. The isomeric enolate is not alkylated under the reaction conditions, for the resulting tricyclic skeleton would be much more strained.

Figure 44 shows a synthesis of another tricyclic skeleton with a seven-membered central ring of the 7_4 type (48).[85] The retrosynthetic analysis follows the lines given in the scheme 7_4b, and the formation of the strategic carbon-carbon bond relies upon electrophilic addition of the carbocationic center to the double bond (47 → 48).

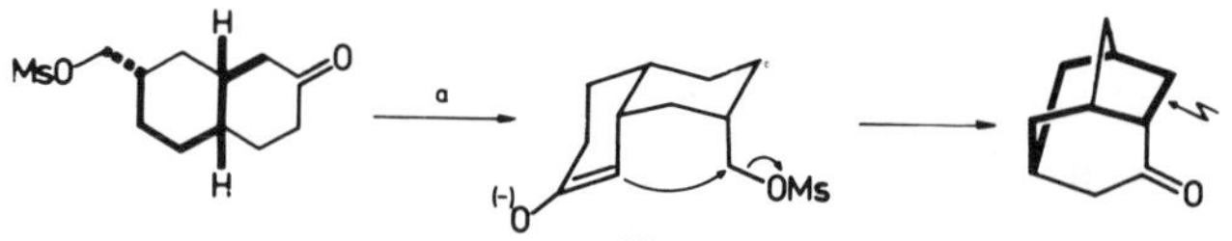

FIGURE 43. (a) NaH.

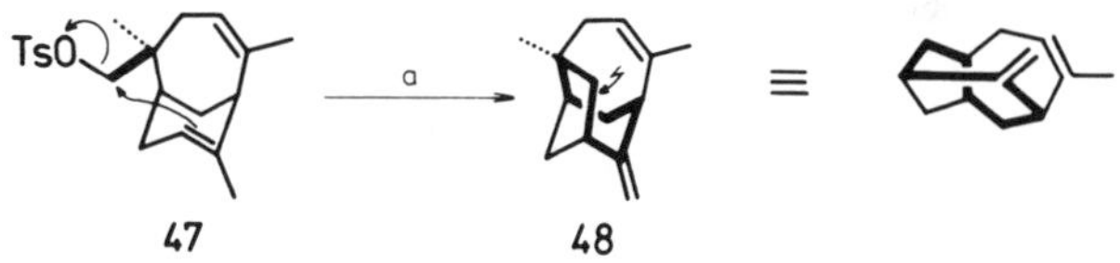

FIGURE 44. (a) 118°C, Me_2SO.

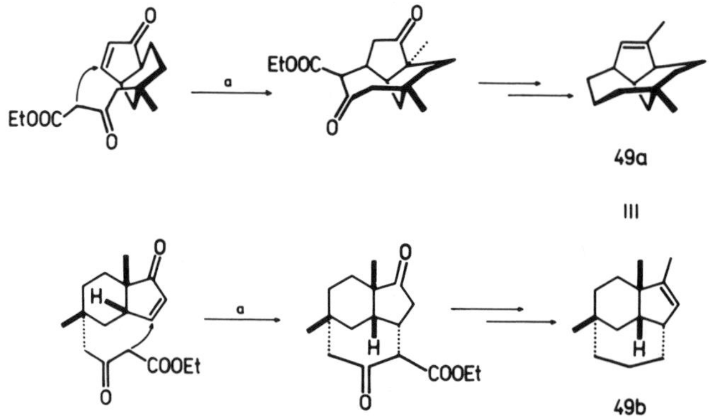

FIGURE 45. (a) Base.

Isoclovene (49) contains a nine-membered central ring (see formula 49a) which is closed up by an intramolecular Michael addition as a key step (Figure 45).[5] However, this tricyclic system can be easily analyzed as a [4.3.0] *ortho*-condensed subsystem bridged by a three-carbon unit (49b). This example shows, that an analysis relying (intellectually) on the location of a "central ring" need not necessarily be the easiest one. Nevertheless, looking for the central ring is a general, though not always the fastest, method of retrosynthesis.

2. Ditopic Cyclization

In the preceding section we have examined the cases of retrosynthetic analysis based on disconnecting one carbon-carbon bond in a monotopic process. An alternative way to tricyclic bridged systems is to disconnect either two bonds of the central ring or one bond of the central ring and one bond of the bridge. Of necessity, simultaneous breaking or formation of two bonds in one reaction step encounters more stringent steric requirements as can be seen on Dreiding models. Some synthons deduced from retrosynthetic analysis could yield isomeric products, the formation of which would require less strain in the transition state than with the desired products. In the following text we will therefore avoid cases of formal analysis and will instead concentrate on those which led to actual synthetic solution.

Of the ditopic reactions which come into account for closing the central ring we shall stress the [2 + 2] and [4 + 2] cycloadditions. These reactions provide an easy access to four- and six-membered rings, respectively, while the products can be further modified. Thus, for instance, the [2 + 2] cycloaddition can be utilized in the synthesis of natural

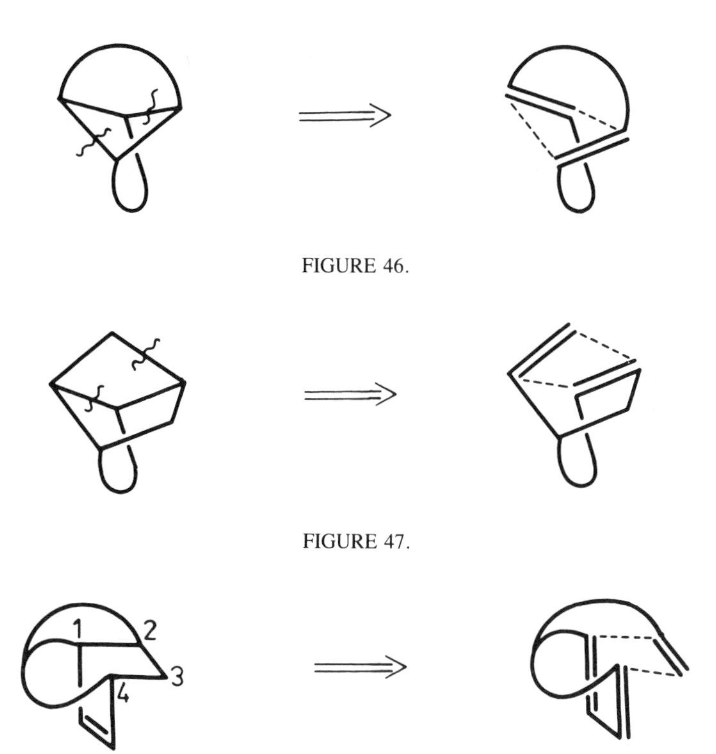

FIGURE 46.

FIGURE 47.

FIGURE 48.

products not containing the cyclobutane ring and the [4 + 2] reaction for compounds which do not contain a six-membered ring.

Figure 46 illustrates the retrosynthetic analysis leading to construction of a cyclobutane central ring and using intramolecular [2 + 2] cycloaddition as the strategic step. Note, however, that this reaction would proceed with proper regiochemistry only if the starting diene has contained a medium ring in which the conformation of the molecule forces a juxtaposition of the double bonds. Otherwise, formation of an *ortho*-condensed system might be a serious side-reaction.

[2 + 2] Cycloaddition can be utilized for a simultaneous construction of a five-membered central ring. Figure 47 shows the retrosynthetic analysis of a skeleton with the cyclopentane central ring which is spanned by a one-carbon bridge. In this case the [2 + 2] cycloaddition closes the central ring while forming the bridge at the same time.

Central (5 + n)-membered rings ($n \geqq 0$) can be formed by [4 + 2] cycloaddition (Figure 48), provided the ring is spanned between C-1 and C-4 by a two-carbon bridge, incorporating thus a cyclohexane ring. This strategy is quite common and often used due to the well-known and predictable features of the [4 + 2] reaction.

If one finds a six-membered ring in the molecule in question, disregarding whether or not the ring is a central one, it is worth the effort to conceive the retrosynthetic analysis so that it incorporates a [4 + 2] cycloaddition. Some possible approaches are depicted in Figures 49 and 50.[56,86,92] Let us finally illustrate the general retrosynthetic analysis with examples of actual (and successful) syntheses.

The analysis shown in Figure 46 may be visualized by the synthesis of α- and β-longipinene (52, Figure 51).[87] The basic skeleton in 51 was constructed by photochemical [2 + 2] cycloaddition of the monocyclic diene system in 50.

The synthesis of a five-membered central ring according to the analysis in Figure 47 is depicted in Figure 52. Irradiation of (−)-carvone affords the tricyclic ketone.[88,89]

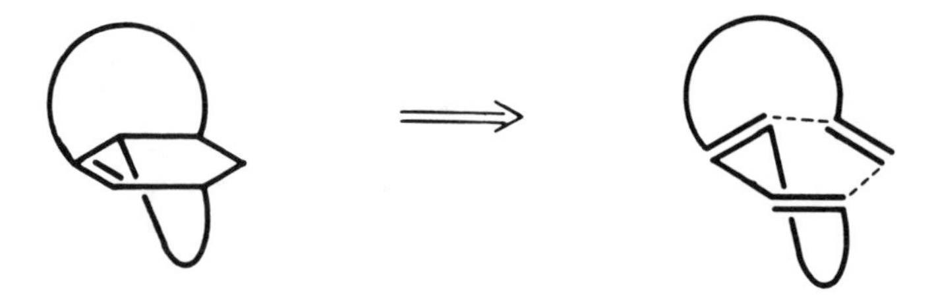

FIGURE 49.

FIGURE 50.

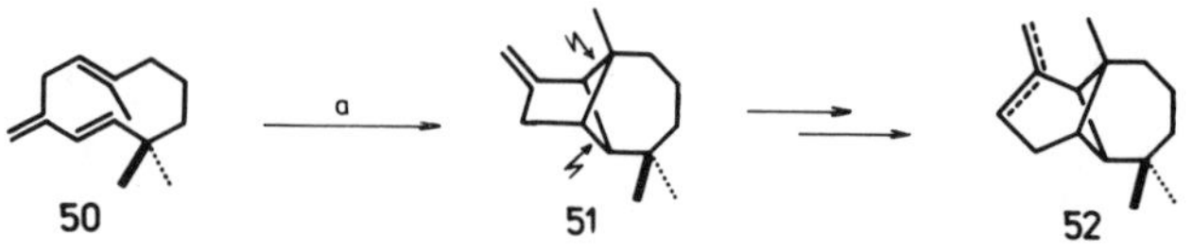

FIGURE 51. (a) hν.

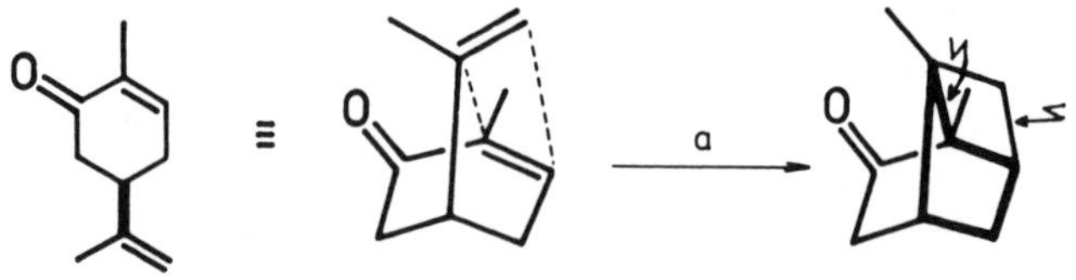

FIGURE 52. (a) hν.

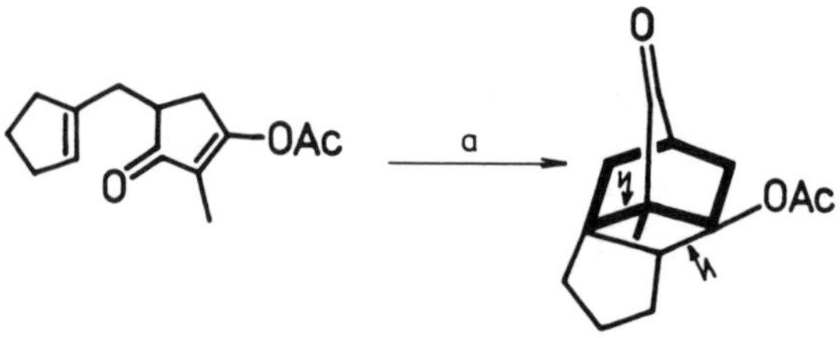

FIGURE 53. (a) hν.

Finally, an application of the [2 + 2] cycloaddition to the synthesis of a six-membered central ring is shown in Figure 53.[90,91]

The [4 + 2] cycloaddition has been employed in syntheses of tricyclic systems much more often then the [2 + 2] cycloaddition. This may be illustrated with the synthesis of 9-isocyanopupukenane (55) in which intramolecular Diels-Alder reaction of triene 53 (Figure 54) represented the key step in constructing the tricyclic skeleton (e.g. in 54) of the 6_2 type.[93] A similar synthesis has been reported.[94]

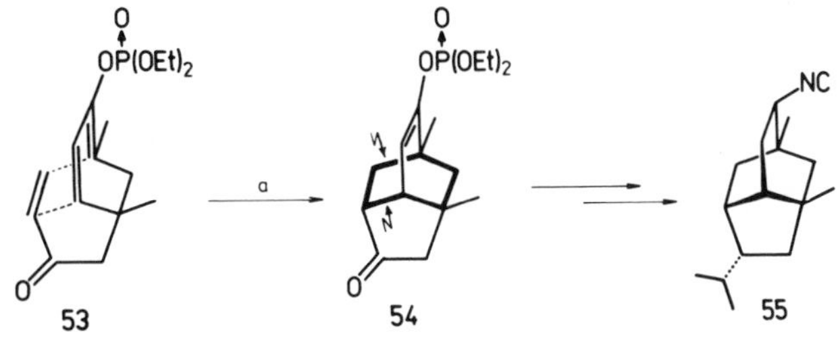

FIGURE 54. (a) reflux in xylene.

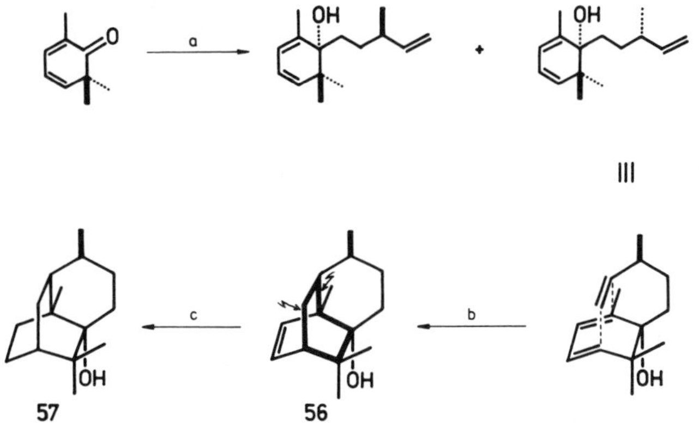

FIGURE 55. (a) $LiCH_2CH_2CH_2(CH_3)CH=CH_2$; (b) 280°C, t-BuOK, decaline or xylene; (c) H_2, Pd.

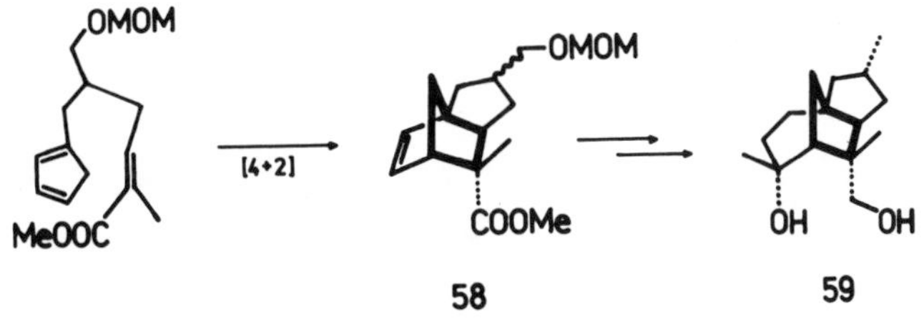

FIGURE 56.

The synthesis of patchouli alcohol (57) by Swiss authors[95,96] was also based on an intramolecular [4 + 2] cycloaddition (Figure 55). The six-membered central ring in 56 was closed in one step with the formation of the cyclohexene ring, affording the basic skeleton of the 6_2 type.

Following the Figure 48, a five-membered central ring may be also created by an intramolecular Diels-Alder reaction (Figure 56).[97] In this instance, one of the bridges of the resulting system 58 was elongated by one-carbon insertion and then converted in several steps to cedranediol 59. For further examples of this strategy see Reference 98 and a review.[99]

B. Syntheses by Bridging a Central Ring

An alternative way to tricyclic skeletons comprises bridging the completed central ring (see Figure 29). The bridging may be, in principle, accomplished by bond formation at or outside the annulation sites (3 and 4, respectively in Figure 29), and may be realized via monotopic or ditopic processes. The retrosynthetic analysis is reduced to finding the way

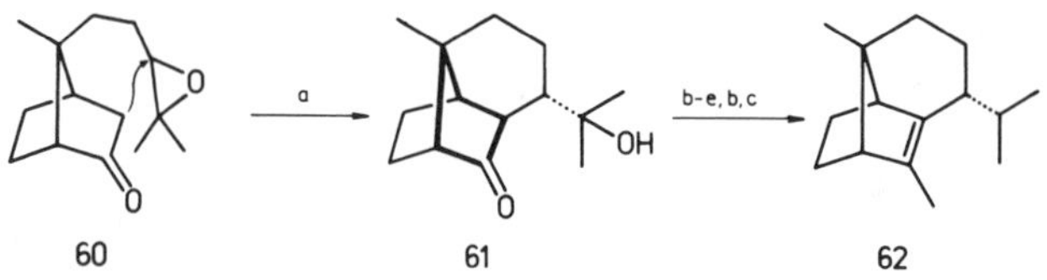

FIGURE 57. (a) t-BuOK; (b) CH_3COCl, $PhNMe_2$; (c) heat; (d) H_2, Pt; (e) CH_3Li.

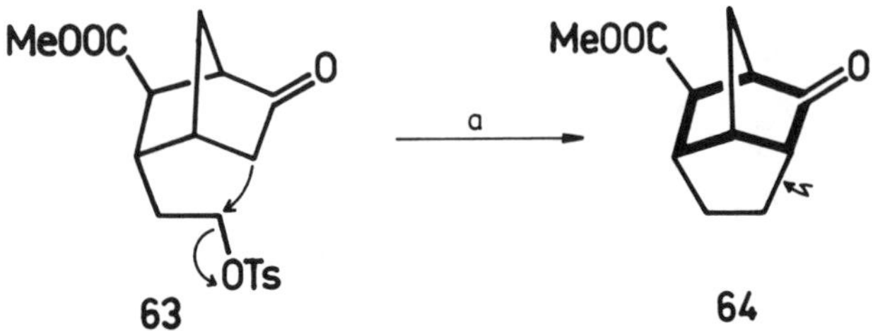

FIGURE 58. (a) t-BuOK, t-BuOH.

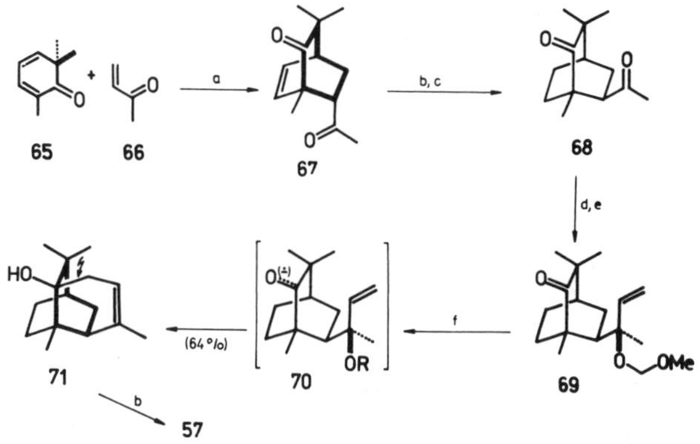

FIGURE 59. (a) Diels-Alder; (b) H_2, Pd – C; (c) CH_3ONa, CH_3OH; (d) CH_2=CH.Li, THF; (e) $CH_3OCH_2N^+Et_3$, CH_3CN; (f) NaH, THF.

to the corresponding substituted bicyclic subsystem. Indeed, numerous approaches to tricyclic systems have their basis in synthesis of bridged bicyclic skeletons which have been described in Section I. In the following text we shall present several methods that constituted the key steps in the synthesis of tricyclic bridged systems.

One of the key steps in the synthesis of sinularene (62, Figure 57) was the formation of the cyclohexane ring bridging the central five-membered ring.[100,101] The ring-closing reaction was the intramolecular opening of the oxirane ring by the enolate anion (60 → 61).

The popular intramolecular substitution of a tosyloxy group by an enolate was employed in the synthesis of the tricyclic ketone 64 (Figure 58).[102] This tricyclic system again belongs to 6_2 type. Other examples of bridging the cyclohexane ring at or outside the annulation sites can be found in numerous twistane syntheses.[81-83]

The synthesis of patchoulenol (71, Figure 59) started with construction of the central six-membered ring by the Diels-Alder reaction (65 + 66 → 67).[103] The formation of the third ring was achieved by the ketyl radical-anion attack on the allylic system in 70.[103-106] Seychellene[106,107] and patchouli alcohol[108] were synthesized in an analogous manner.

Beside spanning the central ring with an n-membered bridge ($n \geqq 1$) it is sometimes

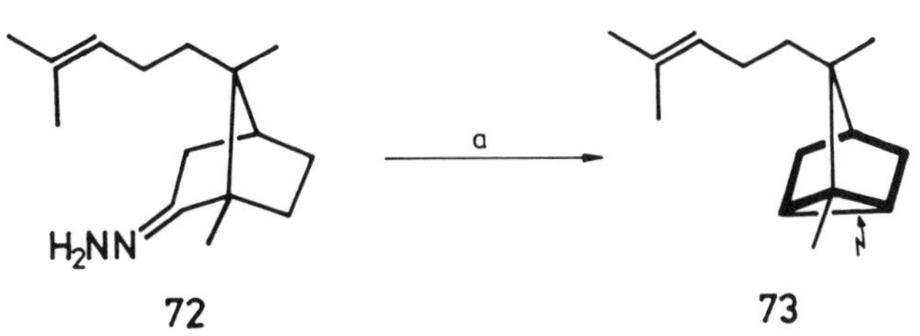

FIGURE 60. (a) HgO, CH_3OH, reflux.

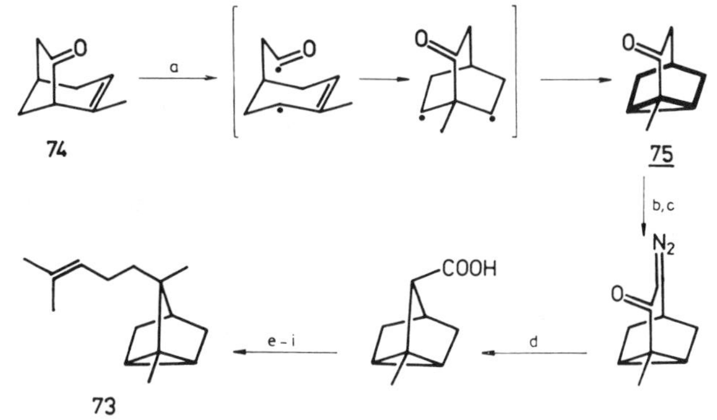

FIGURE 61. (a) hν, Me_2CO; (b) HCO_2Et, NaH; (c) TsN_3, Et_3N; (d) hν, $NaHCO_3$, THF, H_2O; (e) $(i\text{-}Pr)_2NLi$, then BuLi; (f) $Me_2C=CH-CH_2CH_2I$; (g) $LiAlH_4$, THF; (h) TsCl, C_5H_5N; (i) $LiBHEt_3$, THF.

necessary to form a zero-atom bridge, i.e. to connect two positions in the skeleton by a cross-piece bond, with formation of an *ortho*-condensed system. The optimum methodology depends on many factors, namely, the structure of the target molecule, availability of suitable synthons, steric proximity of the positions to be linked and, last but not least, the personality of the author. To illustrate this point we present here two different successful syntheses of α-santalene (73). Figure 60 shows the formation of the cross-piece bond in the [2.2.1] system (72) by oxidation of the hydrazone group with mercuric oxide, whereby the intermediate carbene inserts into the nearest skeletal carbon-hydrogen bond.[109] The same task was also solved via photochemical rearrangement of the ketone 74 accompanied by cyclization to 75. Further transformation of 75, including ring contraction and introduction of the six-carbon side-chain (Figure 61), yielded α-santalene (73).[20] Other methodologies can be found in a review.[26]

C. Rearrangements in the Synthesis of Tricyclic Bridged Systems

Rearrangements often provide an elegant way to cyclic systems which would otherwise be synthesized with difficulty. Of course this also holds for tricyclic bridged systems. Nature itself supplies us with compounds which can be synthetically utilized for rearrangements, e.g. the rearrangement of the pinane skeleton is the basis for the industrial production of synthetic camphor. It is therefore proper to present here at least a few selected examples illustrating the title methodology.

The photochemical rearrangement (a suprafacial [1,3] sigmatropic reaction) of vulgarenone B (76) to vulgarenone A (77) is shown in Figure 62;[110] this transformation, which is a tricyclic analogy of the well-known rearrangement of verbenone to chrysanthenone, converts the starting 4_1 system to a new one corresponding to a 6_1 type.

A more complex example, involving tandem aldol condensation and pinacol rearrange-

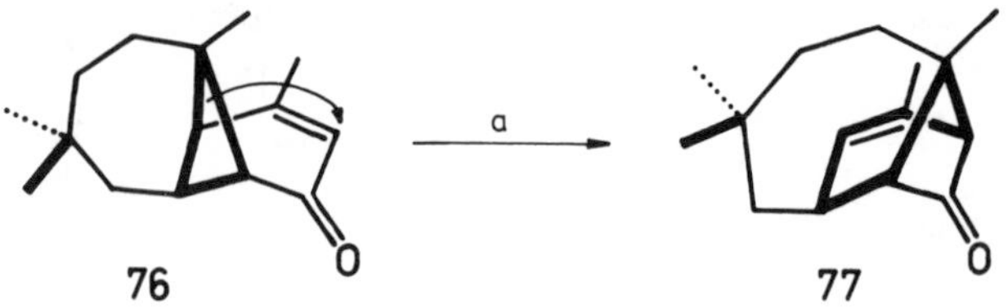

76 → 77 (a)

FIGURE 62. (a) hν.

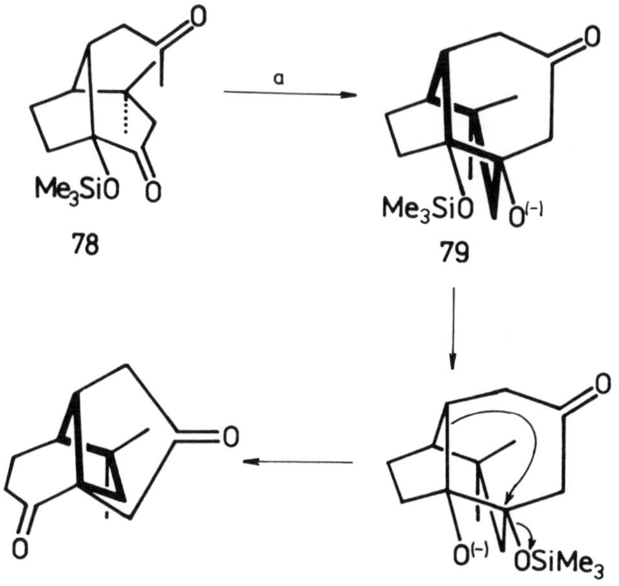

78 → 79 (a) → 80 → 81

FIGURE 63. (a) tBuOK, C_6H_6, 110°C.

ment, is illustrated in Figure 63.[64] Diketone 78 is first converted to the tricyclic intermediate aldol 79 which undergoes migration of the trimethylsilyl group (79 → 80) followed by pinacol rearrangement (80 → 81). In this way the central six-membered ring is contracted to a five-membered one. For excellent reviews on longifolene rearrangements and chemistry see References 111 and 112.

III. POLYCYCLIC BRIDGED SYSTEMS

Retrosynthetic analysis of polycyclic bridged systems leads gradually to simpler subsystems which in turn can be analyzed according to the principles outlined in Chapter 5, Sections I and II. To illustrate this procedure, let us show a few examples of compounds which may be regarded as tricyclic systems having the six-membered central ring, with one cross-piece bond added to make the tetracyclic skeleton.

Cyclosativene (84) contains a six-membered central ring of the 6_2 type, spanned with an additional cross-piece bond. The strategy chosen for the synthesis of 84 involved spanning the central ring with a three-carbon bridge with concomitant formation of the cross-piece bond (Figure 64). This was achieved by an ingenious reaction sequence starting with solvolysis of tosylate 82 accompanied by a two-fold participation of the multiple bonds which formed the desired carbon-carbon bonds (82 → 83).[101,113]

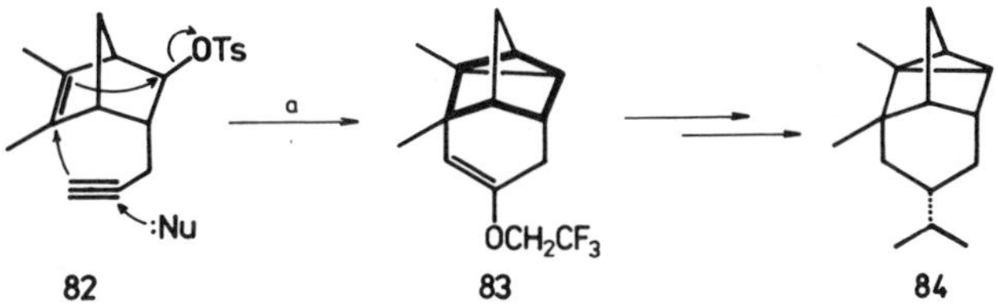

OTs
a
:Nu
82
OCH_2CF_3
83
84

FIGURE 64. (a) CF_3CH_2OH.

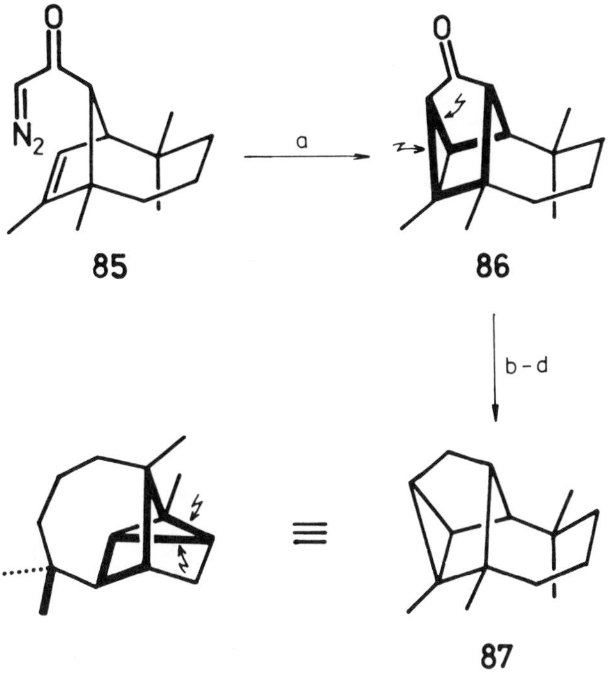

O
N_2
a
85
O
86
b - d
87

FIGURE 65. (a) Cu, THF; (b) $(i\text{-}Bu)_2AlH$, THF; (c) MsCl, Et_3N; (d) $LiAlH_4$.

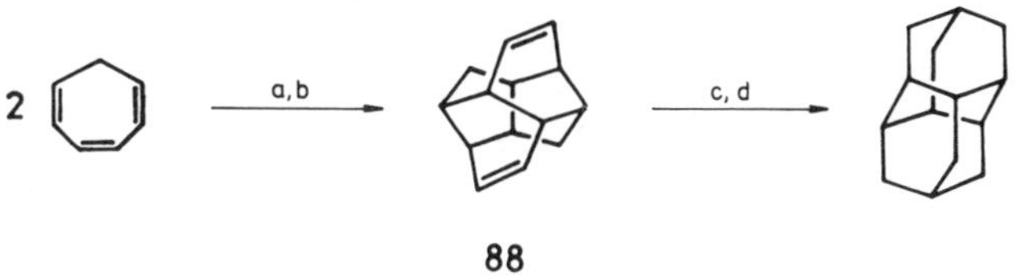

2
a, b
c, d
88

FIGURE 66. (a) $(\eta-C_6H_6)Ti(II)(AlCl_4)_2$, 20°C, 6 hr; (b) 180°C, 6 hr; (c) H_2, Pt, EtOH; (d) $AlCl_3$,CH_2Cl_2, reflux 18 hr.

Longicyclene (87) contains a six-membered central ring spanned by two bridges and stitched by a cross-piece bond. The synthetic strategy (Figure 65)[114] employed ditopic [2 + 1] cycloaddition of a carbene intermediate (85) to form the cyclopropane ring. From the retrosynthetic viewpoint this means forming the central ring plus the cross-piece bond (85 → 86); note that the latter has evolved from the original double bond in 85.

Tandem ditopic annulations offer a variety of approaches to polycyclic bridged systems. For instance the transition-metal catalyzed [6 + 2] dimerization of cycloheptatriene, followed by a spontaneous intramolecular Diels-Alder reaction, afforded the pentacyclic tetradecadiene 88, a starting point of a recent synthesis of diamantane (Figure 66).[115,116]

The chemistry of polycyclic systems is a wide area which continues to challenge the ingenuity of synthetic chemists. We have attempted to outline some fundamental approaches and to select a few recent examples. The interested reader will find further information in References 26 and 117. In contrast to previous chapters, we have stressed retrosynthetic analysis and gave in several instances a detailed description of approaches which are at least theoretically feasible. Some of them would be hampered due to steric or stereoelectronic factors; other might find synthetic applications in the future. Nevertheless, we included all of them into the analytic schemes so as to allow a systematic analysis based on a structure of the target compound.

IV. NOTES ADDED IN PROOF

An intramolecular nonconcerted ketene [2 + 2] addition to an olefinic double bond was employed as the key step in a simple, seven-step synthesis of β-*trans*-bergamotene.[118,119] An earlier synthesis of cantharidin based on the high-pressure Diels-Alder reaction (see Volume II, Chapter 1) was extended to preparative scale.[120] An unusual preference for the boat-like transition state in the intramolecular Diels-Alder addition has been noticed in certain instances of forming a [3.2.1] system.[121]

REFERENCES

1. **Liu, H. J. and Chan, W. H.,** Total synthesis of zizane sesquiterpenes: (−)-Khusimonine, (+)-zizanoic acid, and (−)-epizizanoic acid, *Can. J. Chem.*, 60, 1081, 1982.
2. **Piers, E. and Banville, J.,** Five membered ring annelation via thermal rearrangement of β-chloropropyl-α,β-unsaturated ketones. A new total synthesis of (±)-zizaene, *J. Chem. Soc. Chem. Commun.*, 1138, 1979.
3. **Paquette, L. A., Nitz, T. J., Ross, R. J., and Springer, J. P.,** Ingenane synthetic studies. An expedient approach to highly ogygenated ABC subunits of ingenol via reductive dialkylative annulation and α,β-epoxy ketone photoisomerization, *J. Am. Chem. Soc.*, 106, 1446, 1984.
4. **Oppolzer, W., Pitteloud, R., Bernardinelli, G., and Baettig, K.,** Asymmetric Michael addition of a chiral ester-dienolate: Enantioselective synthesis of (−)-khusimone, *Tetrahed. Lett.*, 24, 4975, 1983.
5. **Baraldi, P. G., Barco, A., Benetti, S., Pollini, G. P., and Simoni, D.,** Total synthesis of (±)-isoclovene, *Tetrahed. Lett.*, 24, 5669, 1983.
6. **House, H. O., Phillips, W. V., Sayer, T. S. B., and Yau, C.-C.,** Chemistry of carbanions. 31. Cyclization of the metal enolates from ω-bromo ketones, *J. Org. Chem.*, 43, 700, 1978.
7. **Wiesner, K., Jay, E. W. K., Tsai, T. Y. R., Demerson, C., Jay, L., Kanno, T., Křepinský, J., Vilím, A., and Wu, C. S.,** The synthesis of delphinine. A stereoselective total synthesis of an optically active advanced relay compound, *Can. J. Chem.*, 50, 1925, 1972.
8. **Wiesner, K., Taylor, W. I., Figdor, S. K., Bartlett, M. F., Armstrong, J. R., and Edwards, J. A.,** Garrya-Alkaloide, II. Mitteil. Weitere Versuche über den Abbau von Garryin und Veatchin, *Ber. Deut. Chem. Ges.*, 86, 800, 1953.
9. **Guthrie, R. W., Henry, W. A., Immer, H., Wong, C. M., Valenta, Z., and Wiesner, K.,** *Collect. Czech. Chem. Commun.*, 31, 602, 1966.
10. **Yamada, K., Suzuki, M., Hayakawa, Y., Aoki, K., Nakamura, H., Nagase, H., and Hirata, Y.,** Total synthesis of dl-dendrobine, *J. Am. Chem. Soc.*, 94, 8278, 1972.
11. **House, H. O., Melillo, D. G., and Sauter, F. J.,** Perhydroindan derivatives. XV. The synthesis of a tetracyclic precursor to epiallogibberic acid, *J. Org. Chem.*, 38, 741, 1973.
12. **Stork, G., Malhotra, S., Thompson, H., and Uchibayashi, M.,** A new cyclization. 2-Methylenecyclopentanols by the chemical reduction of γ-ethinyl ketones, *J. Am. Chem. Soc.*, 87, 1148, 1965.
13. **Becker, K. B.,** The synthesis of cycloalkenes by the intramolecular Wittig reaction, *Helv. Chim. Acta*, 60, 68, 1977.
14. **Becker, K. B.,** The synthesis of strained methylene-bridged bicyclic olefins by the intramolecular Wittig reaction, *Helv. Chim. Acta*, 60, 81, 1977.
15. **Becker, K. B.,** Electrophilic additions to strained bridgehead olefins. Estimation of strain by comparison with the solvolysis of bridgehead bromides, *Helv. Chim. Acta*, 60, 94, 1977.

16. **Becker, K. B. and Chapuis, J. L.,** Synthesis and dimerization of bicyclo [4.4.1] undec-1(11)-ene, a bridged trans-cycloheptane, *Helv. Chim. Acta,* 62, 34, 1979.
17. **House, H. O., Haack, J. L., McDaniel, W. C., and VanDerveer, D.,** Enones with strained double bonds. 8. The bicyclo [3.2.1] octane systems, *J. Org. Chem.,* 48, 1643, 1983.
18. **Kočovský, P.,** Unpublished results.
19. **Larsen, S. D. and Monti, S. A.,** The synthesis of racemic α-trans- and α-cis-bergamotene and α-pinene, *J. Am. Chem. Soc.,* 99, 8015, 1977.
20. **Monti, S. A. and Larsen, S. D.,** Total synthesis of racemic α-santalene and of racemic teresantalic acid, *J. Org. Chem.,* 43, 2282, 1978.
21. **Corey, E. J., Girotra, N. N., and Mathew, C. T.,** Total synthesis of dl-cedrene and dl-cedrol, *J. Am. Chem. Soc.,* 91, 1557, 1969.
22. **Crandall, T. G. and Lawton, R. G.,** A biogenetic-type synthesis of cedrene, *J. Am. Chem. Soc.,* 91, 2127, 1969.
23. **Baird, R. and Winstein, S.,** Neighboring carbon and hydrogen. XLVI. Spiro-(4,5)-deca-1,4-dien-3-one from Ar_1^--5 participation, *J. Am. Chem. Soc.,* 84, 788, 1962.
24. **Masamune, S.,** Total synthesis of diterpenes and diterpene alkaloids. II. A tetracyclic common intermediate. Total syntheses of diterpenes and diterpene alkaloids. IV. Garryine. *J. Am. Chem. Soc.,* 86, 288, 290, 1964.
25. **Mander, L. N., Turner, J. V., and Coombe, B. G.,** Synthetic plant growth regulators. I. The synthesis of (±)-14-nor-helminthosporic acid and related compounds, *Aust. J. Chem.,* 27, 1985, 1974.
26. **ApSimon, J.,** *The Total Synthesis of Natural Products,* Vols. 1—5, Wiley-Interscience, New York, 1973—1983.
27. **Hodgson, G. L., MacSweeney, D. F., and Money, T.,** Cyclisations involving enol acetates: Synthesis of (±)-campherenone, (±)-campherenol and (±)-epicampherenone, *J. Chem. Soc. Chem. Commun.,* 766, 1971.
28. **Hodgson, G. L., MacSweeney, D. F., and Money, T.,** Alternative total synthesis of (±)-β-santalene, (±)-epi-β-santalene, (±)-α-santalene, (±)-copacamphor and (±)-ylangocamphor, *Tetrahed. Lett.,* 3683, 1972.
29. **Hodgson, G. L., MacSweeney, D. F., and Money, T.,** Synthesis of (±)-campherenone, (±)-epicampherenone, (±)-β-santalene, (±)-epi-β-santalene, (±)-α-santalene, (±)-ylangocamphor, and (±) sativene, *J. Chem. Soc. Perkin Trans.,* 1, 2113, 1973.
30. **McMurry, J. E., Andrus, A., Ksander, G. M., Musser, J. H., and Johnson, M. A.,** Stereospecific total synthesis of aphidicolin, *J. Am. Chem. Soc.,* 101, 1330, 1979.
31. **Fleming, I.,** *Frontier Orbitals and Organic Chemical Reactions,* J. Wiley & Sons, Chichester, 1976.
32. **Berson, J. A., Hamlet, Z., and Mueller, W. A.,** The correlation of solvent effect on the stereoselectivities of Diels-Alder reactions by means of linear free energy relationships. A new empirical measure of solvent polarity, *J. Am. Chem. Soc.,* 84, 297, 1962.
33. **Sauer, J. and Kredel, J.,** Optische induktion bei sechsring-cycloadditionen, *Angew. Chem.,* 77, 1037, 1965.
34. **Woodward, R. B. and Katz, T. J.,** The mechanism of Diels-Alder reaction, *Tetrahedron,* 5, 70, 1959.
35. **Scheiner, P., Schmiegel, K. K., Smith, G., and Vaughan, W. R.,** Synthesis of bicyclic nitriles and related compounds. II., *J. Org. Chem.,* 28, 2960, 1963.
36. **Freeman, P. K., Balls, D. M., and Brown, D. J.,** A method for the addition of elements of ketene to some selected dienes in Diels-Alder fashion, *J. Org. Chem.,* 33, 2211, 1968.
37. **Escher, S., Keller, U., and Willhalm, B.,** Neue Phelandren-Derivate aus dem Wurzelöl von *Angelica archangelica* L., *Helv. Chim. Acta,* 62, 2061, 1979.
38. **Ranganathan, S., Ranganathan, D., and Mehrotra, A. K.,** Nitroethylene as a versatile ketene equivalent. Novel one-step preparation of prostaglandin intermediates by reduction and abnormal Nef reaction, *J. Am. Chem. Soc.,* 96, 5261, 1974.
39. **Ranganathan, S., Ranganathan, D., and Iyengar, R.,** A simple and convenient route to 11-deoxyprostaglandins, *Tetrahedron,* 32, 961, 1976.
40. **Ficini, J.,** Ynamine. A versatile tool in organic synthesis, *Tetrahedron,* 32, 1449, 1976.
41. **Rigby, J. H. and Sage, J.-M.,** Short synthesis of bicyclo-[3.2.2] nona-3,6,8-trien-2-one, *J. Org. Chem.,* 48, 3591, 1983.
42. **Ranganathan, S., Ranganathan, D., and Mehrotra, A. K.,** Ketene equivalents, *Synthesis,* 289, 1977.
43. **Oppolzer, W. and Chapuis, C.,** Asymmetric Diels-Alder reaction of a chiral allenic ester. Enantioselective synthesis of (−)-β-santalene, *Tetrahed. Lett.,* 24, 4665, 1983.
44. **Solas, D. and Wolinsky, J.,** Total synthesis of β-santalol, *J. Org. Chem.,* 48, 1988, 1983.
45. **White, K. B. and Reusch, W.,** The synthesis of bicyclo[2.2.2]octan-2-ones by sequential Michael reactions, *Tetrahedron,* 34, 2439, 1978.
46. **Lee, R. A.,** Reactions of ά-dienolates with Michael acceptors. A synthesis of bicyclo [2.2.2] octan-2-ones, *Tetrahed. Lett.,* 3333, 1973.

47. **Cory, R. M. and Chan, D. M. T.,** Bicycloannulation; a one-step synthesis of tricyclo [3.2.1.$0^{2,7}$] octan-6-ones, *Tetrahed. Lett.*, 4441, 1975.
48. **Rubottom, G. M. and Krueger, D. S.,** The synthesis of oxygenated bicyclic systems via the [4 + 2] cyclo-addition reaction of trimethylsilyloxy cyclohexadienes, *Tetrahed. Lett.*, 611, 1977.
49. **Geribaldi, S., Torri, G., and Azzaro, M.,** Synthèses dans la série bicyclo [2.2.2] octanique. II. - bicyclo [2.2.2] octène-5 ones-2 alkylées, *Bull. Soc. Chim. France*, 2836, 1973.
50. **Conia, J. M. and Le Perchec, P.,** The thermal cyclisation of unsaturated carbonyl compounds, *Synthesis*, 1, 1975.
51. **Conia, J. M. and Lange, G. L.,** Thermolysis and photolysis of unsaturated ketones. 26. Preparation of bicyclo [2.2.2] octan-2-ones and bicyclo [2.2.1] heptan -2-ones by thermal cyclization of unsaturated ketones. A facile synthesis of (+)-camphor from (+)-dihydrocarvone, *J. Org. Chem.*, 43, 564, 1978.
52. **Stetter, H., Rämsch, K. D., and Elfert, K.,** 2,2-Bis-(chloromethyl) acetophenon als neues α,α′-Anel-lierungs-reagenz, *Justus Liebigs Ann. Chem.*, 1322, 1974.
53. **Still, W. C.,** A simple synthesis of bicyclo [4.n.1] enones by cyclodialkylation, *Synthesis*, 453, 1976.
54. **Shea, K. J. and Wise, S.,** Intramolecular Diels-Alder reactions. A new entry into bridgehead bicyclo [3.n.1] alkenes, *J. Am. Chem. Soc.*, 100, 6519, 1978.
55. **Shea, K. J. and Wise, S.,** Intramolecular Diels-Alder cycloadditions. Synthesis of substituted derivatives of bicyclo 3.n.1. bridgehead alkenes, *Tetrahed. Lett.*, 1011, 1979.
56. **Fallis, A. G.,** The intramolecular Diels-Alder reaction. Recent advances and synthetic applications, *Can. J. Chem.*, 62, 183, 1984.
57. **Noyori, R., Nishizawa, M., Shimizu, F., Hayakawa, Y., Maruoka, K., Hashimoto, S., Yamamoto, H., and Nozaki, H.,** Intramolecular dibromo ketone-iron carbonyl reaction in terpene synthesis, *J. Am. Chem. Soc.*, 101, 220, 1979.
58. **Baldwin, J. E. and Barden, T. C.,** Discrimination between exo- and endo-2,3-methyl shifts in substituted 2-norbornyl cations on the (+)-camphenilone route to (−)-albene, *J. Am. Chem. Soc.*, 105, 6656, 1983.
59. **Corey, E. J. and Nozoe, S.,** The total synthesis of α-caryophyllene alcohol, *J. Am. Chem. Soc.*, 86, 1652, 1964.
60. **Corey, E. J. and Nozoe, S.,** The total synthesis of α-caryophyllene alcohol, *J. Am. Chem. Soc.*, 87, 5733, 1965.
61. **Ohfune, Y., Shirahama, H., and Matsumoto, T.,** Biogenetic-like synthesis of *d,l*-hirsutene, *Tetrahed. Lett.*, 2795, 1976.
62. **Roberts, S. M.,** Rearrangements of bicyclo [3.2.0] heptan-6-ones. Synthesis of potential prostanoid pre-cursors, *J. Chem. Soc. Chem. Commun.*, 948, 1974.
63. **Yamada, Y., Nagaoka, H., and Kimura, M.,** A convenient synthetic method for the bicyclo [3.2.1] octane ring system, *Synthesis*, 581, 1977.
64. **Monti, S. A. and Dean, T. R.,** An approach to the quadrone skeleton *via* a tandem aldol-pinacol, *J. Org. Chem.*, 47, 2679, 1982.
65. **Breitholle, F. G. and Fallis, A. G.,** Total synthesis of (±)-cedrol and (±)-cedrene *via* an intramolecular Diels-Alder reaction, *J. Org. Chem.*, 43, 1964, 1978.
66. **Trost, B. M. and Hiemstra, H.,** Ion pair effects in an intercalation process. An approach to the bicyclo [5.3.1] undecyl system of taxane, *J. Am. Chem. Soc.*, 104, 886, 1982.
67. **Barker, A. J. and Pattenden, G.,** Intramolecular de Mayo reactions leading to zizaane and related terpenoid systems, *Tetrahed. Lett.*, 21, 3513, 1980.
68. **Oppolzer, W. and Burford, B. S.,** Synthesis of tricyclo [6.2.1.$0^{1,5}$]-undecadiones via intramolecular photoaddition of 5-(1-cyclopentenylmethyl)-3-alkoxy-2-cyclopentenones, *Helv. Chim. Acta*, 63, 788, 1980.
69. **Stork, G. and Clarke, F. H., Jr.,** Cedrol: Stereochemistry and Total synthesis, *J. Am. Chem. Soc.*, 83, 3114, 1961.
70. **Danishefsky, S., Vaughan, K., Gadwood, R. C., and Tsuzuki, K.,** Total synthesis of dl-quadrone, *J. Am. Chem. Soc.*, 102, 4262, 1980.
71. **Danishefsky, S., Vaughan, K., Gadwood, R. C., Tsuzuki, K., and Springer, J. P.,** A novel transfor-mation in the quadrone series, *Tetrahed. Lett.*, 21, 2625, 1980.
72. **Heathcock, C. H.,** The total synthesis of (±)-copaene and (±)-8-isocopaene, *J. Am. Chem. Soc.*, 88, 4111, 1966.
73. **Heathcock, C. H., Badger, R. A., and Patterson, J. W., Jr.,** Total synthesis of (±)-copaene and (±)-ylagene. A general method for the synthesis of tricyclo [4.0.0.$0^{2,7}$] decanes, *J. Am. Chem. Soc.*, 89, 4133, 1967.
74. **Corey, E. J., Ohno, M., Mitra, R. B., and Vatakencherry, P. A.,** Total synthesis of *d,l*-longifolene, *J. Am. Chem. Soc.*, 83, 1251, 1961.
75. **Corey, E. J., Ohno, M., Vatakencherry, P. A., and Mitra, R. B.,** Total synthesis of longifolene, *J. Am. Chem. Soc.*, 86, 478, 1964.
76. **McMurry, J. E. and Isser, S. J.,** Total synthesis of longifolene, *J. Am. Chem. Soc.*, 94, 7132, 1972.

77. **Woodward, R. B., Brutschy, F. J., and Baer, H.,** The structure of santoninic acid, *J. Am. Chem. Soc.*, 70, 4216, 1948.
78. **Jiang, A. Q., Scheffer, J. R., Secco, A. S., Trotter, J., and Wong, Y. F.,** Interannular dehydration as a route to a novel polycyclic ring system, *J. Chem. Soc. Chem. Commun.*, 773, 1983.
79. **Gras, J.-L.,** Stereoselective approach to the norpatchoulene unit. Total synthesis of (±)-iso and deoxy-norpatchoulenol, *Tetrahed. Lett.*, 4117, 1977.
80. **Gibbson, T. W. and Erman, W. F.,** The synthesis of the (−)-α- and (+)-β-cis-bergamotenes, *J. Am. Chem. Soc.*, 91, 4771, 1969.
81. **Tichý, M.,** Absolute configuration of tricyclo [4.4.0.0^{3,8}] decane (twistane), *Tetrahed. Lett.*, 2001, 1972.
82. **Hamon, D. P. G. and Young, R. N.,** The analytical approach to synthesis: Twistane, *Aust. J. Chem.*, 29, 145, 1976.
83. **Schubert, W., and Ugi, I.,** Constitutional symmetry and unique descriptors of molecules, *J. Am. Chem. Soc.*, 100, 37, 1978.
84. **Takaishi, N., Inamoto, Y., Fujikura, Y., and Agami, K.,** Regiospecific intramolecular cyclization in 7-keto-endo-2-cis-decalylcarbinyl methanesulfonate, *J. Org. Chem.*, 44, 650, 1979.
85. **Fráter, G.,** Umwandlung des Tricyclo [5.4.0.0^{3,9}] undecadien-gerüstes in das neue Octahydro-2,5-methano-azulenegerüst, *Helv. Chim. Acta*, 59, 164, 1976.
86. **Snowden, R. L.,** The intramolecular [4 + 2]π-cycloadditions of (3-alkenyl) cyclopentadienes, *Tetrahed. Lett.*, 22, 97, 1981.
87. **Miyashita, M. and Yoshikoshi, A.,** Total synthesis of racemic α- and β-longipinenes, *J. Chem. Soc. Chem. Commun.*, 1173, 1972.
88. **Hodgson, G. L., MacSweeney, D. F., and Money, T.,** Synthesis, absolute configuration, and photo-cyclisation of the sesquiterpene cryptomerin, *J. Chem. Soc. Chem. Commun.*, 236, 1973.
89. **Crawford, R. J., Erman, W. F., and Broaddus, C. D.,** Methylation of limonene. A novel method for the synthesis of bisabolene sesquiterpenes, *J. Am. Chem. Soc.*, 94, 4298, 1972.
90. **Barker, A. J. and Pattenden, G.,** Intramolecular de Mayo reactions leading to zizaane and related terpenoid ring systems, *Tetrahed. Lett.*, 21, 3513, 1980.
91. **Oppolzer, W. and Burford, B. S.,** Synthesis of tricyclo [6.2.1.0^{1,5}] undecadiones via intramolecular photoaddition of 5-(1-cyclopentenylmethyl)-3-alkoxy-2-cyclopentenones, *Helv. Chim. Acta*, 63, 788, 1980.
92. **Snowden, R. L.** A stereoselective total synthesis of (±)-sativene, *Tetrahed. Lett.*, 22, 101, 1981.
93. **Piers, E. and Winter, M.,** A total synthesis of (±)-pupukeanone, *Justus Liebigs Ann. Chem.*, 973, 1982.
94. **Schiehser, G. A. and White, J. D.,** Total synthesis of (±)-9-pupukeanone, *J. Org. Chem.*, 45, 1864, 1980.
95. **Näf, F. and Ohloff, G.,** A short stereoselective total synthesis of racemic patchouli alcohol, *Helv. Chim. Acta*, 57, 1868, 1974.
96. **Näf, F., Decorzant, R., Giersch, W., and Ohloff, G.,** A stereocontrolled access to (±)-, (−)-, and (+)-patchouli alcohol, *Helv. Chim. Acta*, 64, 1387, 1981.
97. **Landry, D. W.,** Total synthesis of (±)-8S,14-cedranediol, *Tetrahedron*, 39, 2761, 1983.
98. **Sternbach, D. P., Hughes, J. W., Burdi, D. F., and Forstot, R. M.,** Synthesis of polyquinanes. I. Intramolecular Diels-Alder reaction, *Tetrahed. Lett.*, 24, 3295, 1983.
99. **Paquette, L. A.,** Recent synthetic developments in polyquinane chemistry, *Topics in Current Chemistry*, 119, 1, 1984.
100. **Collins, P. A. and Wege, D.,** The total synthesis of sinularene, a sesquiterpene hydrocarbon from the soft coral *Sinularia mayi, Aust. J. Chem.*, 32, 1819, 1979.
101. **Baldwin, S. W. and Tomesch, J. C.,** Total synthesis of cyclosativene by cationic olefinic and acetylenic cyclizations *J. Org. Chem.*, 45, 1455, 1980.
102. **Furuichi, K. and Miwa, T.,** Total synthesis of (±)-forsythide aglucone dimethyl ester, *Tetrahed. Lett.*, 3689, 1974.
103. **Bertrand, M., Teisseire, P., and Pélerin, G.,** Une possible solution de rechange a la cyclisation des ϵ-halogéncétones. Application a la synthèse du dérivés du patchoulane, *Nouv. J. Chim.*, 7, 61, 1983.
104. **Bertrand, M., Teisseire, P., and Pélerin, G.,** Sur une nouve synthese du (±) norpatchoulenol, *Tetrahed. Lett.*, 2051, 1980.
105. **Bertrand, M., Teissiere, P., and Pélerin, G.,** Sur une solution de rechange a la cyclisation de ϵ-halogenceetones-application a la synthese du (±)-patchoulol, *Tetrahed. Lett.*, 21, 2055, 1980.
106. **Jung, M. E. and Pan, Y.-G.,** Direct stereoselective total synthesis of (±)-seychellene, *Tetrahed. Lett.*, 21, 3127, 1980.
107. **Mirrington, R. N. and Schmalzl, K. J.,** Studies with bicyclo [2.2.2] octenes. The total synthesis of (±)-seychellene, *J. Org. Chem.*, 37, 2877, 1972.
108. **Mirrington, R. N. and Schmalzl, K. J.,** Studies with bicyclo [2.2.2] octenes. The total synthesis of (+)-patchouli alcohol, *J. Org. Chem.*, 37, 2871, 1972.

109. **Hodgson, G. L., MacSweeney, D. F., and Money, T.,** Synthesis of (±)-campherenone, (±)-epicampherenone, (±)-β-santalene, (±)-β-episantalene, (±)-α-santalene, (±)-ylangocamphor, (±)-copacamphor, and (±)-sativene, *J. Chem. Soc. Perkin Trans.*, 1, 2113, 1973.
110. **Uchio, Y., Matsuo, A., Eguchi, S., Nakayama, M., and Hayashi, S.,** Vulgarenone B, a novel sesquiterpene ketone from *Chrysanthemum vulgare* and its photochemical transformation to vulgarenone A, *Tetrahed. Lett.*, 1191, 1977.
111. **Dev, S.,** Aspects of longifolene chemistry. An example of another facet of natural products chemistry, *Accounts Chem. Res.*, 14, 82, 1981.
112. **Dev, S.,** The chemistry of longifolene and its derivatives, *Fortschr. Chem. Org. Naturst.*, 40, 49, 1981.
113. **Baldwin, S. W. and Tomesch, J. C.,** Stereospecific synthesis of *d,l*-cyclosativene. Alkyne capture of an homoallylic carbonium ion, *Tetrahed. Lett.*, 1055, 1975.
114. **Welch, S. C. and Walters, R. L.,** Total synthesis of (±)-longicyclene, *Synth. Commun.*, 3, 15, 1973.
115. **Mach, K., Antropiusová, H., Tureček, F., Hanuš, V., and Sedmera, P.,** Zwei Neue Pentacyklische Dimere des Cycloheptatriens, *Tetrahed. Lett.*, 21, 4879, 1980.
116. **Tureček, F., Hanuš, V., Sedmera, P., Antropiusová, H., and Mach, K.,** Cycloheptatriene dimers. New precursors of diamantane, *Collect. Czech. Chem. Commun.*, 46, 1474, 1981.
117. **Shea, K. J.,** Recent developments in the synthesis, structure and chemistry of bridgehead alkenes, *Tetrahedron*, 36, 1683, 1980.
118. **Corey, E. J. and Desai, M. C.,** Simple synthesis of (±)-β-*trans*-bergamotene, *Tetrahed. Lett.*, 26, 3535, 1985.
119. **Kulkarni, Y. S. and Snider, B. B.,** Intramolecular [2 + 2] cycloadditions of ketenes. 2. Synthesis of chrysanthenone, β-pinene, β-*cis*-bergamotene, and β-*trans*-bergamotene, *J. Org. Chem.*, 50, 2809, 1985.
120. **Dauben, W. D., Gerdes, J. M., and Smith, D. B.,** Organic reactions at high pressure. The preparative scale synthesis of cantharidin, *J. Org. Chem.*, 50, 2577, 1985.
121. **Koreeda, M. and Luengo, J. I.,** A novel type of intramolecular Diels-Alder reaction involving dienol ethers. An unusual preference for a boat transition state in the incipient ring formation, *J. Org. Chem.*, 49, 2079, 1984.

Chapter 6

MEDIUM AND LARGE RINGS

Beginning with eight-membered rings, both sp^3 and sp^2 carbon centers can play the role of stereogenous elements because both configurations of the double bond (*E* and *Z*) are stable. The synthetic strategy leading to compounds with carbocyclic medium or large rings may be divided into two classes (Figure 1).

The first approach involves a ring-forming cyclization of a suitable precursor, performed as a rule under high or medium dilution conditions (A). The second approach rests on the fragmentation of a cross-piece bond in a system composed of two or more smaller rings. It should be noted here that the procedure A is more often used for synthesis of large-ring compounds ($n \geqq 12$), while procedure B has found application in synthesis of medium-ring compounds ($n = 8 - 12$). Besides these methods, ring enlargement (mostly by [3,3] sigmatropic shift or by "zip" reaction) is also used in making large rings.

I. CYCLIZATION REACTIONS

The synthetic methods for ring formation usually make use of dipolar synthons containing one electrophilic and one nucleophilic center. In order to suppress undesirable bimolecular coupling, it is necessary to work under conditions of medium or high dilution. The classical cyclization methods (Figure 2) are represented by the acyloin (1) and Thorpe reactions (2) (for reviews see References 1 to 4). In addition to these, a number of new methods have been reported in which the electrophilic center is represented by a carbon atom of the oxirane ring, while the nucleophile, usually a carbanion, is generated by proton abstraction from a carbon center adjacent to an electron-withdrawing group (EWG) (3).[5-7] The electrophile may also be realized as a stabilized allylic cation (4).[8,9] Intramolecular addition of the acyl halide moiety to the double bond, catalyzed by Lewis acids, (5), has been successfully employed as a ring-forming reaction.[10] Other methods are exemplified by McMurry dicarbonyl coupling (6),[11-17] cyclization of diols induced by potassium hexachlorotungstenate,[18] intramolecular coupling of allyl bromides (7) with nickel tetracarbonyl[19-22] or tosylmethyl isocyanide,[23,24] by palladium mediated intramolecular coupling of vinyl iodides and vinyl ketones,[25] and the Wittig reaction.[26,27] Methods for closing lactone macrolide rings will be treated in Chapter 1 in Volume II; for stereoselective substitution of large rings see Chapter 7.

Figure 3 shows a cyclization employing a stabilized allylic cation as an electrophilic species.[8] The nucleophilic counterpart (10 or 11) is generated first by abstraction of the labile proton from 8 or 9. The electrophilic center is generated from the allylacetate moiety by a palladium (0) reagent anchored to a polystyrene support (10,11 → 12). The reaction between the two centers results in the formation of the ring (12 → 13 + 14). The purpose of the polymer-bound catalyst is to suppress the bimolecular side reaction. This technique is referred to as "pseudodilution", for the low surface concentration of the catalyst on the polymeric support makes it possible to work with relatively high concentration of the substrate (0.5 to 1.0 *M*)! This advantage is somewhat lessened by the fact that the salt, formed by proton abstraction (8,9 → 10,11) penetrates the polymer very slowly to reach the catalyst. This technical difficulty was removed by an ingenious variation in which the order of generating the reactive centers was reversed (Figure 4).[9] Vinyl epoxide 15 reacts with the polymer-bound catalyst (15 → 16) with concomitant opening of the oxirane ring, whereby the alkoxide anion remains in the molecule. The latter then serves as an internal base, abstracting the proton from the nucleophilic center (16 → 17), which induces the catalyzed cyclization (17 → 18 + 19). The reaction is reportedly very easy to perform due to the pseudodilution principle: a concentrated solution of the substrate (e.g. 15) is simply run

FIGURE 1.

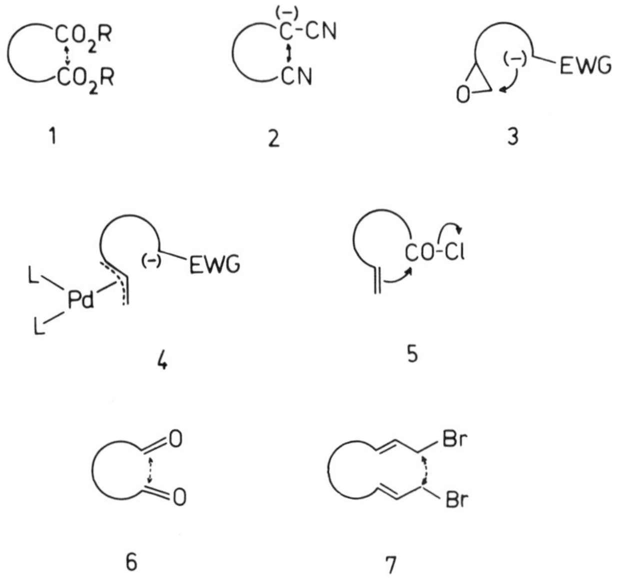

FIGURE 2.

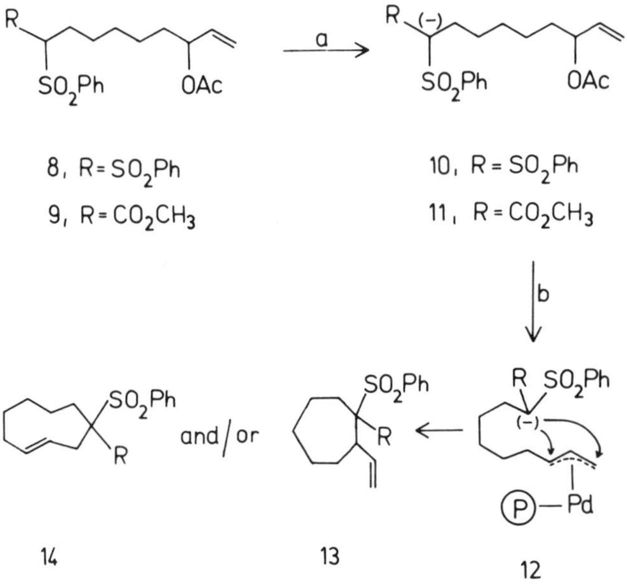

FIGURE 3. (a) Base; (b) Pd —Ⓟ.

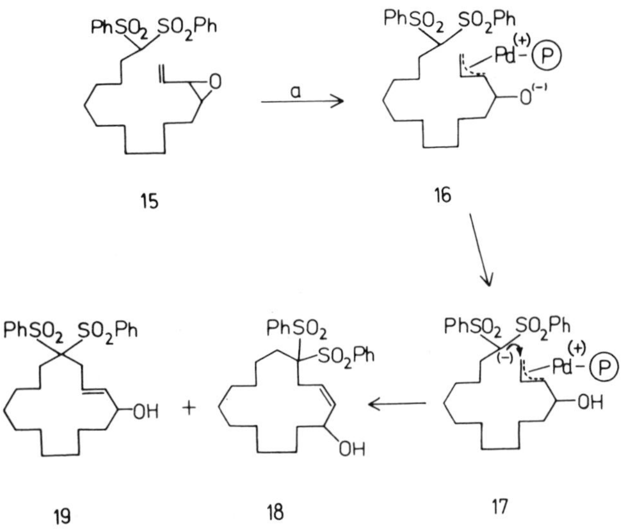

FIGURE 4. (a) Pd —Ⓟ, 18-crown-6.

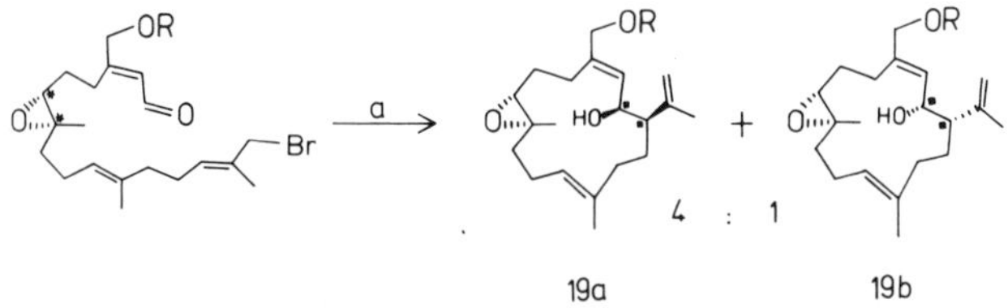

FIGURE 5. (a) $CrCl_2$, THF, 25°C 6 hr.

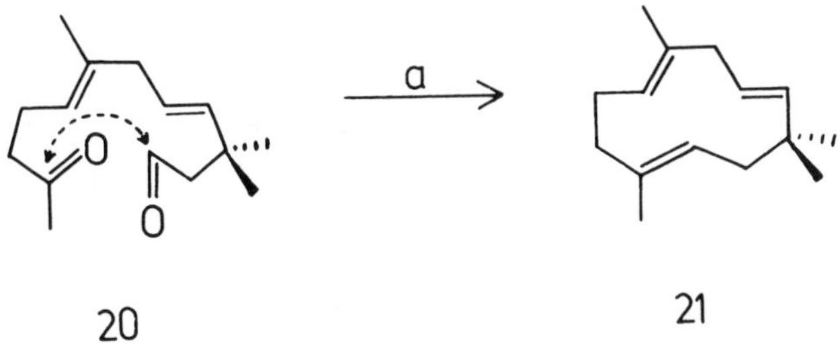

FIGURE 6. (a) $TiCl_3$, Zn – Cu, reflux 18 hr.

through a column packed with the polymer containing the catalyst, and the products (18 and 19) are collected as the eluate. Final oxidation of the hydroxyl group in 18 or 19 affords a single enone with a *trans* double bond.

Another ring closure method encompasses intramolecular reductive coupling of an allyl bromide moiety with an aldehydo group (Figure 5).[28] Note, that in this case the reaction is highly stereoselective giving rise to the *threo*-isomer 19a. Moreover, molecular geometry favors this diastereoisomer (with respect to the chiral centers of oxirane ring) over the other *threo*-isomer (19b) in a 4:1 ratio.

The McMurry synthesis of humulene (21, Figure 6)[17] was based on titanium-induced dicarbonyl coupling as the ring-forming reaction. The coupling gives preferentially isomers with a *trans* double bond. The original version of the method (using titanium trichloride and lithium aluminum hydride) was modified, using instead a zinc-copper couple for the reduction (for a review see Reference 11).

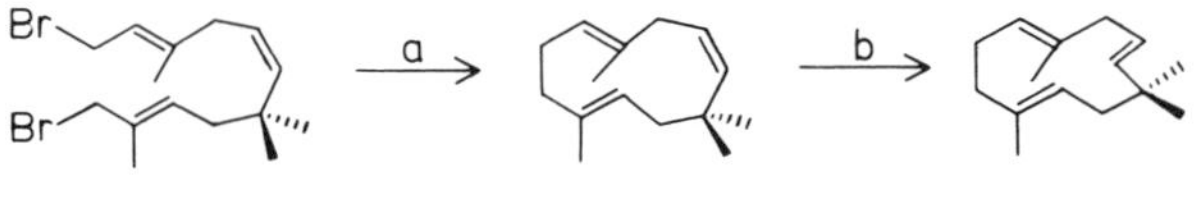

FIGURE 7. (a) $Ni(CO)_4$; (b) hν, $(PhS)_2$.

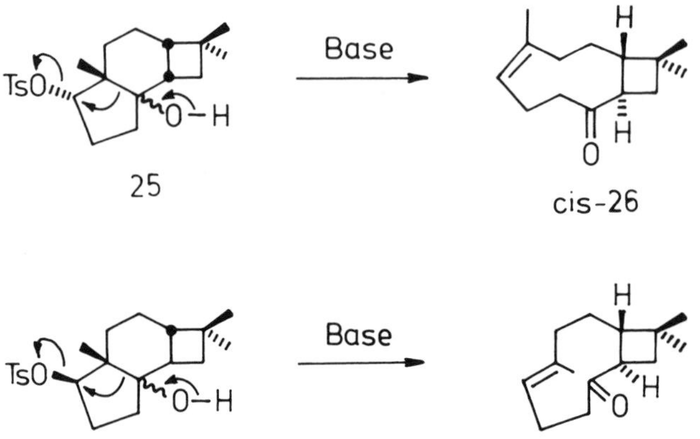

FIGURE 8. Base = t-BuOK, Me_2SO.

The last example shows a coupling of the *bis*-allylbromide 22 (Figure 7)[19,20] mediated by nickel tetracarbonyl. The cyclization, designed to produce *cis*-humulene (23), occurs at the allylic positions. The required *trans*-isomer 24 was obtained by radical *cis-trans* isomerization (23 → 24) (see also Reference 23).

II. FRAGMENTATION REACTIONS

Instead of closing a new ring we can make use of peripheral skeletal bonds in a bicyclic system in which the cross-piece bond is broken. The annulation of rings in the starting bicyclic system determines the configuration of the newly formed double bond. Since bicyclic compounds are often easily accessible and chemical transformations on these systems are well defined, the fragmentation strategy displays high flexibility and offers many tactical variations.

A frequently used method for disconnecting cross-piece bonds is the Grob fragmentation of 1,3-diol monotosylates (Figure 8).[29-35] The configuration of the double bond formed is determined by mutual orientation of the leaving tosyloxy group and the adjacent methyl, while the configuration of the hydroxyl group does not matter. The monotosylate 25 affords the *cis*-isomer 26, whereas *trans*-isomer is formed from the epimeric tosylate 27. Note that the bridgehead position α- to the carbonyl group is affected by base-catalyzed equilibration.

Another stereospecific fragmentation of hydroxy tosylates with a decalin system is shown in Figure 9.[34] The Eschenmoser fragmentation of tosylhydrazones derived from α-epoxy ketones[36] is also applicable to synthesis of medium and large carbocyclic rings. Further examples can be found in Reference 37.

The de Mayo retro-aldolization[38-40] (Figure 10)[41,42] is analogous to the Grob method. The ketones 34 and 35, prepared by intramolecular [2 + 2] photoaddition (cf. Figure 10), undergo retro-aldolization, giving rise to cyclooctanedione derivatives 36 and 37. This strategy has

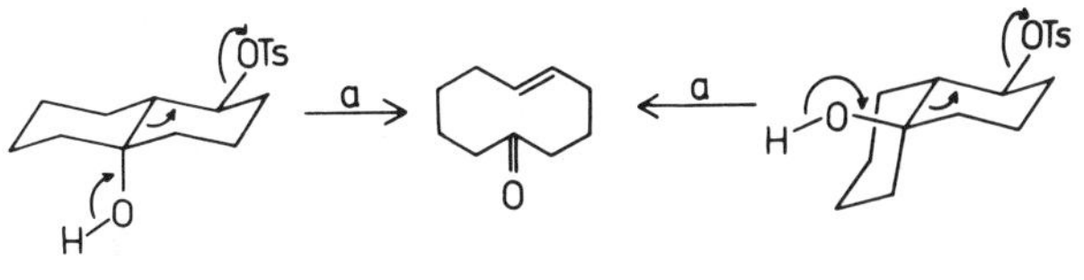

28 trans-29 30

31 cis-29

FIGURE 9. (a) t-BuOK.

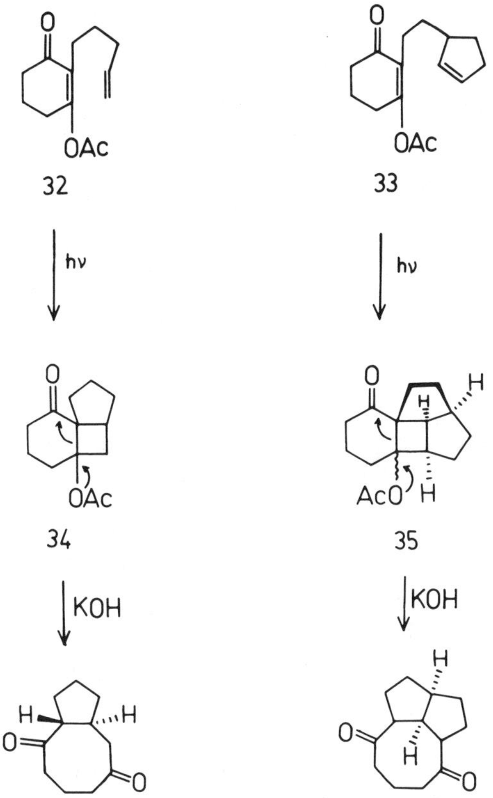

32 33 34 35 36 37

FIGURE 10.

been employed by Oppolzer and Pattenden in their syntheses of numerous natural products (for a review see Reference 40).

In addition to the above-mentioned polar fragmentations, one can use the Cope rearrangement or electrocyclic retroadditions. Figure 11 displays a sequential removal of the cross-piece bond in a [4.4.0] system by means of Cope rearrangement.[43] The double bond in 38

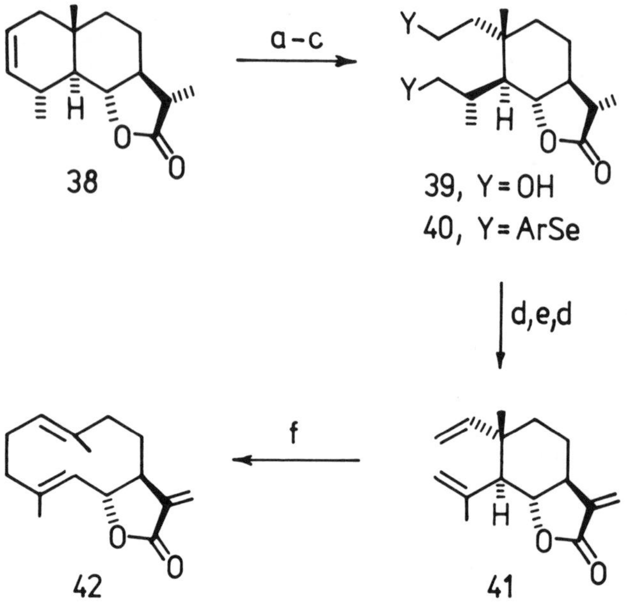

FIGURE 11. (a) O_3, CH_2Cl_2, CH_3OH; (b) $NaBH_4$; (c) ArSeCN, Bu_3P, THF, C_5H_5N; (d) H_2O_2, THF; (e) $(i\text{-}Pr)_2NLi$, $(PhSe)_2$, THF, HMPA; (f) 200°C.

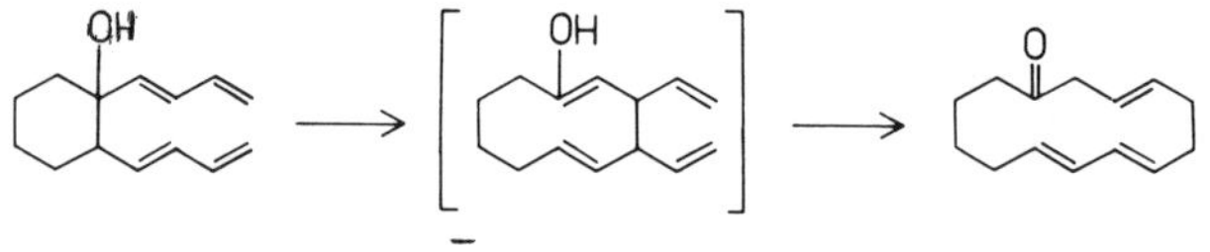

FIGURE 12.

was first cleaved by ozonolysis (38 → 39) to form a dialdehyde which was converted to the requisite diene 41 in several steps. Cope rearrangement in 41 restored the peripheral bond while cleaving the skeletal cross-piece (41 → 42). For similar strategy see Reference 44. Tandem use of two successive Cope rearrangements has been reported as a means of macroexpansion methodology[45] (Figure 12). As a heteroanalog, Claisen rearrangement has also been used for ring-enlargement, namely in the synthesis of muscone (Figure 13).[46]

The photochemical retrocyclization of the diene 43 (Figure 14)[47] to triene 44 formed a strategic basis of the Corey synthesis of dihydrocostunolide (45). The conrotatory mechanism, predicted by Woodward-Hoffmann rules, enforces the E configuration of both double bonds being formed.

Figure 15 illustrates[48] the steric course of a thermal retroaddition in the tricyclic derivative 48. According to the symmetry rules, thermal (ground state) [2 + 2] retroaddition should proceed via a supra-antara mechanism. The experimental results agree well with the prediction: *cis-anti-cis* isomer 48 is transformed to E,Z-diene 49. By fixing the configuration of the future α,β-enone bond (as in the lactone 50), the thermal retro-$[2_s + 2_a]$ addition proceeds through an alternative transition state, yielding the Z,E isomer 51. The lactone ring here plays the role of a control element. For similar methodologies cf. References 49 to 51.

III. "ZIP" REACTIONS

In contrast to smaller rings, where (n + 1) enlargement is a typical procedure (see Chapter

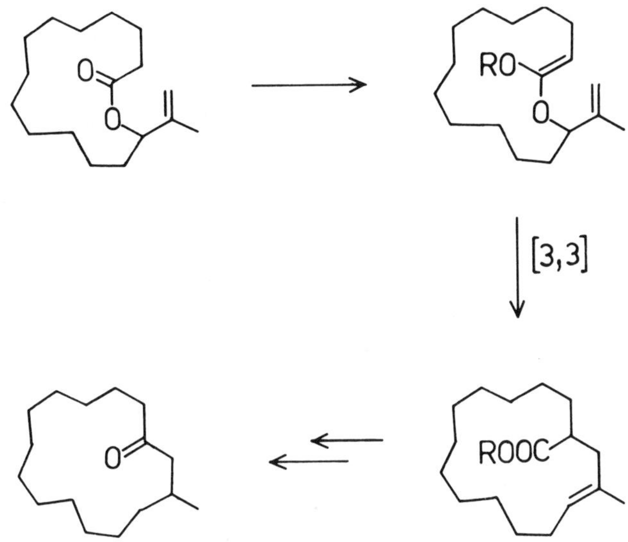

FIGURE 13. R = Et_3Si.

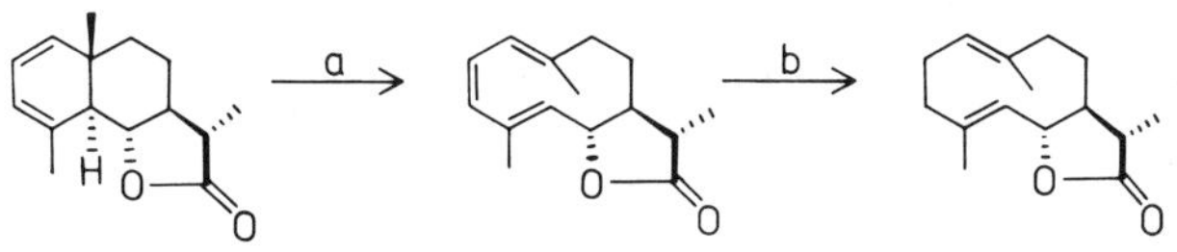

43 44 45

FIGURE 14. (a) hν, CH_3OH, −18°C; (b) H_2, Ni, −18°C.

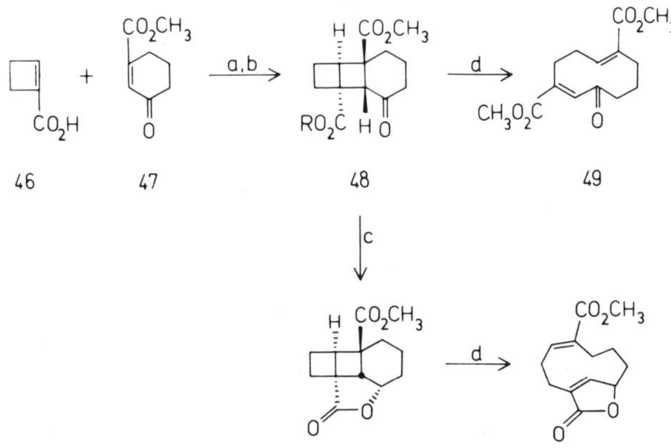

46 47 48 49

50 51

FIGURE 15. (a) hν (350 nm), C_6H_6; (b) CH_2N_2; (c) $NaBH_3CN$; (d) 164°C.

3), methods have been developed for selective (n + m) enlargements of pre-existing rings. The ''zip'' transamidation reaction will be discussed in detail in Chapter 1, Volume II. Here we present a carbocyclic modification resulting in (n + 2) or (n + 4) ring enlargement (Figure 16).[52]

FIGURE 16. (a) $CH_2{=}CHCOCH_3$, Ph_3P, THF; (b) t-BuOK, THF.

IV. NOTES ADDED IN PROOF

Further application of the McMurry dicarbonyl coupling has led to an efficient synthesis of four sesquiterpenes with a 10-membered ring, namely bicyclogermacrene,[53] lepidozene,[53] helminthogermacrene,[54] and germacrene A.[54] It has been found that the latter undergoes the Cope rearrangement under the reaction conditions to give β-elemene.[54] Cyclodimerization of 2,4-pentadienoid acid methyl ester catalyzed by nickel(O) and stereodirected by diethylaluminum ethoxide has been shown to produce stereoselectively the dimethyl ester of *trans*-1,5-cyclooctadiene-3,4-dicarboxylic acid in an 83% yield.[55]

The Grob-type fragmentation has been implemented in a total synthesis of sericenine.[56]

REFERENCES

1. **Finley, K. T.,** The acyloin condensation as a cyclization method, *Chem. Rev.*, 64, 573, 1964.
2. **Strating, J., Reiffers, S., and Wynberg, H.,** An improved method for the preparation of non-enolisable vic-diketones *via* acyloin condensation, *Synthesis*, 209, 1971.
3. **Rühlmann, K.,** Die Umsetzung von Carbonsäureestern mit Natrium in Gegenwart von Trimethylchlorsilan, *Synthesis*, 236, 1971.
4. **Schaefer, J. P. and Bloomfield, J. J.,** The Dieckmann condensation (including the Thorpe-Ziegler condensation), *Org. Reactions*, 15, 1, 1967.
5. **Kodama, M., Matsuki, Y., and Itô, S.,** Syntheses of macrocyclic terpenoids by intramolecular cyclization I. (±)-Cembrene-A, a termite pheromone, and (±)-nephthenol, *Tetrahed. Lett.*, 3065, 1975.
6. **Kodama, M., Matsuki, Y., and Itô, S.,** Synthesis of macrocyclic terpenoids by intramolecular cyclization II. Germacrane-type sesquiterpenes, *Tetrahed. Lett.*, 1121, 1976.
7. **Corbel, B. and Durst, T.,** Control of ring size resulting from γ-epoxysulfone and γ-epoxynitrite cyclization. Formation of either cyclopropyl or cyclobutyl derivatives. *J. Org. Chem.*, 41, 3648, 1976.
8. **Trost, B. M.,** Transition metal templates for selectivity in organic synthesis, *Pure Appl. Chem.*, 53, 2357, 1981.
9. **Trost, B. M. and Warner, R. W.,** Macrocyclization via an isomerization reaction at high concentrations promoted by palladium templates, *J. Am. Chem. Soc.*, 104, 6112, 1982.
10. **Kato, T., Kobayashi, T., and Kitahara, Y.,** Cyclization of polyenes XVI. Biogenetic type synthesis of cembrene type compounds, *Tetrahed. Lett.*, 3299, 1975.
11. **McMurry, J. E.,** Titanium-induced dicarbonyl coupling reactions, *Accounts Chem. Res.*, 16, 405, 1983.
12. **McMurry, J. E. and Fleming, M. P.,** A new method for the reductive coupling of carbonyls to olefins. Synthesis of β-carotene, *J. Am. Chem. Soc.*, 96, 4708, 1974.
13. **McMurry, J. E. and Fleming, M. P.,** Improved procedure for the reductive coupling of carbonyls to olefins and for the reduction of diols to olefins, *J. Org. Chem.*, 41, 896, 1976.
14. **McMurry, J. E. and Krepski, L. R.,** Synthesis of unsymetrical olefins by titanium (0) induced mixed carbonyl coupling. Some comments on the mechanism of the pinacol reactions, *J. Org. Chem.*, 41, 3929, 1976.
15. **McMurry, J. E. and Kees, K. L.,** Synthesis of cycloalkanes by intramolecular titanium induced dicarbonyl coupling, *J. Org. Chem.*, 42, 2655, 1977.

16. **McMurry, J. E., Fleming, M. P., Kees, K. L., and Krepski, L. R.,** Titanium-induced reductive coupling of carbonyls to olefins, *J. Org. Chem.*, 43, 3255, 1978.
17. **McMurry, J. E. and Matz, J. R.,** Stereospecific synthesis of humulene by titanium-induced dicarbonyl coupling, *Tetrahed. Lett.*, 23, 2723, 1982.
18. **Sharples, K. B. and Flood, T. C.,** Direct deoxygenation of vicinal diols with Tungsten (IV). A new olefin synthesis, *J. Chem. Soc. Chem. Commun.*, 370, 1972.
19. **Corey, E. J. and Hamanaka, E.,** A new synthetic approach to medium-size carbocyclic systems, *J. Am. Chem. Soc.*, 86, 1641, 1964.
20. **Corey, E. J. and Hamanaka, E.,** Total synthesis of humulene, *J. Am. Chem. Soc.*, 89, 2758, 1967.
21. **Dauben, G. H., Beasly, M. D., Broadhurst, M. D., Muller, B., Peppard, D. J., Pesnelle, P., and Sutter, S.,** A synthesis of cembrene. A 14-membered ring diterpene, *J. Am. Chem. Soc.*, 96, 4724, 1974.
22. **Corey, E. J. and Wat, E. K. W.,** The synthesis of large-ring 1,5-dienes by cyclization of allylic dibromides with nickel carbonyl, *J. Am. Chem. Soc.*, 89, 2757, 1967.
23. **Frazza, M. S. and Roberts, B. W.,** Synthesis of 4-substituted cis,cis-1,6-cyclodecadienes via cycloalkylation, *Tetrahed. Lett.*, 22, 4193, 1981.
24. **van Leusen, D. and van Leusen, A. M.,** A simple synthesis of cyclobutanone, *Synthesis*, 325, 1980.
25. **Ziegler, F. E., Chakraborty, U. R., and Weisenfeld, R. B.,** A palladium-catalyzed carbon-carbon bond formation of conjugated dienones. A macrocyclic dienone lactone model for the carbomycins, *Tetrahedron*, 37, 4035, 1981.
26. **Stork, G. and Nakamura, E.,** Large-ring lactones by internal ketophosphonate cyclizations, *J. Org. Chem.*, 44, 4010, 1979.
27. **Still, W. C. and Novack, V. J.,** Macrocyclic stereocontrol. Total synthesis of (±)-3-deoxy-rosaranolide, *J. Am. Chem. Soc.*, 106, 1148, 1984.
28. **Still, W. C. and Mobilio, D.,** Synthesis of asperdiol, *J. Org. Chem.*, 48, 4785, 1983.
29. **Grob, C. A., Kiefer, H. R., Lutz, H., and Wilkins, H.,** The stereochemistry of synchronous fragmentation, *Tetrahed. Lett.*, 2901, 1964.
30. **Corey, E. J., Mitra, R. B., and Uda, H.,** Total synthesis of *d,l*-caryophyllene and *d,l*-isocaryophyllene, *J. Am. Chem. Soc.*, 85, 362, 1963.
31. **Corey, E. J., Mitra, R. B., and Uda, H.,** Total synthesis of *d,l*-caryophyllene and *d,l*-isocaryophyllene, *J. Am. Chem. Soc.*, 86, 485, 1964.
32. **ApSimon, J.,** *The Total Synthesis of Natural Products*, Vol. 1—5, Wiley-Interscience, New York, 1973—1983.
33. **Patel, H. A. and Dev, S.,** Products active on arthropod - III. Insect juvenile hormone mimics (Part 3). Cyclononane analogue of *Cecropia* juvenile hormone, *Tetrahedron*, 37, 1577, 1981.
34. **Wharton, P. S. and Hiegel, G. A.,** Fragmentation of 1,10-decalindiol monotosylates, *J. Org. Chem.*, 30, 3254, 1965.
35. **Sternbach, D., Shibuya, M., Jaisli, F., Bonetti, M., and Eschenmoser, A.,** Ein fragmentatiever zugang zu Makroliden: (5-E,8-Z)-6-methyl-5,8-undecadien-11-olid, *Angew. Chem.*, 91, 670, 1979.
36. **Felix, D., Schreiber, J., Ohlhoff, G., and Eschenmoser, A.,** α,β-Epoxyketon → alkynon-Fragmentierung I. Synthese von Exalton und *rac* -muscon aus cyclodecanon, *Helv. Chim. Acta*, 54, 2896, 1971.
37. **Oppolzer, W. and Wylie, R. D.,** Total synthesis of β-bulnesene and 1-*epi*-β-bulnesene by intramolecular photoaddition, *Helv. Chim. Acta*, 63, 1198, 1980.
38. **De Mayo, P.,** Enone Photoannelation, *Acc. Chem. Res.*, 4, 41, 1971.
39. **Loutfy, R. O. and de Mayo, P.,** On the mechanism of enone photoannelation. Activation energies and the role of exciplexes, *J. Am. Chem. Soc.*, 99, 3559, 1977.
40. **Oppolzer, W.,** Intramolecular (2 + 2) photoaddition/cyclobutane-fragmentation sequence in organic synthesis, *Acc. Chem. Res.*, 15, 135, 1982.
41. **Begley, M. J., Mellor, M., and Pattenden, G.,** A new approach to fused carbocycles. Intramolecular photocyclisations of 1,3-dione enol acetates, *J. Chem. Soc. Chem. Commun.*, 235, 1979.
42. **Oppolzer, W. and Bird, T. G. C.,** Intramolecular de Mayo reactions of 3-acetoxy-2-alkenyl-2-cyclohexenones, *Helv. Chim. Acta*, 62, 1199, 1979.
43. **Grieco, P. A. and Nishizawa, M.,** Total synthesis of (+)-costunolide, *J. Org. Chem.*, 42, 1717, 1977.
44. **Ando, M., Tajima, K., and Takase, K.,** Studies on the synthesis of sesquiterpene lactones. 8. Synthesis of saussurea lactone, 8-deoxymelitensin, and 11,12-dehydro-8-deoxymelitensin via a novel fragmentation reaction, *J. Org. Chem.*, 48, 1211, 1983.
45. **Wander, P. A., Sieburth, S. McN., Petraitis, J. J., and Singh, S. K.,** Macroexpansion methodology. Medium ring synthesis based on an eight unit ring expansion process, *Tetrahedron*, 37, 3967, 1981.
46. **Brunner, R. K. and Borschberg, H.-J.,** Construction of large carbocyclic rings by Ireland-Claisen rearrangement of O-silylated lactone enolates. Synthesis of (±)-muscone, *Helv. Chim. Acta*, 66, 2608, 1983.
47. **Corey, E. J. and Hortmann, G. F.,** The total synthesis of dihydrocostunolide, *J. Am. Chem. Soc.*, 87, 5736, 1965.

48. **Lane, G. L. and Otulakowski, J. A.,** Improved preparation of methyl-3-oxo-1-cyclohexene-1-carboxylate and its use in the synthesis of substituted 1,5-cyclodecadienes, *J. Org. Chem.*, 47, 5093, 1982.
49. **Wender, P. A. and Lechleiter, J. C.,** Total synthesis of (±)-isabelin, *J. Am. Chem. Soc.*, 102, 6340, 1980.
50. **Williams, J. R. and Cleary, T. P.,** A short synthesis of the 6-*epi*-arteannuin B skeleton, *J. Chem. Soc. Chem. Commun.*, 626, 1982.
51. **Lange, G. L., Huggins, M.-A., and Neidert, E.,** Facile synthesis of substituted 1,5-cyclodecadienes. Employing a photoaddition-thermolysis sequence, *Tetrahed. Lett.*, 4409, 1976.
52. **Nakashita, Y. and Hesse, M.,** The carbon zip reaction. A method for expanding carbocycles, *Helv. Chim. Acta*, 66, 845, 1983.
53. **McMurry, J. E. and Bosch, G. K.,** Synthesis of bicyclogermacrene and lepidozene, *Tetrahed. Lett.*, 26, 2167, 1985.
54. **McMurry, J. E. and Kočovský, P.,** Synthesis of helminthogermacrene and β-elemene, *Tetrahed. Lett.*, 26, 2171, 1985.
55. **Brun, P., Tenaglia, A., and Waegell, B.,** Nickel zero catalyzed cyclodimerization of 2,4-pentadienoic acid methyl ester. A remarkable directing effect of Et_2AlOEt, *Tetrahed. Lett.*, 26, 5685, 1985.
56. **Honan, M. C., Balasuryia, A., and Cresp, T. M.,** Total synthesis of sericenine, *J. Org. Chem.*, 50, 4326, 1985.

Chapter 7

STEREOSELECTIVE SUBSTITUTION IN CYCLIC SYSTEMS

Almost any natural product possessing a cyclic skeleton contains a corollary of substituents arranged in a stereospecific manner and its biological activity may be very sensitive to subtle variations of relative configuration of substituents. Perhaps this is why organic chemists have spent so much effort on developing methodologies that would enable one to introduce various substituents into cyclic systems stereoselectively. In this Chapter we will outline only the principal strategic and tactical routes leading to substituted ring systems, illustrated with selected examples. According to the methodology used, the topic can be divided into three subsections dealing with *de novo* formation of substituted rings, stereoselective attachment of substituents to existing rings, and finally, configurational changes of substituents already present in the ring system. Other means of construction of cycles, e.g. contract and expansion of pre-existing rings, have already been discussed in Chapter 3, Section III (see also References 1 and 2).

I. *De novo* CONSTRUCTION OF A STEREOSPECIFICALLY SUBSTITUTED RING

In the preceding chapters we have discussed numerous synthetic ways leading to stereoselective construction of various ring systems. In this section we will focus on the selection of suitable synthons and cyclization methods with the aim to achieve a required geometric arrangement of substituents on the skeleton being built.

As already mentioned in the previous text, a suitable and reliable means of *de novo* construction of a ring is a cycloaddition. For instance, rings bearing two *cis*-oriented vicinal substituents can be formed in a stereoselective manner by means of various cycloaddition reactions, as generally shown in Figure 1. The corresponding *trans*-isomers could be obtained similarly, starting from *trans*-disubstituted olefins. Other ring-closure methods rely on intramolecular Michael-type or S_N2 cyclizations. However, it is of eminent importance to consider carefully and in advance the stereoelectronic factors which may hamper these reactions[3-6] (for a detailed discussion see Chapters 1 and 2 in Volume II). Some of the methods are specific to different ring systems, and these will be outlined in the following paragraphs.

A. Cyclopropropane Derivatives

One of the means of forming a stereospecifically substituted cyclopropane ring is the reaction of carbanions with conjugated esters (Figure 2). The carbanion 1 which arises by a conjugate addition of the sulfone carbanion to ethyl 3-methylbutenoate, undergoes intramolecular nucleophilic substitution[7] to give the more stable *trans*-isomer of ethyl chrysanthemate (2). Another excellent method is based on ring closure during intramolecular cleavage of the oxirane ring by a carbanion generated at the α-position to a nitrile group (Figure 3).[8-10] This method is generally applicable to construction of rings of various sizes. If two rings can arise by attack on the closer or further oxirane carbon atom, the formation of the smaller ring will be preferred (Figure 3).[8]

Another route to stereospecifically substituted cyclopropanes is the contraction of suitable derivatives of cyclobutane. For some recent methods see References 11 and 12.

B. Cyclobutane Derivatives

Photochemical [2 + 2] cycloaddition is probably the most popular synthetic route to substituted cyclobutanes.[1] Of few other methods, we stress here the cyclization of epoxy-

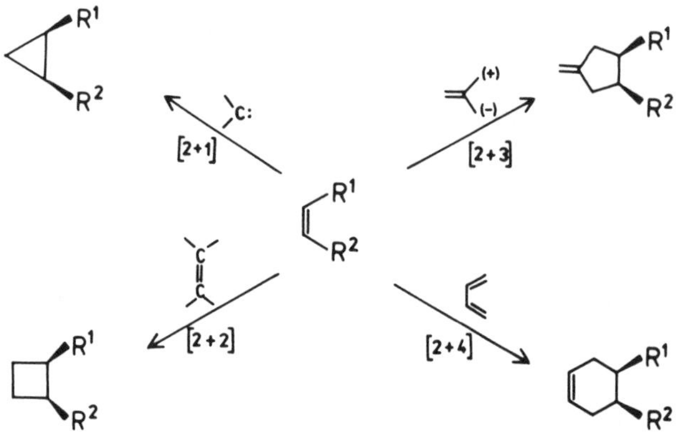

FIGURE 1.

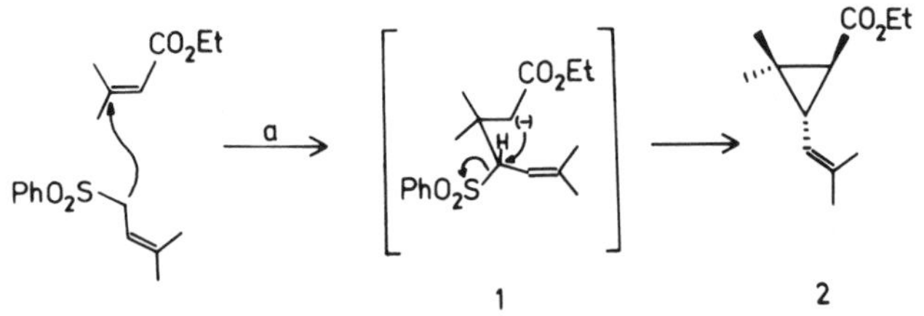

FIGURE 2. t-BuOK, THF.

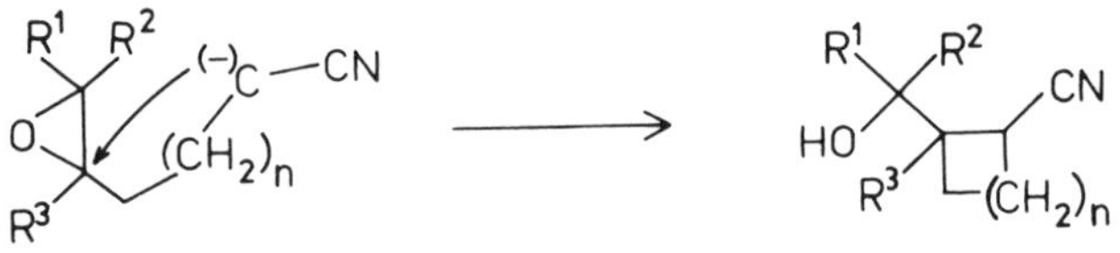

FIGURE 3.

nitriles which is exemplified by the synthesis of grandisol (3, Figure 4).[8] The high stereoselectivity of the cyclization may be explained by a preferential formation of the less-strained transition state A (Figure 4).[8]

C. Cyclopentane Derivatives

Beside the cyclization of epoxy nitriles, cyclopentane derivatives can be prepared by intramolecular substitution of a suitable leaving group by a carbanion.[13,14] This tactic has been employed particularly in syntheses of prostaglandins[15,16] and brefeldin A.[17-19]

Intramolecular Michael addition (Figure 5) proceeds preferentially through the transition state A which gives rise to a *trans*-disubstituted product. The method works excellently in closing six-membered rings as well.[20]

Another method formally consists of a [4 + 1] addition, which is executed in two steps (Figure 6,a):[21] The first step comprises a carbene addition to a diene followed by a butyl lithium induced [1,3] shift. This represents a fancy method of stereoselective formation of a five-membered ring with a hydroxy group. For a related method and synthetic applications see References 22 to 24.

D. Cyclohexane Derivatives

Natural products containing the cyclohexane ring are ubiquitous, and hence the majority

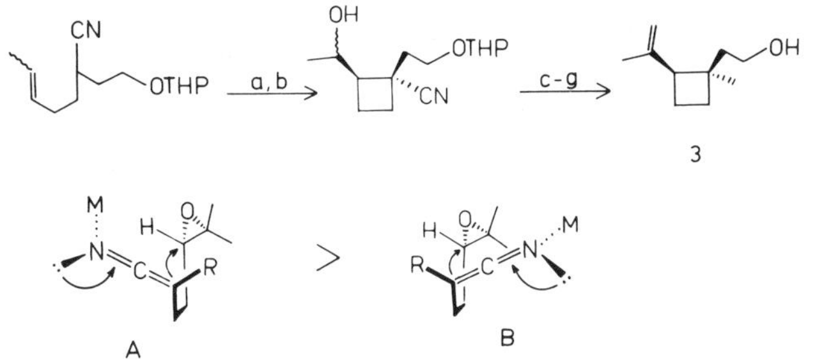

CN OTHP a, b OH OTHP CN c-g OH

3

M H O N C R > H O M R C N

A B

FIGURE 4. (a) m – Cl – C_6H_4 – CO_3H; (b) $(Me_3Si)_2NLi$; (c) $(i\text{-}Bu)_2AlH$; (d) N_2H_4, KOH; (e) CrO_3, H_2SO_4, Me_2CO; (f) $Ph_3P{\cdot}CH_3I$, BuLi, THF; (g) $0.05M$ – $HClO_4$, THF, H_2O.

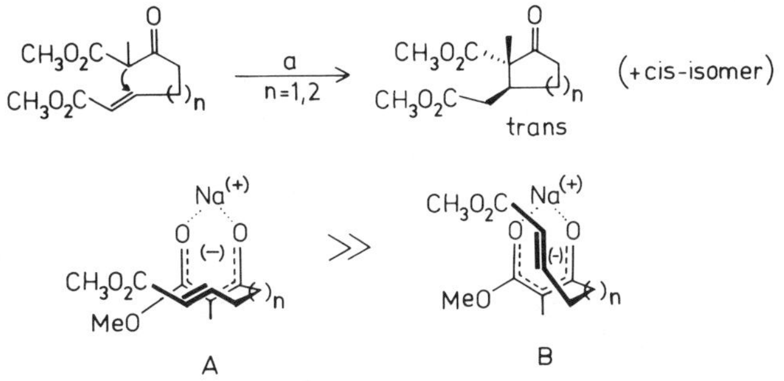

CH_3O_2C O CH_3O_2C ( )n a n=1,2 CH_3O_2C O CH_3O_2C ( )n trans (+cis-isomer)

$Na^{(+)}$ O O (–) CH_3O_2C MeO ( )n ≫ CH_3O_2C $Na^{(+)}$ O O (–) MeO ( )n

A B

FIGURE 5. (a) NaH, r.t., 15 min.

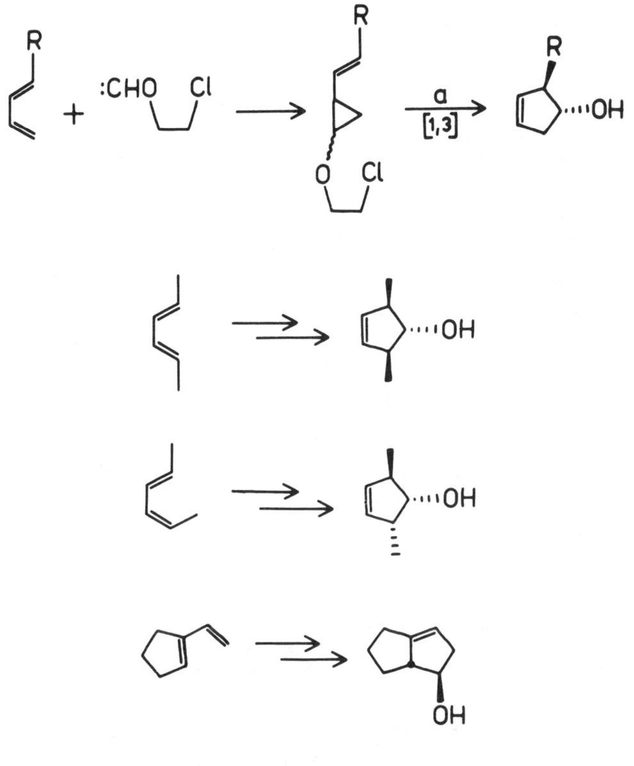

R + :CHO Cl → R O Cl a [1,3] → R OH

→ OH

→ OH

→ OH

FIGURE 6. BuLi.

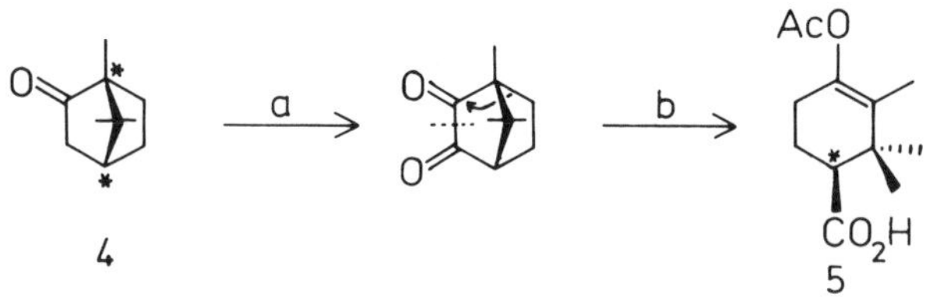

FIGURE 7. (a) SeO_2; (b) $BF_3 \cdot Et_2O$, Ac_2O.

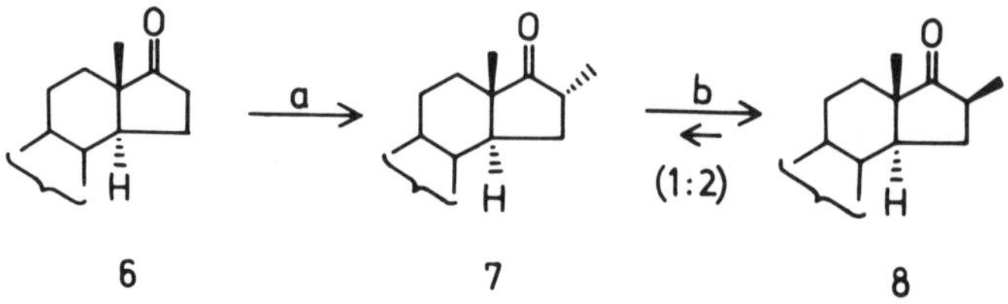

FIGURE 8. (a) $(i\text{-}Pr)_2NLi$, then CH_3I; (b) NaOH, CH_3OH, reflux.

of synthetic strategies takes advantage of compounds that already contain a six-membered ring and that can be stereoselectively functionalized. Except for the Diels-Alder reaction and Robinson annulation, there are only a few methods based on direct cyclization of a suitably substituted precursor. One of these is the intramolecular Michael addition described in the preceding paragraph (Figure 5).[20]

An alternative way to substituted cyclohexanes is expansion of the five-membered ring in the corresponding cyclopentane derivative (see Chapter 3, Section III). This approach is illustrated in Figure 7, where the chiral camphor skeleton 4 is transformed to optically active substituted cyclohexanecarboxylic acid 5.[25-27] This procedure involves a ring fragmentation accompanied by skeletal rearrangement.

II. STEREOSELECTIVE ATTACHMENT OF SUBSTITUENTS

Aside from cycloaddition reactions, a common strategy leading to substituted five- and six-membered rings is to build the ring first and then attach the substituents. The synthetic arsenal for the latter step is very rich, affording various methods of functionalization, substituent transposition, etc. In all cases the synthetic plan must take into account the kinetic or thermodynamic control of the key reaction steps as shown in Figure 8.[28,29] Methylation of steroidal ketone 6 gave the 16α-methyl ketone 7 (kinetic control) which was transformed in alkaline medium to a mixture of both isomers in which the more stable 16β-isomer 8 prevailed.

A. Stereoselective Introduction of Substituents by Addition to Double Bonds

Methods based on substitution reactions are not completely reliable for a stereoselective introduction of a substituent into the ring, for they may or may not proceed via S_N2 mechanism. It is more reliable to utilize the addition reactions to multiple bonds in which the hybridization of the reaction center changes from sp^2 to sp^3. The stereochemical course of the addition depends on the accessibility of the double bond from one or the other side of the cyclic system (under kinetic control) and can be influenced by control elements.

The stereoselective introduction of heteroatoms mostly relies upon electrophilic additions,[30] or the analogous cleavage of the oxirane ring.[31,32] These reactions may be affected by competing electronic, steric, or stereoelectronic factors. The actual course of the addition, especially with six-membered rings,[33] can be directed by modifying the structure of the

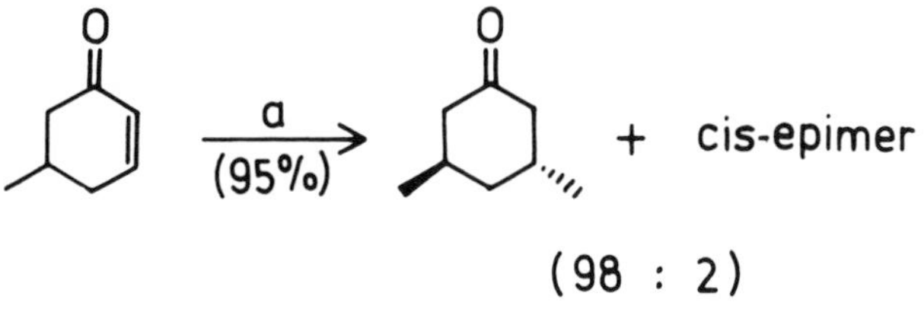

FIGURE 9. (a) $(CH_3)_2CuLi$.

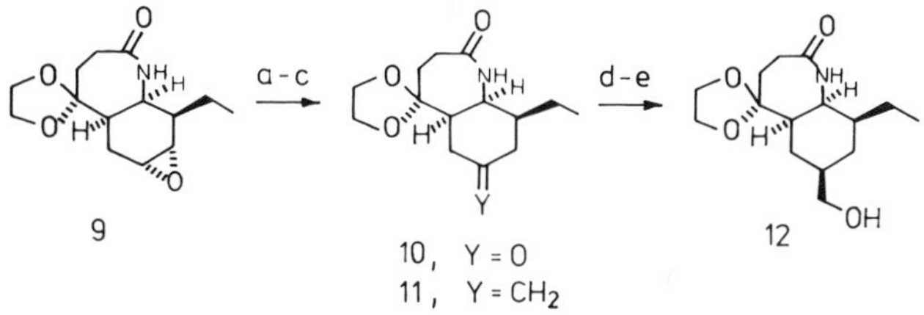

FIGURE 10. (a) $LiAlH_4$; (b) CrO_3, C_5H_5N; (c) Wittig react.; (d) B_2H_6; (e) H_2O_2, OH^-.

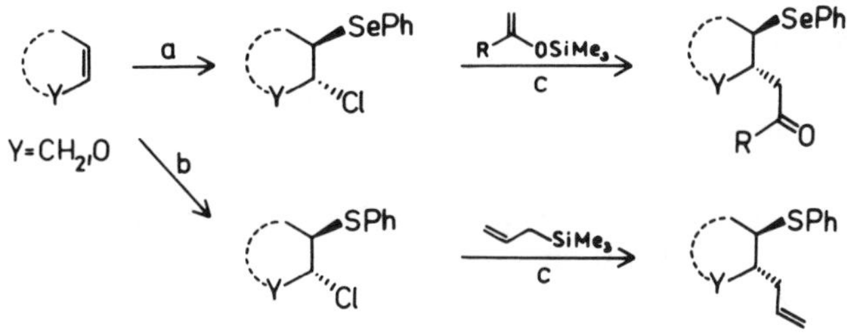

FIGURE 11. (a) PhSeCl; (b) PhSCl; (c) $ZnBr_2$.

starting compound and the reaction conditions. In this way, the addition can be designed to proceed either according or contrary to the Markovnikov or Fürst-Plattner rules (see for example Figure 16 in Chapter 1).

The introduction of carbon substituents (e.g. alkyl groups) is usually achieved through 1,4-addition of carbanions to a conjugated double bond of an enone or α,β-unsaturated ester system (Figure 9).[34-37] Asymmetric additions to cyclopentenones have been reported to provide high optical yields.[38]

While nucleophilic attack to activated endocyclic double bonds creates the asymmetric center directly, even an unactivated endocyclic double bond can be transformed to alkyl substituents by an indirect procedure (Figure 10).[39] The exomethylene derivative 11, obtained from the ketone 10, was converted to the alcohol 12 by stereoselective hydroboration. Since the synthesis started from the epoxide 9 (Figure 10), which in turn was prepared from the corresponding olefin, the overall sequence represents an alternative tactic for the regio- and stereoselective introduction of a functionalized substituent on the endocyclic double bond.

A shorter method of a stereoselective introduction of an alkyl substituent to an unactivated double bond is illustrated in (Figure 11)[40] (see also References 2, 41, and 42). Other methods employ cleavage of cyclopropane rings, which in turn are usually obtained by a [2 + 1] addition to a nonactivated double bond (see Chapter 7, Section II.E.1).

Stereoselective hydrogenation of an exocyclic or trisubstituted endocyclic double bond can be effectively used for synthetic purposes.[1] Very common are the methods utilizing the

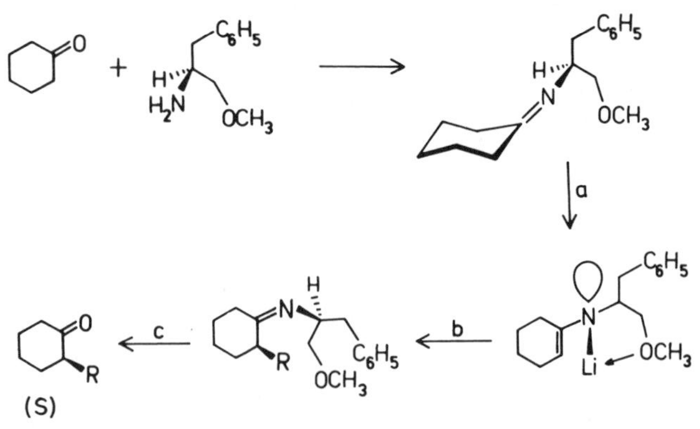

FIGURE 12. (a) $(i\text{-}Pr)_2NLi$, THF, −20°C; (b) RX, −78°C; (c) H_3O^+.

$sp^2 \rightarrow sp^3$ change of bonds between carbon atoms and heteroatoms, such as stereoselective reduction of ketones, oximes, etc.[1,43,44]

Another common alkylation method creating a new chiral center is based on reactions of electrophilic reagents with enamines[45,46] or α-metallated imines. If the enamine or imine is prepared from an optically active amine, the alkylation often affords chiral products in high optical yields[47-54] (Figure 12).[53] Since the optically active amine can be recycled, the method provides a cost-efficient route to chiral compounds. In some cases, the method allows for the steric control within the alkyl substituent being introduced.[55] (For a detailed discussion see Chapter 2 in Volume II.)

B. Stereoselective Introduction of Substituents in Vicinal Positions

Compounds containing vicinal substituents arranged with a defined geometry can be essentially synthesized by alkylation or addition methods as discussed in the preceding paragraph. Nevertheless, if the groups are not implanted in one reaction step, the substituents already present in the molecule would affect the stereochemistry of the next steps so that the optimum tactics leading to vicinally substituted cyclic compounds may differ from that leading to compounds with other substitution patterns (1,3-, 1,4-, etc.). The methods creating carbon-heteroatom bonds, mostly based on electrophilic additions, have been reviewed in detail.[30-32] We shall focus on the formation of the carbon-carbon bonds in compounds, where the synthon of the vicinal system is represented by a double bond or an oxirane ring. Finally, several cases of introducing vicinal hydroxyl and/or amino groups will be presented as well.

The first approach (Figure 13)[56] employs as the initial step the 1,4-addition of an organocuprate reagent (13 → 14) to a latent cyclopentadienone system in which one of the double bonds has been protected as a Diels-Alder adduct. The addition takes place from the less-hindered *exo*- side of the skeleton. The ketone 14 is further alkylated at the α-position, but this step is controlled by the β-methyl group and proceeds from the α-side (15). The double bond was liberated by the retro-Diels-Alder reaction yielding the vicinally substituted cyclopentanone 16.

An ingenious synthetic procedure introducing four vicinal, *cis*-oriented methyl groups was developed by Jommi et al.[57,58] in their synthesis of (−)-pinguisone (25, Figure 14). 1,4-Addition of lithium dimethylcuprate to optically active enedione 17 afforded as expected the *cis*-annulated product 18. The system was reactivated by introducing two enone double bonds (19), but the nucleophilic 1,4-addition to 19 was found to proceed at the more reactive cyclopentenone double bond from the less-hindered α-side, i.e. with "incorrect" stereochemistry. This unfavorable reactivity was utilized, however, for protection of the cyclopentenone double bond as sulfides 20, and 21. The latter compound reacted with another

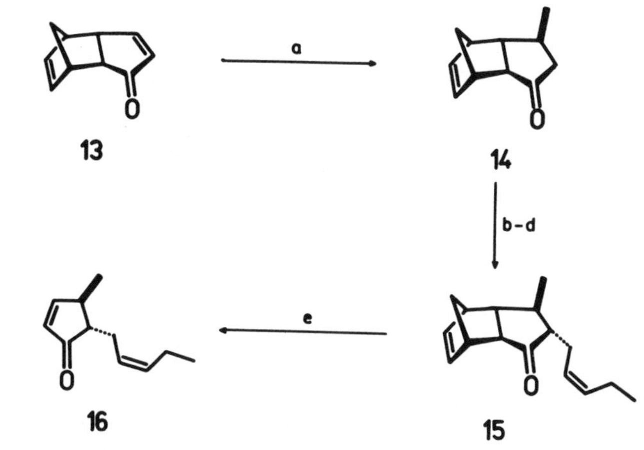

FIGURE 13. (a) CH_3MgX, Cu^+; (b) $C_6H_{11}NH_2$; (c) $(i\text{-}Pr)_2NLi$; (d) $Cl-CH_2-CH{=}CH-CH_2CH_3$; (e) 600°C.

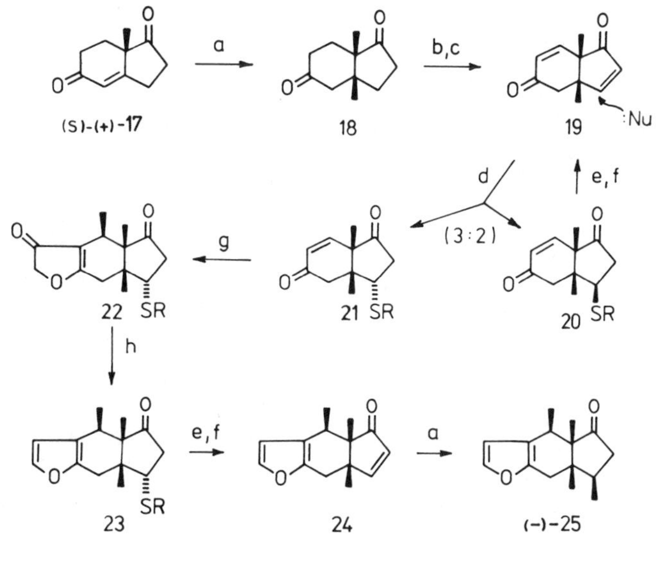

FIGURE 14. (a) Me_2CuLi, −25°C; (b) 2 eq. Br_2, AcOH; (c) $CaCO_3$, CH_3CONMe_2; (d) 1 eq. i-PrSH; (e) $NaIO_4$; (f) $CaCO_3$, CCl_4, heat; (g) Me_2CuLi, then quenching with $ClCH_2COCl$; (h) 9-BBN.

equivalent of the cuprate reagent to give a trimethyl derivative with correct relative configuration of all three methyl groups. The sulfide 20 reacts in the other way but can be recycled back to the ketone 19. The auxiliary sulfide group in 21 thus works as a control element which imposes a certain conformation of the flexible skeleton and in this way directs the approach of the nucleophile to the cyclohexenone double bond. The enolate intermediate obtained after the 1,4-addition in 21 was quenched with chloroacetyl chloride to furnish the heterocyclic compound 22 which was converted by the Jommi method to the furan derivative 23. The cyclopentenone double bond was recreated by sulfoxide elimination (23 → 24) and the product was converted to (−)-pinguisone by the final 1,4-addition of lithium dimethylcuprate. The β-attack of the cuprate in 24 is made possible by favorable conformation of the bicyclic system imposed by the planar furan ring. The latter structural unit thus

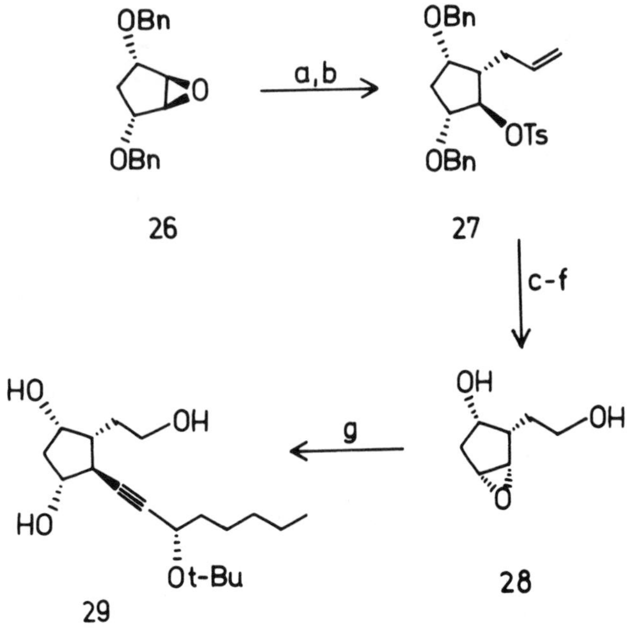

FIGURE 15. (a) $(CH_2{=}CH-CH_2)_2CuLi$, Et_2O; (b) TsCl, C_5H_5N; (c) O_3, CH_2Cl_2; (d) $NaBH_4$; (e) H_2, Pd/C; (f) KOH; (g) $(MeO)_2Al-C{\equiv}C-CH(C_5H_{11})-O-t-Bu$.

functions as a control element in conformation kinetic control of the reaction stereochemistry. Moreover, the control element constitutes a part of the target molecule and therefore it need not be removed. It should be noted that the total synthesis of the optically active product has helped to determine the absolute configuration of natural (−)-pinguisone.

The second method of introducing the vicinal alkyl group is based on cleavage of the oxirane ring in the α,β-epoxy alcohols (see Figure 15).[59,60] The reaction of the epoxide 26 with the organocopper reagent affords the corresponding alcohol, which is converted via a tosylate 27 to the epoxide 28 with a transposed oxirane ring. Cleavage of the latter with the organoaluminum reagent introduces stereoselectively the second alkyl group (29). A great number of similar examples can be found in prostaglandin syntheses (for review see References 1, and 61 to 63). For stereoselective alkylation of complexes with transition metals see References 64 to 66.

Stereoselective introduction of two or three vicinal hydroxyl groups can be illustrated with Ganem's elegant syntheses of chorismic (34) and shikimic acid (36) (Figure 16).[67,68] Dihydrobenzoic acid was first converted to the bridged bromolactone 30. Functionalization of the allylic position in 30 yielded the bromide 31 which on acetolysis afforded two isomeric acetates 32a, and 33a (~1:1). The hydroxy lactone 33b (from 33a) was hydrolyzed to furnish chorismic acid (34),[67] while the second isomer 32 could be transformed to 33. The synthesis of shikimic acid (36) required introduction of another hydroxyl group, which was achieved by successive epoxidation (33b → 35a), debromination (35a → 35b) and hydrolysis in which the carbon-oxygen bond of the oxirane ring served as a leaving group to make the double bond (35b → 36). For a different approach to chorismic acid see Reference 69.

Generally, *cis*-1,2-diols are easily prepared by osmylation of a double bond, whereas their *trans*-isomers may be obtained by a two-step process, i.e. epoxidation followed by cleavage of the oxirane ring. Similar routes exist for a stereoselective preparation of vicinal aminoalcohols: amino osmylation[70,71] and cleavage of epoxides.[72] A different strategy, however, has been employed in the synthesis of daunosamine, the essential component of several

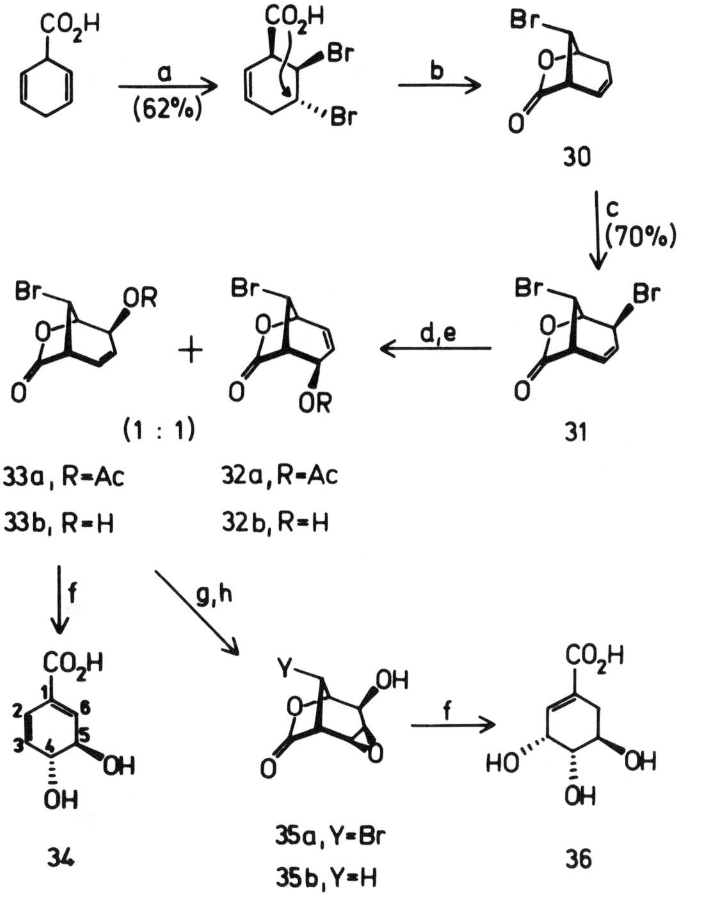

FIGURE 16. (a) Br_2, CH_2Cl_2; (b) $NaHCO_3$; (c) NBS, $(PhCO)_2O_2$, CCl_4, reflux 18 hr; (d) AcONa, HMPA; (e) 10% H_2SO_4, H_2O, THF, reflux; (f) KOH; (g) CF_3CO_3H, $ClCH_2CH_2Cl$, reflux 23 hr; (h) Bu_3SnH, AIBN, toluene, reflux 2 hr.

antitumor antibiotics (Figure 17).[73] In this case, a vicinal *cis*-amino group was introduced by an electrophilic addition to an allylic alcohol: the unfavorable equatorial attack by nitrogen on an intermediate iodonium ion was facilitated by neighboring group participation (see also Section II.D).

A spectacular example of successive introduction of six vicinal O- and N-containing groups into all available positions in a cyclohexane ring is given by the synthesis of the antibiotic fortamine (Figure 18).[74] Starting from cyclohexadiene, Knapp et al. introduced the first *trans*-1,2-unit by tandem oxirane formation and opening. The second oxygen was selectively attached with the help of neighboring group participation (37 → 38). The next oxygen was again planted via tandem epoxidation and cleavage (38 → 39 → 40) as was the last pair of O- and N-substituents.

Aside from electrophilic additions and related processes transforming sp^2 to sp^3 centers, also S_N2' reactions with neighboring group participation may be considered as a potentially useful methodology for the stereoselective introduction of substituents (Figure 19).[75]

C. [3 + 3] Sigmatropic Rearrangements

The Cope and Claisen rearrangements,[76,77] or a tandem version thereof, represent a powerful synthetic tool for the stereoselective formation of carbon-carbon bonds. We have already

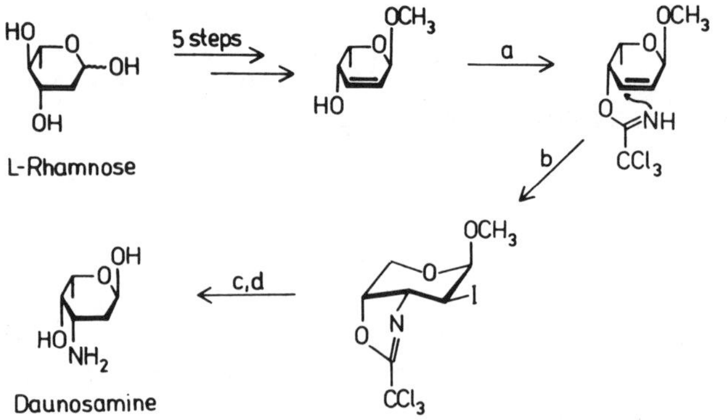

L-Rhamnose → (5 steps) → (a) → (b) → (c,d) → Daunosamine

FIGURE 17. (a) Cl_3CCN, NaH, CH_2Cl_2, 0° → 20°C, 3 hr; (b) $(Collidine)_2I \cdot ClO_4$, CH_3CN, 20°C, 3—5 d; (c) Bu_3SnH, C_6H_6, AIBN, reflux 3 hr; (d) TsOH, C_5H_5N, H_2O, 100°C, 2.5 hr.

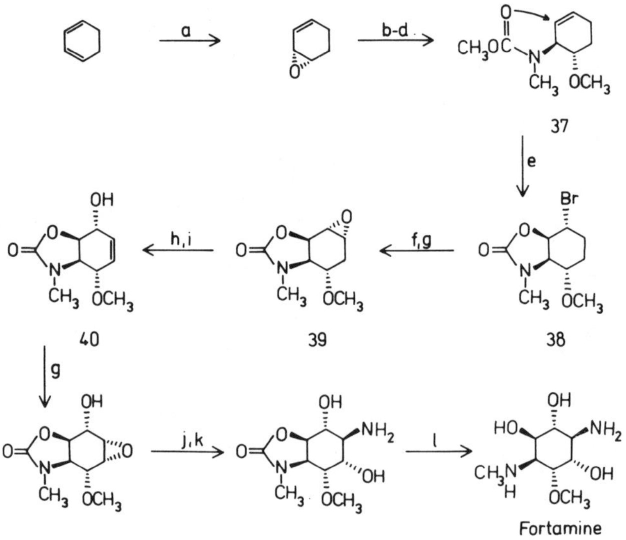

(a) → (b-d) → 37 → (e) → 38 → (f,g) → 39 → (h,i) → 40 → (g) → (j,k) → (l) → Fortamine

FIGURE 18. (a) CH_3CO_3H, Na_2CO_3, CH_2Cl_2, H_2O; (b) CH_3NH_2, CH_3OH, 70°C; (c) $ClCO_2CH_3$, CH_3OH, Na_2CO_3; (d) CH_3I, NaH, THF; (e) $(Collidine)_2Br \cdot ClO_4$, CH_2Cl_2, −78°C; (f) DBU, toluene, 85°C; (g) $(CF_3CO)_2O$, 90%H_2O_2, CH_2Cl_2, 0°C; (h) PhSeNa, EtOH; (i) m−Cl−$C_6H_4CO_3H$; (j) NaN_3, NH_4Cl, 65°C; (k) H_2, Pd/C, CH_3OH; (l) 4*M*-HCl, 100°C.

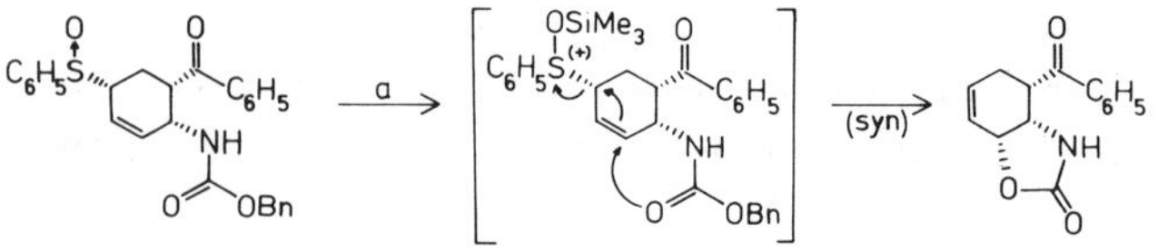

(a) → (syn)

FIGURE 19. (a) $Me_3SiO_3SCF_3$, CH_2Cl_2, 25°C

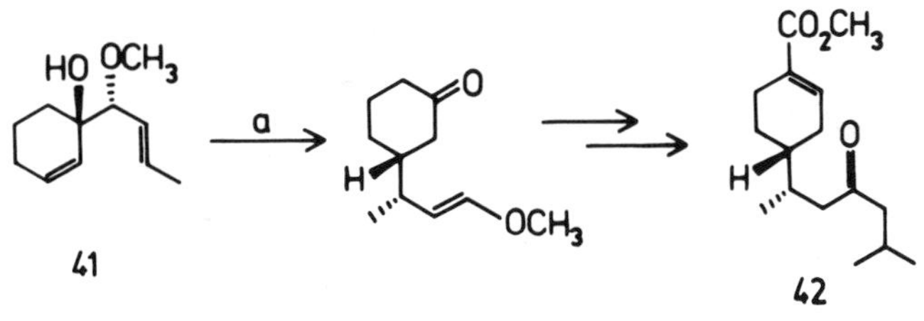

FIGURE 20. (a) KH, 110°C.

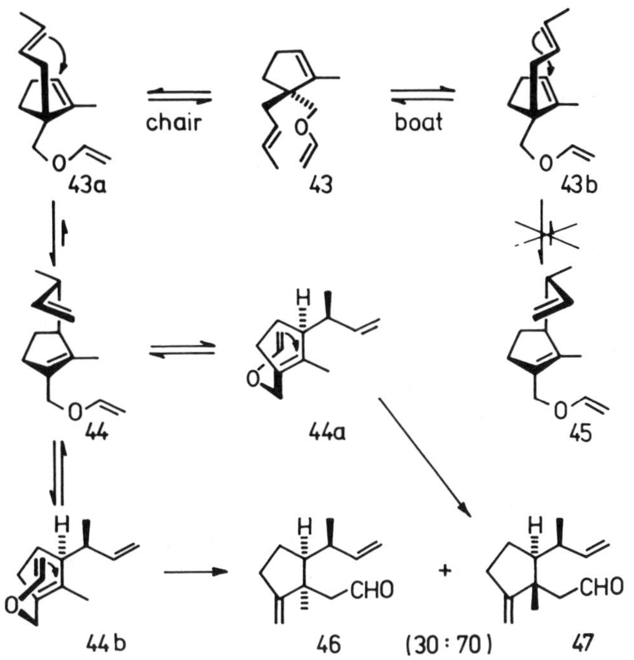

FIGURE 21.

outlined the use of these reactions when discussing the syntheses of *ortho*-condensed and bridged systems (Chapters 3 and 5). This section will deal with stereoselective introduction of substituents via [3,3] rearrangements.

Oxy-Cope rearrangement of diene 41, which proceeds in a concerted fashion via a chair-like transition state, was utilized in the synthesis of juvabione 42 (Figure 20).[78] Note that the reaction, beside transcribing the chirality center onto the ring, also creates a new center in the side chain. The configuration of the latter corresponds to the (E)-configuration of the side-chain double bond in 41.

Another synthesis widely employing [3,3] sigmatropic rearrangements is illustrated in Figure 21.[79,80] The synthetic sequence consists of a tandem Cope-Claisen rearrangement which, compared to a simple Cope reaction, is more favorable in that the irreversible Claisen reaction traps the kinetic products of the Cope rearrangement and thus prevents them from isomerizing to the thermodynamically more stable products. Depending on the reacting conformation of 43 (chair-like 43a or boat-like 43b), the Cope rearrangement could lead to a kinetic product 44 or to the more stable isomer 45 which would eventually prevail due to the reversibility of the Cope reaction. However, the kinetically preferred isomer 44 is consumed in the subsequent Claisen rearrangement, giving a mixture of aldehydes 46, 47,

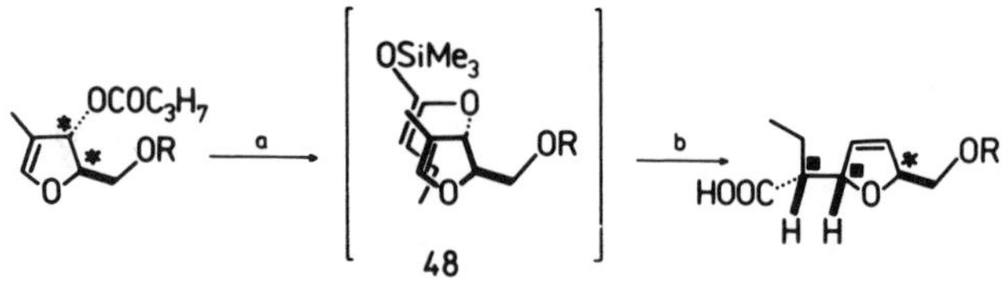

FIGURE 22. (a) $(i\text{-}Pr)_2NLi$, THF, HMPA, then Me_3SiCl; (b) OH^-, H_2O, 25°C.

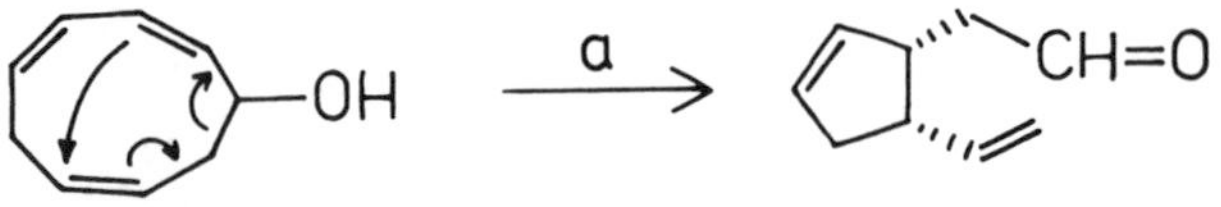

FIGURE 23. (a) KH, THF.

which drives the overall reaction in the desired direction. The relative configuration of the vicinal substituents in 46 and 47 is due to involvement of two transition states in the Claisen rearrangement. Note that the less-strained geometry with both large alkyl groups being on the opposite side of the ring is preferred (44a → 47). The configuration of the chirality center created by the Cope reaction in the side chain is largely determined by the (E)-configuration of the double bond in 43 and by the chair-like geometry of the preferred transition state 43a.

One of the double bonds that are necessary for the Claisen rearrangement to occur can be generated by enolizing an ester group (Figure 22) followed by stabilizing the enol system as a trimethylsilyl enolether.[81] Configuration of the enol double bond in 48 depends on the reaction conditions (for a detailed discussion see Chapter 2 in Volume II).[76] The chirality center on the ring (*) is transduced to the new position (■).

The Cope rearrangement in cyclononatrienol (Figure 23)[83] follows a different strategic line aimed at building a cross-piece bond with concomitant cleavage of the nine-membered ring. *cis*-Configuration of the substituents in the product follows from the disrotatory formation of the cross-piece bond and (Z)-configuration of the double bonds in the starting triene. A similar strategy has been used by Crabbé et al.[82] in a synthesis of prostaglandin.

D. Neighboring Groups as Control Elements

The role of control elements was discussed in general in Chapter 1, Section III.B. This section will focus on cases in which the control element participates in the reaction by forming a covalent bond.

Halolactonization[84-86] and related methods[87-89] are perhaps most often used for stereoselective introduction of an oxygen function (hydroxyl or ether). The methods have been studied in detail, especially in connection with prostaglandin syntheses[1,61-63] (see also Chapter 1 in Volume II).

One of the pivotal steps in prostaglandin synthesis,[90,91] the regio- and stereoselective introduction of the hydroxyl group at C-9α, is depicted in Figure 24. Iodolactonization of the unsaturated acid 49 afforded the lactone 50 which was converted in further steps to the prostaglandin intermediate 51.[1,63]

Participation by the formate group in hypobromous acid addition to the steroid derivative 52 was utilized for the regio- and stereoselective formation of the 5β-hydroxyl group in the synthesis of 14-deoxystrophantidin (Figure 25).[92] The bromine atom and the formate group were removed in the next step (53 → 54), and after oxidation of the 19-hydroxyl, the 3β-hydroxyl group was liberated by hydrolysis (54 → 55).

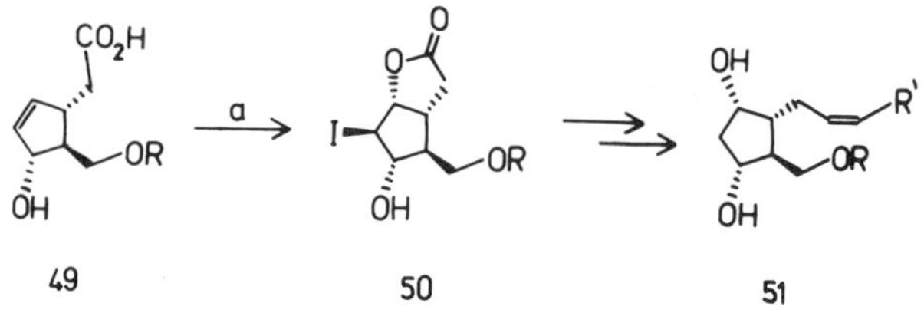

FIGURE 24. (a) KI, I_2.

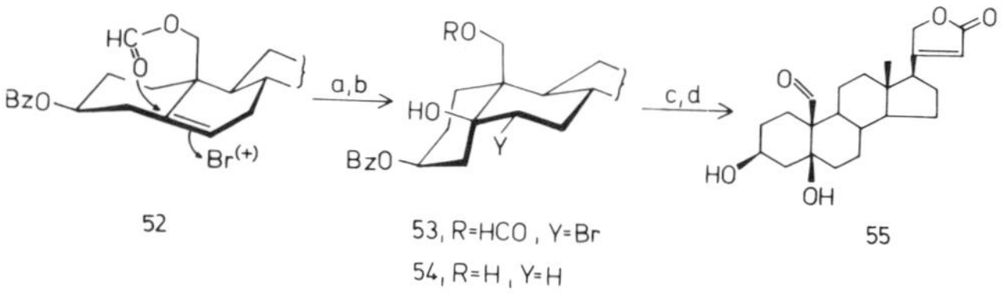

FIGURE 25. (a) NBA, $HClO_4$, dioxane, H_2O, r.t.; (b) Ra – Ni, EtOH, reflux; (c) CrO_3, H_2SO_4, Me_2CO; (d) K_2CO_3, CH_3OH.

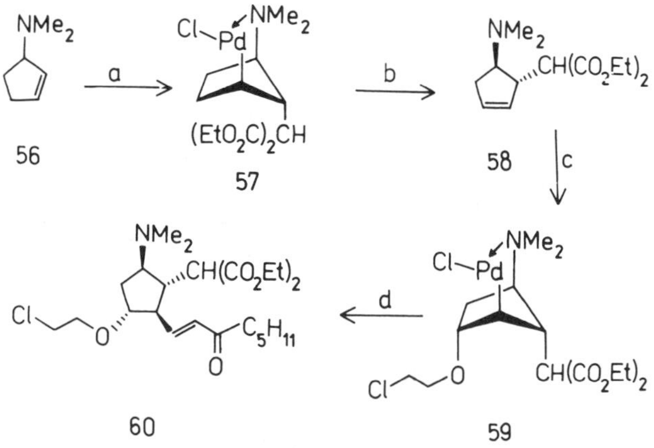

FIGURE 26. (a) Li_2PdCl_4, Na^+ $^-CH(CO_2Et)_2$, THF, 0°C, 2 hr; (b) (i-Pr)$_2$NEt, heat; (c) Li_2PdCl_4, Cl – CH_2CH_2 – OH, (i-Pr)$_2$NEt, Me_2SO; (d) CH_2=CH–COC_5H_{11}.

Certain functional groups can be employed to chelate transition metal cations (e.g. palladium) and thus control the stereochemistry of the subsequent synthetic reactions. Such a strategy is shown in Figure 26, where it effects stereoselective substitution in a five-membered ring.[93] The complexation of the dimethylamino group with the palladium electrophile directs the attack of the latter to the double bond (57), so that the nucleophile (malonate ester) approaches from the opposite side of the ring, giving rise to the *trans*-disubstituted cyclopentene 58. The second nucleophile is introduced in the same way, which results in a *cis*-arrangement of the two groups in 59 and 60.

E. Auxiliary Rings as Precursors of Substituents

Disconnecting a ring in a bicyclosystem is an excellent means of forming two substituents with a defined stereochemistry. Conceptually, this is a reversal of the ring synthesis via cyclization at the annulation site (cf. Chapter 3), and it would be used in cases where the ring system is easily available by some other synthetic methodology. Ring fission in an

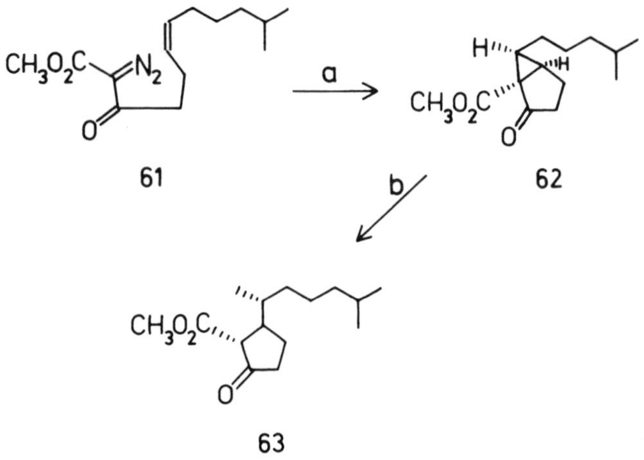

FIGURE 27. (a) Cu, toluene, reflux; (b) $(CH_3)_2CuLi$.

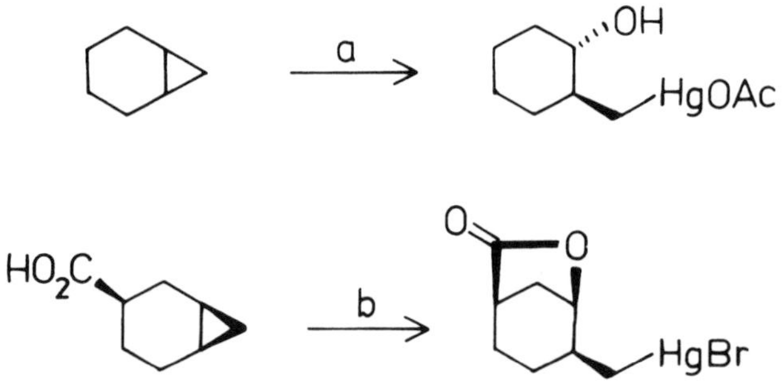

FIGURE 28. (a) $(AcO)_2Hg$, H_2O, 100°C; (b) $Hg(ClO_4)_2$, KBr, H_2O.

ortho-condensed system will produce a pair of vicinal substituents, while substitution patterns with more remote groups (1,3-, 1,4-, etc.) could be obtained by cleaving bridged ring systems. Needless to say, these methods may serve for introduction of carbon or heteroatom substituents and they may be combined with insertion reactions (vide infra).

1. Introduction of Substituents via Cleavage of ortho-Condensed Systems

The cyclopropane ring can be a synthon of a substituent as shown in Figure 27.[94] Intramolecular addition of a diazoketone[95] to the double bond in 61 gives rise to the *cis*-annulated [3.1.0] system 62. The relative configuration of the three chirality centers in 62 is due to the *cis*-stereochemistry of the addition and the original configuration of the double bond in 61. The cyclopropane ring is stereoselectively cleaved with an organocuprate reagent at the outside position, preserving the cyclopentane ring. The synthetic strategy that uses a three-membered ring as a synthon of a substituent[96] has been widely utilized in prostaglandin syntheses.[1,61-63]

Oxymercuration, too, is a good method for converting a cyclo-propane ring into two vicinal substituents (Figure 28).[97,98] The stereochemical outcome depends on a mechanism, which may in turn be dependent on the substrate structure and other factors (for a detailed discussion see Reference 98). α-Cyclopropyl ketones[99] and α-methylenecyclopropanes[100] are easily cleaved by acids. Furthermore, β-cyclopropyl ketones undergo smooth splitting when brominated.[101,102]

The four-membered ring obtained by the addition of ketene derivatives to a double bond

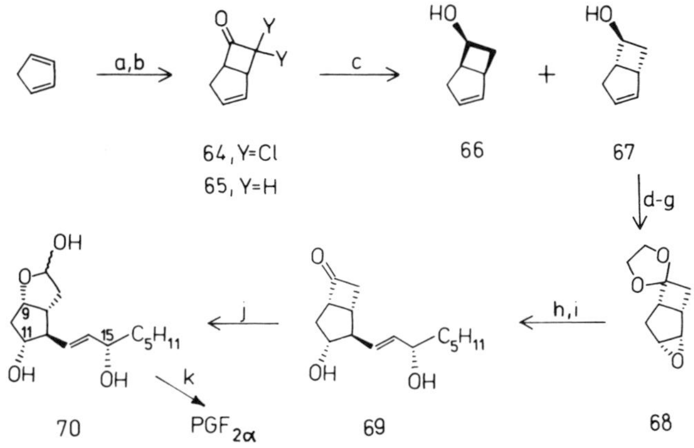

FIGURE 29. $Cl_2C{=}C{=}O$; (b) Zn, NH_4Cl; (c) yeast, separation; (d) CrO_3: H_2SO_4, Me_2CO; (e) ketalisation; (g) 1. HOBr, 2. base; (h) R_2CuLi; (i) deketalization; (j) hν, H_2O; (k) Wittig react.

has become a popular synthon of both the "upper" side chain of prostaglandins[1,61-63] and the 9α-hydroxyl group (Figure 29).[61] The bicyclic dichloroketone 64, prepared by the addition of dichloroketene to cyclopentadiene was reduced with zinc to the racemic ketone 65. The latter compound is a starting point of the synthesis of racemic prostaglandins. Nevertheless, it turned out more advantageous to resolve the racemate 65 to enantiomers and conduct the synthesis with an optically active compound.[61] Reduction of 65 with baker's yeast proceeds with high (S)-stereoselectivity, affording diastereoisomeric alcohols 66 and 67 which can be separated by chromatography or careful distillation on a spinning-band column. The alcohol 67 was oxidized (d) to the optically active (homochiral) ketone 65. The next synthetic steps involved protection of the oxo group as a dioxolane, epoxidation (68), and cleavage of the epoxide with an organocopper reagent. Note that the epoxide ring in 68 has to have the α-configuration, which would have required that the reagent attack from the more hindered side of the ring. This is the reason why the epoxide ring was formed in two steps, i.e. by hypobromous acid addition (the nucleophilic hydroxyl group is forced to approach from the α-side) followed by ring closure of the intermediate bromohydrin. After deprotection of the ketone group (69) the cyclobutanone ring was enlarged by a photochemical rearrangement and the resulting lactol 70 was coupled with a Wittig reagent with simultaneous deprotection of the 9α-hydroxyl group. The synthesis starting from 65 thus yielded the natural enantiomer of prostaglandin $F_{2\alpha}$.

It is remarkable that even the second enantiomer of 65 can be used to prepare the natural $PGF_{2\alpha}$, though via a different synthetic route (cf. Figure 36). Such syntheses, where both enantiomers of the starting compound can be converted (by different routes, of course) to a single enantiomer of the target compound are called enantioconvergent.

A slightly different way of preparing and modifying a four-membered ring as a synthon of a substitutent relies upon addition of ynamines to a double bond (for a review cf. Reference 103). In contrast to ketenes, ynamines can function as ${}_{\pi}2_s$ components in both [2 + 2] and [4 + 2] cycloadditions. Here we present an example of [2 + 2] cycloaddition (Figure 30) while the second possibility will be dealt with later in this Chapter.

The [2 + 2] addition of the ynamine to the cyclopentenone double bond gives rise to the *cis*-annulated system 73 with a reactive enamine function in the four-membered ring (Figure 30). Hydrolysis of the enamine is followed by a retro-Dieckmann fission of the cyclobutanone ring and affords substituted cyclopentanones 77 and 78. However, the stereochemistry of the hydrolysis depends on the reagent used.[103,104] Acidic hydrolysis begins with proton

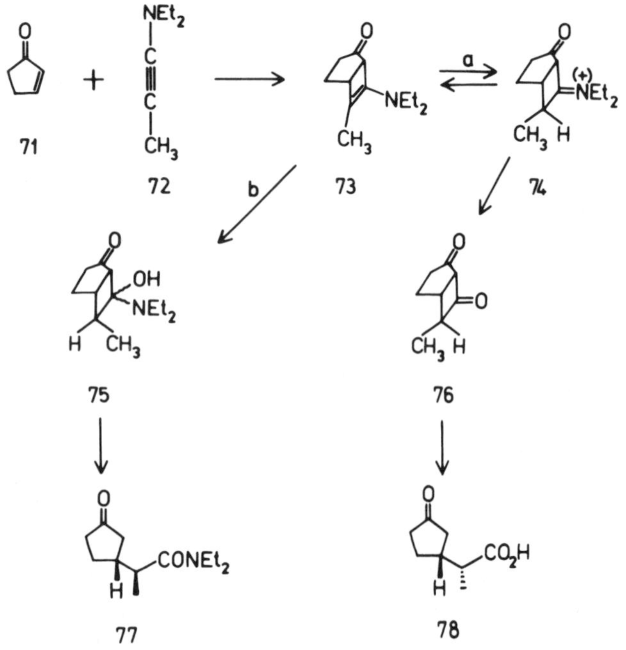

FIGURE 30. (a) 1%-HCl; (b) cat. NaOH.

attachment from the more accessible exo-side of the skeleton, and the following hydrolysis of the iminium species 74 gives the endo-methyl cyclobutanone 76, which is further cleaved under the reaction conditions to yield the final acid 78. By contrast, hydrolysis in a neutral or basic medium goes through the more stable exo-methyl derivative 75, yielding eventually the amide 77. The overall sequence (71 → 77, 78) formally corresponds to a 1,4-addition to the enone system in 71. However, the ynamine route makes it possible to control the stereochemistry of the side chain which would be hardly feasible with simple conjugate additions.

Vicinal substituents on the cyclopropane ring can be formed from a five-membered ring as a corresponding synthon (Figure 31).[105] Intramolecular addition of a diazoketone function to the double bond in 79 gives rise to the [3.1.0] system (80) which is further substituted in the α-position to the keto group (81,82). The following fragmentation of the mesylate 82 cleaves the five-membered ring and the intermediate *cis*-chrysanthemic acid (83) isomerises to the *trans*-isomer 84.[105]

The six-membered ring making a part of an *ortho*-condensed [4.x.0] system is a suitable synthon of two vicinal substituents because of its ready availability. On numerous examples of this kind we have selected two that illustrate the synthetic strategy based on successive [2 + 2] or [4 + 2] annulation and ring cleavage.

The cyclobutane ring of grandisol (87, Figure 32)[106] was constructed by the photochemical addition of ethylene to the unsaturated lactone 85. The lactone ring in the adduct 86 was transformed in several steps to the two grandisol substituents. Here the *syn*-stereochemistry of the [2 + 2] addition determined the relative configuration of the cyclobutane substituents. This strategy has been used in other syntheses, too.[1]

Stereoselective substitution in a cyclohexane derivative can be achieved through the Diels-Alder reaction using a cyclic dienophile (Figure 33).[107] The enedione system in the adduct 88 served as a synthon of the aldehyde and carboxymethyl groups, while the double bond of the cyclohexene ring in 88 was converted to a protected *trans*-1,2-diol system. This synthetic route, based on the oxidative cleavage of the auxiliary six-membered ring, afforded

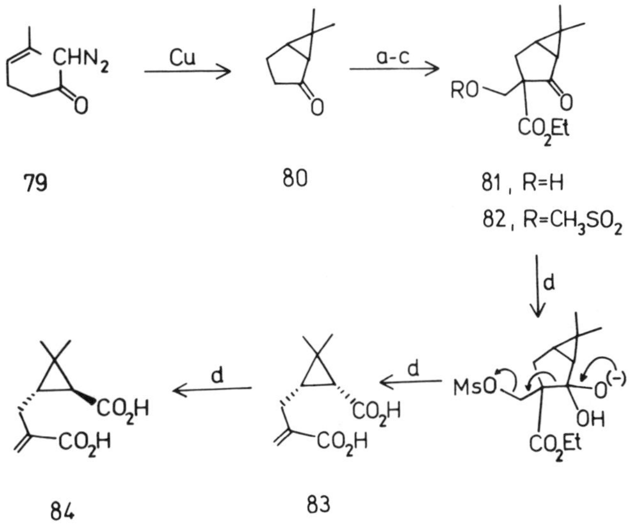

FIGURE 31. (a) $(EtO)_2CO$; (b) CH_2O; (c) CH_3SO_2Cl; (d) KOH,CH_3OH.

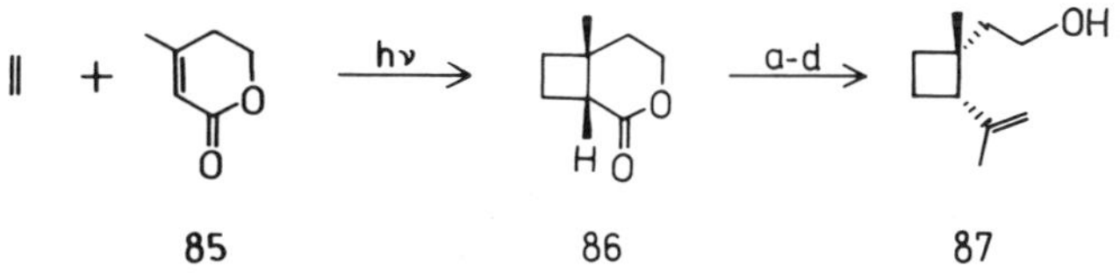

FIGURE 32. (a) CH_3Li; (b) Ac_2O; (c) heat; (d) OH^-.

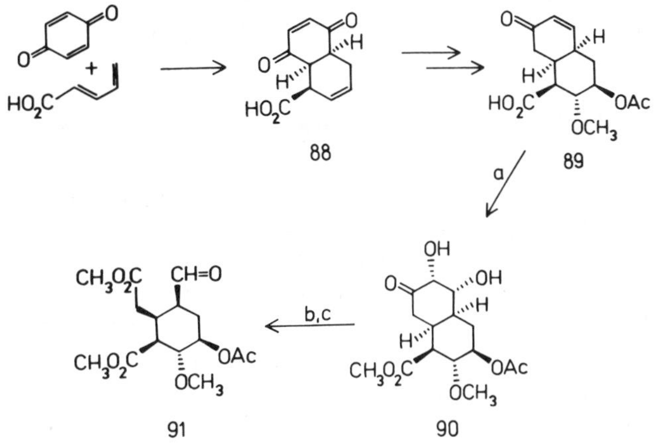

FIGURE 33. (a) OsO_4; (b) HIO_4; (c) CH_2N_2.

cis-disubstituted derivative 91 which would be inaccessible by a direct Diels-Alder reaction.

An aromatic ring with a suitably placed methoxy group can be converted to a cyclohexenone (analogous to enone 89) by Birch reduction. The enone system then can be cleaved as shown in Figure 33.[1,25,108]

2. *Introduction of Substituents via Cleavage of Bridged Systems*

In the preceding paragraph we have shown that an auxiliary *ortho*-condensed ring can be employed as a synthon of two vicinal substituents. If a 1,3-disubstituted derivative is needed,

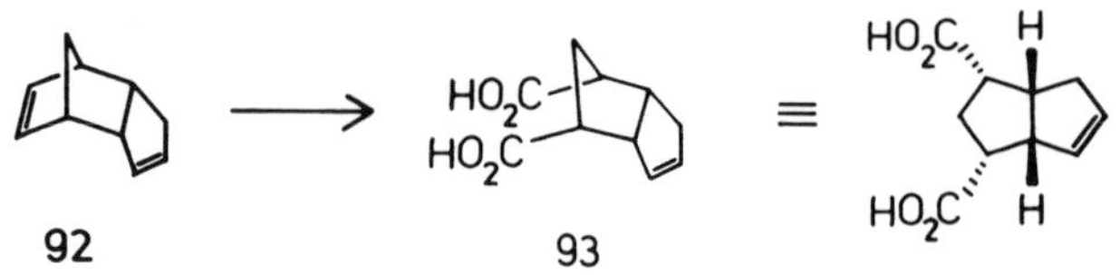

FIGURE 34.

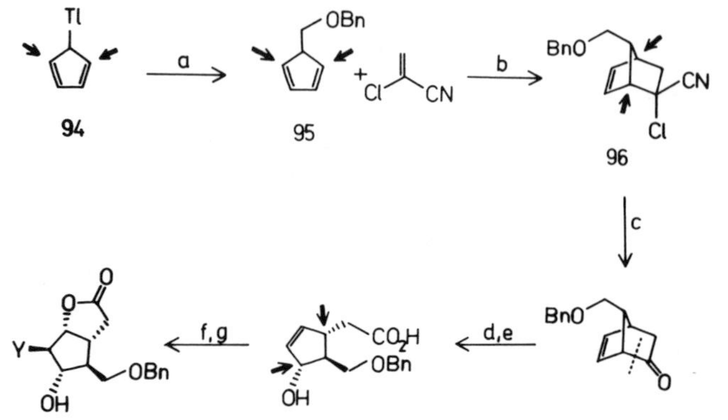

FIGURE 35. (a) $BrCH_2OCH_2Ph$; (b) Diels-Alder; (c) KOH, Me_2SO; (d) $m-Cl-C_6H_4CO_3H$; (e) NaOH; (f) KI, I_2; (g) Bu_3SnH.

then we can use a bridged [x.y.l] system containing either carbocyclic or heterocyclic rings. Because of the bridging, the ring would afford primarily *cis*-substituted compounds. *trans*-Disubstituted derivatives are prepared either by modifying the *cis*-isomers or by using a different strategy.

Figure 34 shows an example of oxidative cleavage of the more reactive double bond in dicyclopentadiene 92, yielding *cis*-dicarboxylic acid 93.[109] Stereoselective 1,3-substitution of the cyclopentane ring via the oxidative cleavage of a norbornene derivative has been brought to perfection in the now-classical Corey synthesis of prostaglandins (Figure 35).[1,59-63,90,91] In this case, alkylation of cyclopentadienyl thallium (94) afforded the substituted cyclopentadiene 95 which was converted to norbornenone 97. The Diels-Alder step (95 → 96) introduced two *cis*-oriented substituents of the future prostanoid system (indicated by arrows), while the benzyloxymethylene group was a synthon of the C-2 substituent. Cleavage of the norbornene system was achieved through the Bayer-Villiger reaction to give the acid 98. The substituent at C-7 in 97 protects the norbornene double bond by hindering access of the reagent from the exo-side. Note that the relative configuration of the three substituents in 98 is due to the stereochemistry of the Diels-Alder reaction (95 → 96) which proceeds from the more accessible reverse side. The fourth substituent was introduced by iodolactonization followed by reductive removal of iodine (98 → 99 → 100).

Another way to a bridged [2.2.1] system is shown in Figure 36.[61] The starting optically active ketone 65 (cf. Figure 29) is a waste product of the enantioselective synthesis of prostaglandin $F_{2\alpha}$. Hypobromous acid addition to 65 gave rise to bromohydrin 101 which could be internally alkylated to close the highly strained [$3.2.0.0^{2,7}$] system (102). After having protected the hydroxyl group, the cyclobutanone ring was cleaved by an organocopper reagent designed to introduce a synthon of the prostaglandin ''upper'' chain (104). The hydroxyl group in 101 constitutes the 9α-hydroxyl group of the target $PGF_{2\alpha}$. The overall

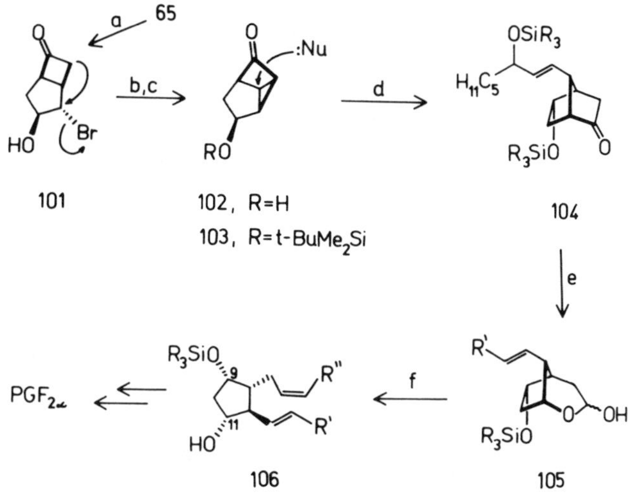

FIGURE 36. (a) $CH_3CONHBr$, H_2O, Me_2CO; (b) KOH; (c) t-Bu-Me_2SiCl; (d) $[C_5H_{11}(R_3SiO)CH-CH{=}CH-Cu-C{\equiv}C-C_3H_7]Li$; (e) CH_3CO_3H, buffer, $-20°C$; (f) Wittig react.

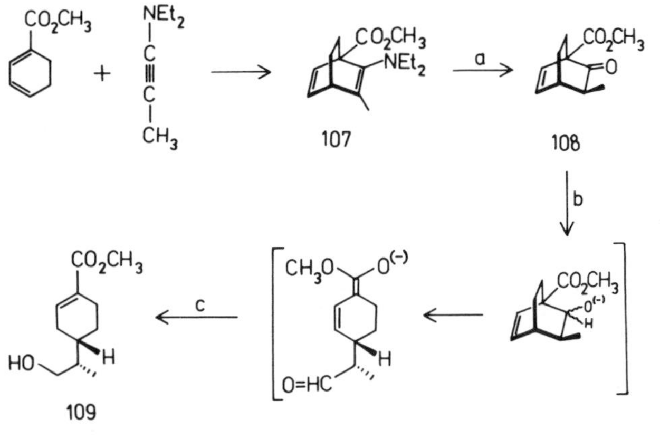

FIGURE 37. (a) H^+; (b) $NaBH_4$; (c) H_2O.

synthetic procedure completes the enantioconvergent synthesis of $PGF_{2\alpha}$ by utilizing the second enantiomer of the starting ketone 65 (cf. Figure 29).

As we have seen in the previous section, reactive ynamines can be employed as synthons of functionalized alkyl substituents. Beside the [2 + 2] cycloaddition, ynamines can react as $_{\pi}2_s$ components in a [4 + 2] reaction as depicted in Figure 37.[103] The Diels-Alder reaction of the ynamine with methyl dihydrobenzoate provided the bridged [2.2.2] system (107). Acidic hydrolysis of the enamine function in 107 is surprisingly stereoselective. The other epimer arises in only minor amounts. The β-hydroxy ester, obtained from 108 upon reduction with sodium borohydride, undergoes ring opening followed by the reduction of the newly formed aldehyde group, eventually yielding the ester 109. Since the [2 + 2] ynamine addition formally resulted in a 1,4-addition to the conjugated double bond (see Figure 30), the present [4 + 2] addition may be regarded as a homologous (1,6) process, i.e. addition to a dienester group.

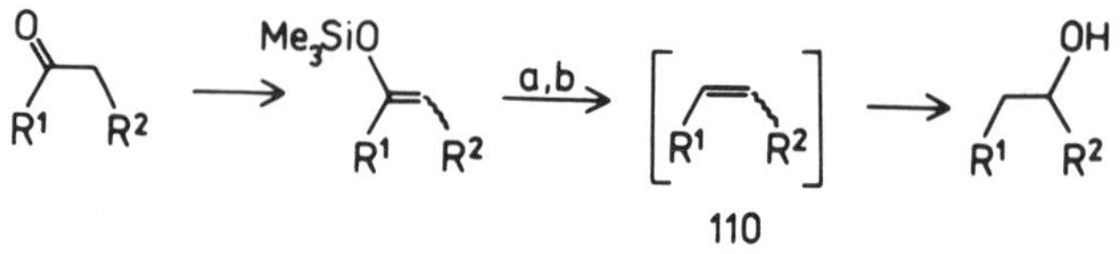

FIGURE 38. (a) 9-BBN, THF, reflux 16 hr; (b) H_2O_2, NaOH.

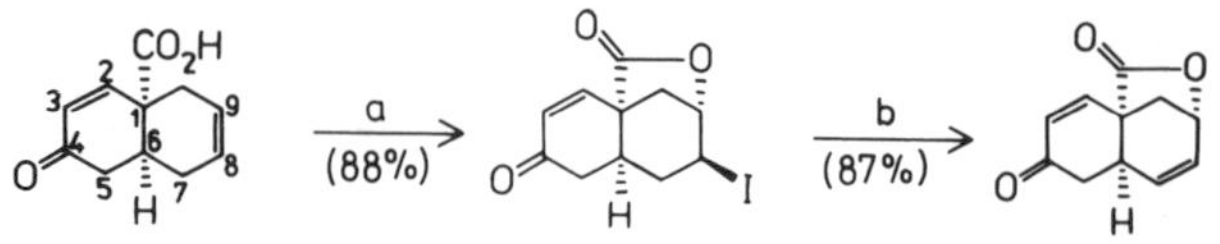

FIGURE 39. (a) KI_3, $NaHCO_3$, H_2O, r.t., 48 hr; (b) DBU.

F. Transposition of Substituents

Transposition of substituents in a synthetic intermediate is often a necessary step relating two parts of the overall strategy. Of a variety of synthetic methods resulting in the transposition of substituents or functional groups, only some are inherently stereoselective. In the next paragraphs we will deal with 1,2- and 1,3-transpositions.

1. 1,2-Transpositions

By definition, a 1,2-transposition interchanges the position of two vicinal substituents or functional groups. However, a majority of methods developed to this purpose, e.g. Meakins transposition[110] and others,[111-114] involve an sp^2 intermediate, usually a ketone, so that the reaction stereochemistry depends on factors that are not connected with the process itself. A potentially stereoselective transposition method,[115] a shift of an oxygen function, is shown in Figure 38. Due to the high stereoselectivity of 9-borabicyclononane, the hydroboration of the trimethylsilylenolether could proceed preferentially from one side of the system and afford the corresponding stereoisomer. However, the success of the reaction is based on the regioselectivity of the hydroboration of the intermediate olefin 110.

Even if a 1,2-transposition does not result in formation of a new stereogenous group, the stereochemistry of the individual reaction steps may be crucial for the regioselectivity. This points to the role of control elements that can influence the steric course of transposition reactions. For instance, a transposition of the 8,9-double bond in the bicyclic acid (Figure 39)[115a] was based on the stereoselective course of iodolactonization which introduced the axial iodine atom to be regioselectively eliminated in the next step.

A more complex case of successive transpositions of a double bond is shown in Figure 40.[116] The synthesis was aimed at converting the easily accessible $\Delta^{5,6}$-steroid 110a to a $\Delta^{3,4}$-derivative 110 g. Hypobromous acid addition to 110a is controlled by participation of the 19-acetoxy group and proceeds to give the 5β-alcohol 115b. The 6α-bromine atom was removed by a tin hydride reduction (b) and the equatorial 5β-hydroxyl was eliminated. Since the elimination proceeds mostly via an *anti*-mechanism (110c), the double bond is created between C-4 and C-5 (110d) by abstraction of the only available antiperiplanar proton (4α). The olefin 110d was converted in two steps to a mixture of 5α-H and 5β-H ketones which was equilibrated to give the more stable 5α-H isomer 110e. Hydride reduction of the latter furnished stereoselectively an axial 4β-alcohol which was converted to benzoate 110f. Pyrolytic elimination of benzoic acid in 110f proceeds via a *syn*-mechanism for which there is only one hydrogen available (3β) and hence the reaction yields selectively the 3,4-unsaturated compound 110 g.

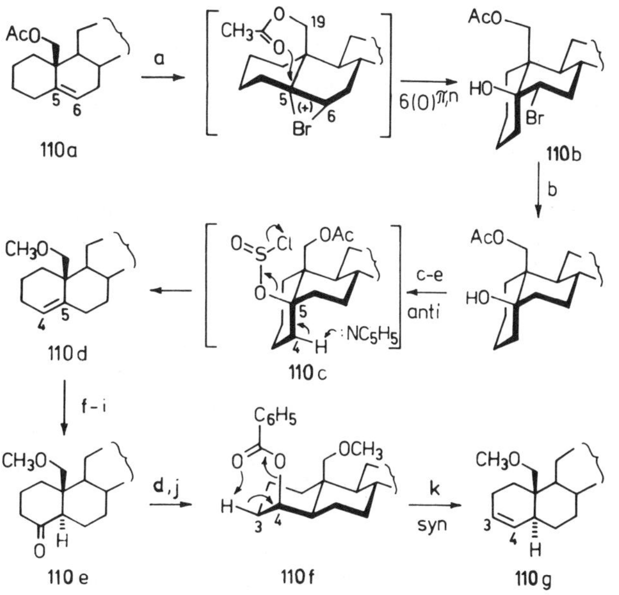

FIGURE 40. (a) NBA, 10% $HClO_4$, dioxane, H_2O, 20°C, 15 min.; (b) Bu_3SnH, AIBN, C_6H_6, reflux, 30 min; (c) $SOCl_2$, C_5H_5N, 0°C, 10 min; (d) $LiAlH_4$, Et_2O; (e) CH_3I, NaH, DME; (f) BH_3, THF, 0°C, 2 hr; (g) H_2O_2, OH^-, THF, H_2O; (h) CrO_3, H_2SO_4, Me_2CO; (i) 1% KOH, CH_3OH (equilibration); (j) PhCOCl, C_5H_5N; (k) 345°C, 10 min.

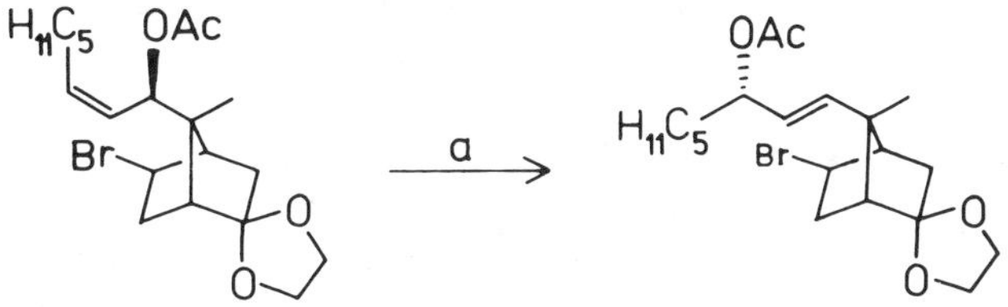

FIGURE 41. (a) $PdCl_2(CH_3CN)_2$.

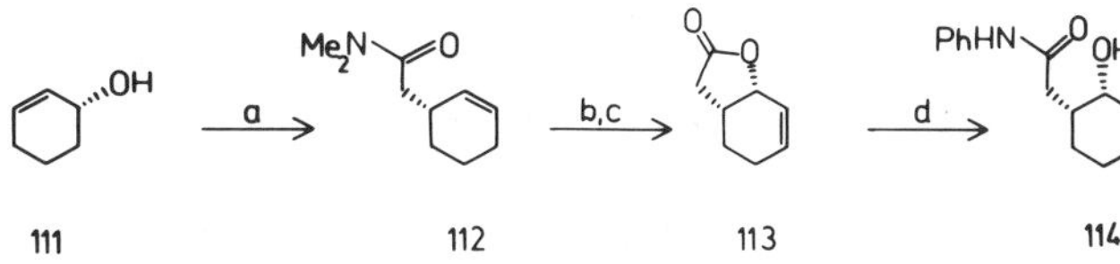

FIGURE 42. (a) $CH_3C(OMe)_2NMe_2$; (b) J_2; (c) DBU; (d) PhNHLi.

2. 1,3-Transpositions

1,3-Transposition can be accomplished by allylic rearrangement or S_N2' substitution[117-121a] (for a review see References 122 to 125). A significant improvement in the stereoselectivity of 1,3-transpositions was achieved by using palladium complexes to stabilize[126-128] the transient allyl cation (Figure 41)[129] (for a review see References 130 and 131).

An elegant and often used method of 1,3-transposition is based on the Claisen rearrangement. Such a stereoselective rearrangement is shown in Figure 42.[132] The product of the Claisen rearrangement in 111 (112) was converted to 113 by an iodolactonization-elimination sequence. The synthesis may be regarded as a tandem 1,2-1,3-transposition of the hydroxyl

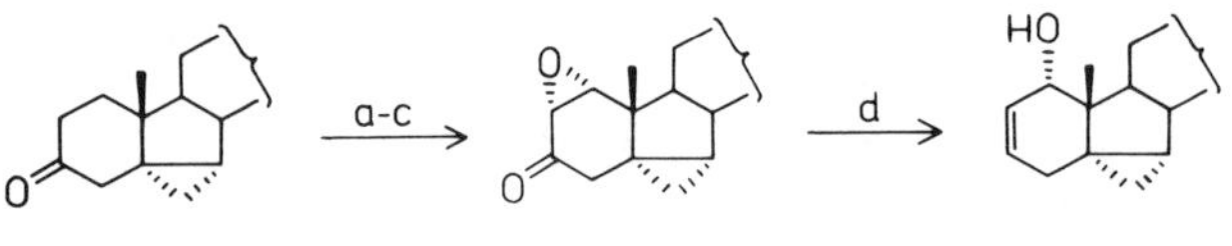

115 → (a-c) → 116 → (d) → 117

FIGURE 43. (a) $PhN^+Me_3{\cdot}Br_3^-$, THF; (b) Collidine, reflux; (c) H_2O_2, NaOH, dioxane, H_2O, r.t.; (d) $N_2H_4{\cdot}H_2O$, reflux, 15 min.

FIGURE 44.

and the double bond accompanied by the introduction of a new substituent. The lactone 113 was further converted to the unsaturated amide 114 which, when treated with acetamide dimethylacetal, underwent a second Claisen rearrangement resulting in a stereoselective introduction of another carboxymethyl group.[132] For other methods of isomerisation of allylic alcohols see References 133 and 134. Homoallylic processes are also known.[135]

Another stereoselective 1,3-transposition of the oxygen atom is shown in Figure 43.[136] Ketone 115 is first transformed to an α,β-unsaturated ketone that reacts with alkaline hydrogen peroxide to give the 1α,2α-epoxy ketone 116. Reduction of 116 with hydrazine provides the allylic alcohol 117 in which the configuration of the hydroxyl group corresponds to that of the epoxide ring.[137]

G. Inversion of Configuration

Although one of the principal purposes of a stereoselective synthesis is to provide compounds with desired stereochemistry, it may happen that one substituent in an intermediate or in the final compound has a reverse, "wrong" relative configuration. Instead of designing a new synthetic plan to correct this flaw, we may attempt to change the configuration at the given chirality center by inversion. A similar case occurs if we start the synthesis from a natural product possessing one of the chirality centers with an opposite configuration.

The inversion of configuration can be realized via an S_N2 reaction[138-143] (for reviews see References 141 to 144) (Figure 44), or as a reaction sequence accompanied with a change of hybridization $sp^3 \rightarrow sp^2 \rightarrow sp^3$. Although as the S_N2 attack accompanied by the Walden inversion is a textbook example, it may fail due to side reactions or intervention of other functional groups in the substrate molecule. For instance, the S_N2 reaction at C-3 in 3β-tosyloxy-5,6-unsaturated steroids[32] proceeds with retention of configuration because of participation by the homoallylic double bond. The inversion can be accomplished, however, provided the double bond has been protected by suitable groups. This is shown in Figure 45[145] where the *in situ* formed 5α-acetoxy group in 121 not only protects the double bond but simultaneously functions as an internal nucleophile in the substitution of the mesyloxy group.

The second means of inversion of configuration is mostly employed for reverting configuration of a hydroxyl group. Of many examples of this kind we have chosen one from the chemistry of steroids (Figure 46).[146] The steroid alcohol 123 with a natural 3β-hydroxyl group was oxidized to ketone 124 which cleanly afforded the axial alcohol 125 on reduction with bulky lithium tri-sec-butylborohydride. The choice of the reducing agent enables one to control the reaction to yield either an axial or equatorial alcohol.[43,44] Since ketones are

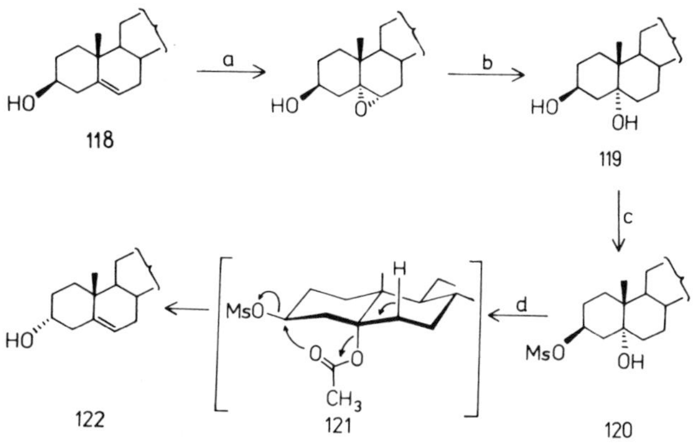

FIGURE 45. (a) RCO_3H; (b) H_2, Pt, AcOH; (c) CH_3SO_2Cl; (d) CH_3COCl, $C_6H_5NMe_2$, reflux.

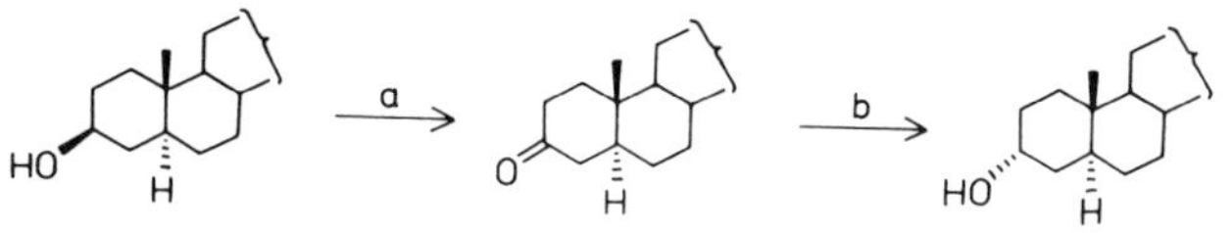

FIGURE 46. (a) CrO_3, H_2SO_4, Me_2CO; (b) LiB(sec.$C_4H_9)_3$H.

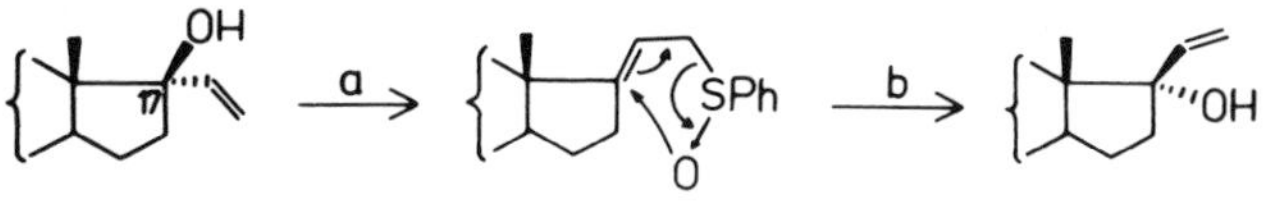

FIGURE 47. (a) PhSCl; (b) $(MeO)_3P$.

usually sufficiently reactive, stereoselective reductions can be often carried out in the presence of other functional groups.[43,44] The steric course of the reaction can be further affected by using a control element that would force the substrate molecule to a conformation needed for the desired reaction course (see Figure 12 in Chapter 1).

The inversion methods are mostly used to change configuration at secondary carbon atoms. Inversion at primary centers comes into consideration e.g., with deuterium-labeled derivatives only.[147] On the other hand, inversion at tertiary centers poses a serious problem (for a review see Reference 148) since eliminations[141] and rearrangements compete with S_N2 reactions. The change of hybridization cannot be considered for structural reasons.

An elegant method of changing configuration at a tertiary center is shown in Figure 47.[149] The steroidal 17α-vinyl derivative 126, which is readily available from a 17-ketone, rearranges to the sulfoxide 127 when treated with benzensulfenylchloride. The sulfoxide acts with trimethylphosphite (b) via a [3.2] rearrangement (for a review see Reference 150), giving rise to the allylic alcohol 128 with a reversed configuration at C-17. This alcohol is a useful starting compound in the synthesis of corticoids. Other means of configurational inversion have been discussed in Chapter 1.

H. Stereoselective Substitution in Medium and Large Rings

Stereochemical control in the synthesis of compounds containing rings larger than six- or

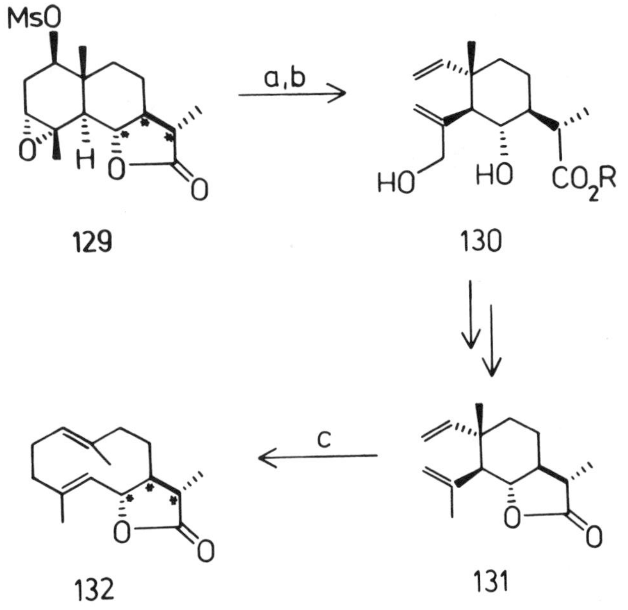

FIGURE 48. (a) $(RO)_3Al$; (b) H_3O^+; (c) 210°C, 5 min.

seven-membered represents a special challenge. As with smaller rings (3- to 6-membered), medium and large rings may also be synthesized by two principal strategies: (1) By *de novo* formation of rings from stereohomogenous precursors and (2) by a stereoselective introduction of substituents into the pre-existing rings. The first approach has been treated in Chapter 6. It relies on one of the means illustrated in Figure 10 in Chapter 1, particularly on connection of enantiomerically pure synthons (4 in Figure 10 in Chapter 1). Vicinal centers, of course, may be effectively controlled by asymmetric induction. (For remote control see for example References 151 to 154 and Chapter 1). In this subsection we describe the second strategy.

One of the possible means of the synthesis is stitching the target ring with cross-piece bonds. Obviously, the stereocontrol problem is then reduced to stereoselective substitution of smaller rings, preferably six-membered, which is relatively well documented. Figure 48 illustrates one of the examples.[155] A ten-membered ring of the lactone 132 is derived from a [4.4.0] precursor (containing already all stereogeneous elements) by a several step fragmentation of a cross-piece bond. In this instance, all three chiral centers of the lactone ring have been synthesized in the [4.4.0] system 129, whereas *trans*-annulation in the compound 129 served as a synthon of E-double bonds of the final product 132.

Aside from this rather classical approach, the systematic study of medium and large rings in recent years brought much deeper insight and flourished into new avenues of synthetic methodology. Although medium- and large-ring compounds are usually capable of existence in a number of stable conformations, NMR,[156-161] X-ray studies,[162,163] and semiempirical calculations[164-171] revealed that only a few of these conformations are low enough in energy to be appreciably populated at normal temperatures.[172] Thus, conformational near-homogeneity has been seen in even-membered cycloalkanes up to cyclohexadecane.[173,174] In contrast, odd-membered cycles tend to assume less symmetrical forms with smaller energy differences between stable conformations.[175-177] Generally, macrocycles tend to exist in conformations having as few *trans*-annular nonbonded repulsions and high-energy torsional arrangements as possible. For example, cyclooctane largely prefers a boat-chair (BC) conformation. Cyclodecane seems to be most stable in the boat-chair-boat (BCB) conformation (Figure 49).

FIGURE 49.

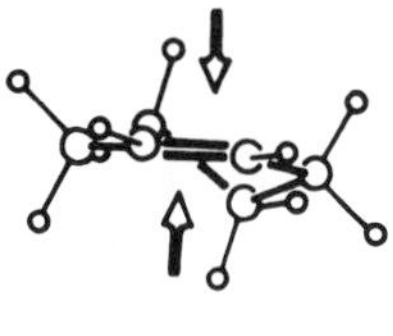

FIGURE 50.

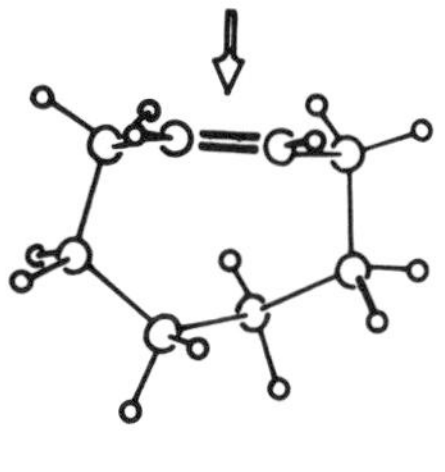

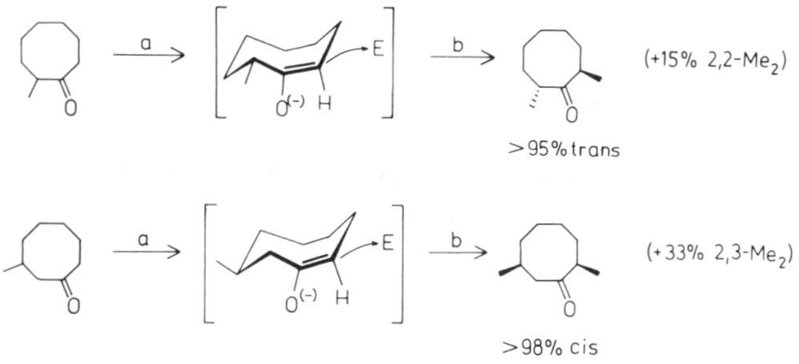

FIGURE 51. (a) $(i\text{-}Pr)_2NLi$, THF, $-78°C$; (b) CH_3I, $-60°C$.

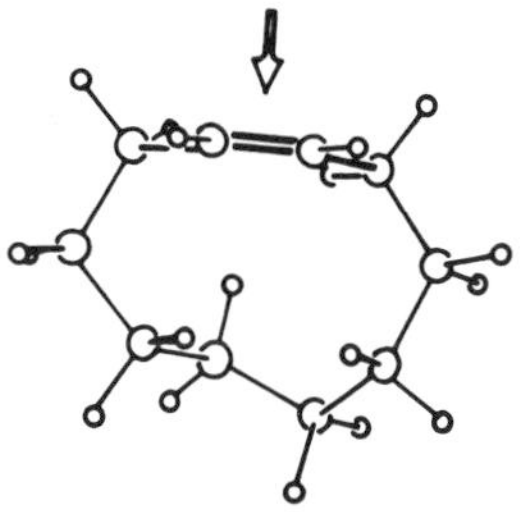

If a double bond is introduced into a ring, the molecule still tends to assume a preferred conformation, e.g. in cyclooctene and cyclodecene (Figure 50). Note, however, that in contrast to cyclohexene (133), in both cyclooctene (134) and cyclodecene (135) the two faces of the π-system are sterically very different! Since one face is severely hindered, it would be expected that a variety of additional reactions would occur preferentially, or perhaps exclusively, from the less hindered peripheral face.[178,179] Actually, for example, alkylation of kinetically generated enolates of methylcyclooctanones turned out to be highly stereoselective (Figure 51).[178] The high diastereoselectivity here stands in a marked contrast to the 60:40 and 80:20 mixtures obtained in alkylation of six-membered analogs.[178] Also cuprate addition to 8-methylcycloct-2-en-1-one results in a highly stereoselective formation of a single adduct (Figure 52).[178] Whereas α-alkylation of the next higher homolog, i.e. 2-methylcyclononan-1-one, lacks stereoselectivity,[178] the cuprate addition to methylcyclono-

Nu:
a
(⟝99:1)

FIGURE 52. (a) Me_2CuLi, Et_2O, 0°C.

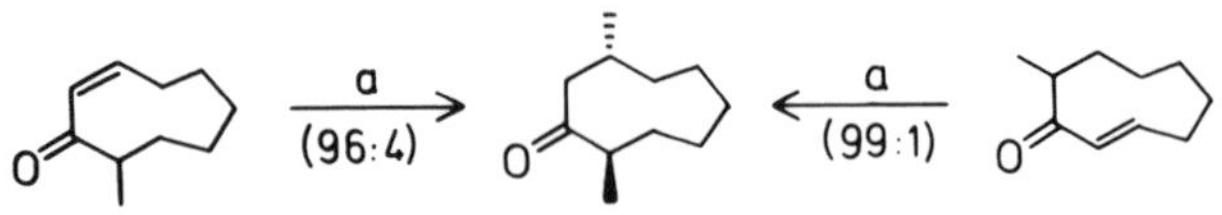

a
(96:4)
a
(99:1)
136 (cis)
(trans)
137 (trans)

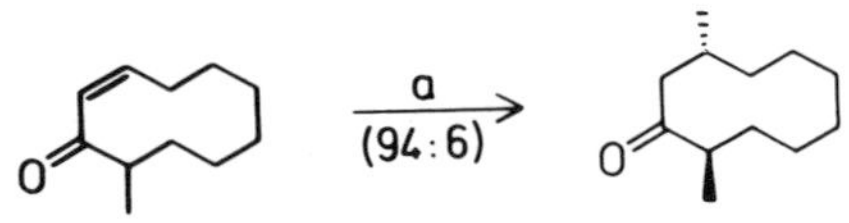

a
(94:6)
138 (cis)
(trans)

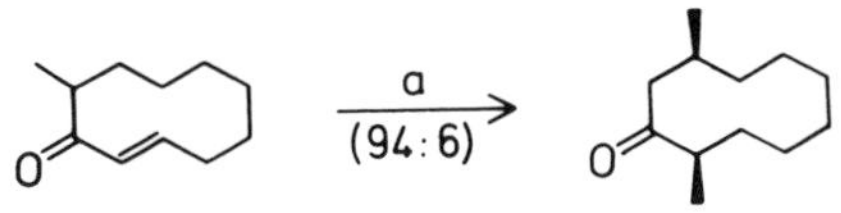

a
(94:6)
139 (trans)
(cis)

FIGURE 53. (a) Me_2CuLi, Et_2O, 0°C.

nenones 136 and 137 is again highly stereoselective as are additions to methylcyclodecenones 138 and 139 (Figure 53).[178]

These results have been rationalized[178] by conformational analysis based on force-field[180] calculations. It has been found that it is possible to predict product distribution based on starting material energy (early transition state) or product energies (late transition state).[178,181] For a detailed discussion see Reference 178.

In larger rings it is not easy to predict which particular conformation(s) will take part in a given reaction because of small energy differences between distinct conformations. Nevertheless, it is believed that local conformational control[178,182] is responsible for the high stereoselectivity of reactions in these systems regardless of the conformation of the rest of the molecule. Thus, even in alkenes with 15-membered rings, reactions like epoxidation and osmylation proceed with high stereoselectivity (Figures 54 and 55).[182]

The described type of stereocontrol has been recently used in several synthesis of macrocyclic natural products of their analogs, e.g. 3-deoxyrosaranolide,[183] asperdiol,[184,185] and others.[179,186]

III. NOTES ADDED IN PROOF

A novel method of enantioselective *de novo* ring construction has been developed: upon alkylation with aliphatic α,ω-dihalogenides, doubly enolized dimenthyl succinate gives the corresponding cyclic products with vicinal *trans*-oriented carboxylic groups in high optical purity.[187] Thus, for instance, alkylation with bromochloromethane results in the formation

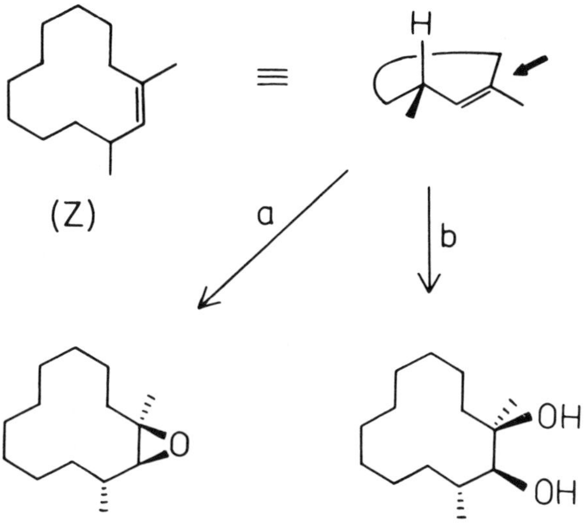

FIGURE 54. (a) $m-Cl-C_6H_4CO_3H$; (b) OsO_4.

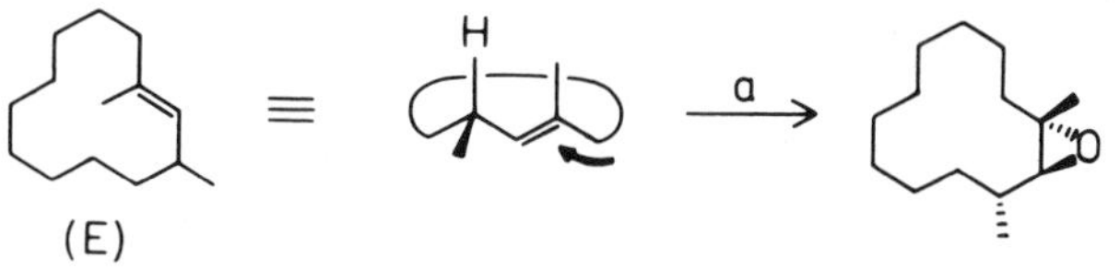

FIGURE 55. (a) $m-Cl-C_6H_4CO_3H$.

of cyclopropane-*trans*-dicarboxylate in 99% ee.[187] 6-Bromo-4-hexenoates cyclize on base treatment via an S_N2' process to *trans*-2-vinyl cyclopropane carboxylates.[188] A novel, enantioselective, palladium(O) catalyzed synthesis of substituted cyclopropanes has been developed.[189] Cyclopropanation of olefins with trialkylaluminum - alkylidene iodide represents an alternative to the Simmons-Smith reaction with different chemoselectivity.[190] Intramolecular nitrile oxide cycloaddition provided a simple stereoselective route to substituted cyclopentanes[191] suited for prostaglandin synthesis. Further progress has been achieved in palladium-mediated cyclization of α,ω-dienes or enynes to substituted cyclopentanes.[192,193] The tandem Michael addition — intramolecular alkylation allows the synthesis of carbocyclic compounds with concomitant stereocontrol of the extracyclic chiral centers.[194] Further development in Danheiser cyclization appeared.[195] The intramolecular Prins reaction catalyzed by chiral Lewis acid gives up to 90% ee. of the cyclized products.[196]

Highly enantioselective α-alkylation of ketones with chiral phase-transfer catalysis gives up to 92% ee.[197] The origin of the stereoselectivity of cyclohexanone hydrazone alkylations has been thoroughly studied.[198,199] A palladium-catalyzed stereospecific equivalent of Michael addition has been developed.[200] (−)-Methyljasmonate was synthesized by the Posner chiral sulfoxide methodology.[201] For a different approach to the same target molecule see Reference 202. A tandem organocopper conjugate addition — enolate trapping reactions have been reviewed.[203] A highly stereospecific addition of alkyl and acetylene units into the vicinal positions has been achieved by the sequential addition of alkylmercury compound to an olefinic double bond followed by transmetalation with palladium and subsequent coupling with alkynyl lithium.[204] Vinylation of cyclohexadiene catalyzed by a $Ni(COD)_2$ — Et_2AlCl — threophos system gives (S)-(+)-3-vinyl cyclohexene in 93% ee.[205] The stereochemistry of palladium-mediated allylic oxidation of menthene seems to be controlled by the side-chain double bond.[206] A full account and further improvements of palladium-catalyzed 1,4-difunctionalization of cyclic 1,3-dienes have been published.[207-212] Palladium (O)-catalyzed opening of vinyl epoxides in the CO_2 atmosphere leads to cyclic carbonates and may thus serve as a cis-hydroxylation equivalent.[213] The versatility of the manganese(III)-mediated oxidation[214] and γ-lactone annulation[215,216] has been demonstrated on several examples. Absolute stereocontrol of two vicinal chiral centers has been achieved by an ingenious application of borane chemistry: B-2-cyclohexen-1-yl diisopinocamphenylborane, produced by treatment of 1,3-cyclohexadiene with diisopinocamphenyl borane, reacts with aldehydes via an allylic rearrangement to afford 1-(2-cyclohexenyl)-1-alkanols in high optical purity.[217] For other examples of application of chiral hydroboration see Reference 218. A stereospecific sequence of S_N2 substitution of allylic alcohols with aryl selenide, followed by a suprafacial [3,2] rearrangement of the corresponding selenoxide, provides a useful route to isomeric allylic alcohols with concomitant inversion of configuration.[219] Some difficulties encountered in S_N2' reactions of vinyl epoxides with organocuprates[220] have now been circumvented.[221] The role of the control element in ring functionalization is well demonstrated by stereodirectivity of epoxidation of allylic[222-224] and homoallylic[116,225,226] alcohols and their derivatives. Note that while $VO(acac)_2$ mediated epoxidation of cyclic allyl alcohols always gives *syn*-products,[222,223] the stereochemistry of peroxyacid oxidation dramatically depends on the ring size.[224] Thus cyclohexenol gives a 95:5 *syn/anti* ratio on MCPBA treatment, cycloheptenol gives 61:39 and cyclooctenol 0.2:99.8![224] This behavior was rationalized by different conformations in transition states[224] (for a review see Reference 227). *syn*-Stereodirectivity effect has also been observed in epoxidation of several allylic ethers.[228] A lithium salt formed by the reaction of methylenetriphenylphosphorane with t-butyl or sec-butyl lithium can split the oxirane ring to produce a Wittig-type reagent that can be used in further steps.[229] Enantiomerically pure aminoalcohols were obtained by the cleavage of achiral epoxides with chiral amine reagents.[230] A simple trick of changing pH may direct halolactonization of an unsaturated hydroxy acid in order to convert one enantiomer to either enantiomer of the product.[231] For further examples of stereoselective halolactonization and related reactions see References 232 to 234. New avenues of stereoselective polyfunctionalization of cyclic dienes and related compounds, provided by development of transition-metal chemistry in last few years,[66,235-246] have flourished into an enantiospecific synthesis of (−)-gabaculin[245] and a new approach to steroid skeleton.[241]

Cyclohexane-*cis*-1,2-dicarboxylic acids may be synthesized in three steps, starting by dichloroketene addition to cyclohexenes.[247] Tetrabutylammonium nitrate has been found to convert tosylates of secondary alcohols into nitrates with inversion of configuration.[248] A new reagent, arising from methylaluminum bis(2,6-di-*tert*-butyl-4-alkylphenoxide) and methyl lithium, was developed for achieving unusual equatorial and anti-Cram selectivity in carbonyl alkylations.[249] Cyclopalladation was used in order to activate methyl groups.[250]

REFERENCES

1. **ApSimon, J.,** *The Total Synthesis of Natural Products,* Vols. 1—5, Wiley-Interscience, New York, 1973—1983.
2. **Snider, B. B. and Phillips, G. B.,** Reaction of enol ethers with formaldehyde and organoaluminium compounds, *J. Org. Chem.,* 48, 2789, 1983.
3. **Lim, M.-I. and Marquez, V. E.,** Total synthesis of (−)-neplanocin A, *Tetrahed. Lett.,* 24, 5559, 1983.
4. **Baldwin, J. E., Kruse, L. I.,** Rules for ring closure. Stereoelectronic control in the endocyclic alkylation of ketone enolates, *J. Chem. Soc. Chem. Commun.,* 233, 1977.
5. **House, H. O., Phillips, W. V., Sayer, T. S. B., and Yau, C.-C.,** Chemistry of carbanions. 31. Cyclization of the metal enolates from ω-bromo ketones, *J. Org. Chem.,* 43, 700, 1978.

6. **Stork, G., Depezay, J. C., and D'Angelo, J.**, Synthesis of small rings *via* the protected cyanohydrin method, *Tetrahed. Lett.*, 389, 1975.
7. **Martel, J. and Huynh, C.**, Synthèse de l'acide chrysanthemique. (Note préliminaire) II. — Accès stérèosèlectif au (±)-*trans* chrysanthémate, *Bull. Soc. Chim. France,* 985, 1967.
8. **Stork, G. and Cohen, J. F.**, Ring size in epoxynitrile cyclization. A general synthesis of functionally substituted cyclobutanes. Application to (±)-grandisol, *J. Am. Chem. Soc.*, 96, 5270, 1974.
9. **Stork, G., Cama, L. D., and Coulson, D. R.**, A general method of ring formation, *J. Am. Chem. Soc.*, 96, 5268, 1974.
10. **Trost, B. M. and Bogdanowicz, M. J.**, New synthetic reactions. A versatile cyclobutanone (Spiroannelation) and γ-butyrolactone (lactone annelation) synthesis, *J. Am. Chem. Soc.*, 95, 5321, 1973.
11. **Shono, T., Matsumura, Y., Tsubata, K., and Sugihara, Y.**, New synthesis of cyclopropanes from 1,3-dicarbonyl compounds utilizing electroreduction of 1,3-dimethanesulfonates, *J. Org. Chem.*, 47, 3090, 1980.
12. **Brookhart, M., Timmers, D., Tucker, J. R., Williams, G. D., Husk, G. R., Brunner, H., and Hammer, B.**, Enantioselective cyclopropane synthesis using the chiral carbene complexes ($S_{Fe}S_C$)- and ($R_{Fe}S_C$)-(C_5H_5)(CO)(Ph_2R*P) Fe = $CHCH_3^+$ (R* = (S)-2-methylbutyl). Role of metal *vs.* ligand chirality in the optical induction, *J. Am. Chem. Soc.*, 105, 6721, 1983.
13. **Kočovský, P.**, Chiral synthons and intermediates in organic synthesis I., *Chem. Listy,* 76, 1147, 1982 (in Czech).
14. **Kočovský, P. and Černý, M.**, Chiral synthons and intermediates in organic synthesis II, *Chem. Listy,* 77, 373, 1983 (in Czech).
15. **Stork, G. and Takahashi, T.**, Chiral synthesis of prostaglandins (PGE_1) from D-glyceraldehyde, *J. Am. Chem. Soc.*, 99, 1275, 1977.
16. **Stork, G., Takahashi, T., Kawamoto, I., and Suzuki, T.**, Total synthesis of prostaglandin $F_{2\alpha}$ by chirality transfer from D-glucose, *J. Am. Chem. Soc.*, 100, 8272, 1978.
17. **Kitahara, T., Mori, K., and Matsui, M.**, A total synthesis of (+)-brefeldin A, *Tetrahed. Lett.*, 3021, 1979.
18. **Corey, E. J. and Wollenberg, R. A.**, Total synthesis of (±)-brefeldin A (Part IV), *Tetrahed. Lett.*, 2243, 1977.
19. **Bartlett, P. A. and Green, F. R., III.**, Total synthesis of brefeldin A, *J. Am. Chem. Soc.*, 100, 4858, 1978.
20. **Stork, G., Winkler, J. D., and Saccomano, N. A.**, Stereochemical control in the construction of vicinally substituted cyclopentanes and cyclohexanes. Intramolecular conjugate addition of β-ketoester anions, *Tetrahed. Lett.*, 465, 1983.
21. **Danheiser, R. L., Davila, C. M., Auchus, R. J., and Kadonaga, J. T.**, A stereoselective synthesis of cyclopentene derivatives from 1,3-dienes, *J. Am. Chem. Soc.*, 103, 1443, 1981.
22. **Short, R. P., Revol, J. M., Ranu, B. C. and Hudlicky, T.**, General method of synthesis of cyclopentanoid terpenic acids. Stereocontrolled total synthesis of (±)-isocomenic acid and (±)-epiisocomenic acid, *J. Org. Chem.*, 48, 4453, 1983.
23. **Hudlický, T., Reddy, D. B., Govindan, S. V., Kulp, T., Still, B., and Sheth, J. P.**, Intramolecular cyclopentene annulation. 3. Synthesis and carbon-13 nuclear magnetic resonance spectroscopy of bicyclic cyclopentene lactones as potential perhydroazulene and/or monoterpene synthons, *J. Org. Chem.*, 48, 3422, 1983.
24. **Govindan, S. V., Hudlický, T., and Koszyk, F. J.**, Two topologically distinct total syntheses of (±)-sarkomycin, *J. Org. Chem.*, 48, 3581, 1983.
25. **Woodward, R. B.**, Recent advances in the chemistry of natural products, *Pure Appl. Chem.*, 17, 519, 1968.
26. **Woodward, R. B.**, Recent advances in the chemistry of natural products, *Pure Appl. Chem.*, 25, 283, 1971.
27. **Woodward, R. B.**, The total synthesis of vitamin B_{12}, *Pure Appl. Chem.*, 33, 145, 1973.
28. **Fouquey, C. J. J., Bouton, M. M., Fortin, M., and Tournemine, C.**, 1- And 16-alkylated derivatives of A-nor(5α)-androstane, *Eur. J. Med. Chem. Chim. Ther.*, 17, 355, 1982; *Chem. Abstr.* 98, 47101x, 1983.
29. **Heathcock, C. H., Tice, C. M., and Germroth, T. C.**, Synthesis of sesquiterpene antitumor lactones. Total synthesis of (±)-parthenin, *J. Am. Chem. Soc.*, 104, 6081, 1982.
30. **de la Mare, P. B. D. and Bolton, P.**, *Electrophilic Additions to Unsaturated Systems,* Elsevier, 1982.
31. **Berti, G.**, Stereochemical aspects of the synthesis of 1,2-epoxides, *Topics Stereochem.*, 7, 93, 1973.
32. **Kirk, D. N. and Hartshorn, M. P.**, *Steroid Reaction Mechanisms,* Elsevier, Amsterdam, 1968.
33. **Zefirov, N. S., Samoshin, V. V., and Zemlyanova, T. G.**, The influence of remote substituents on the reactivity of the cyclohexene double bond, *Tetrahed. Lett.*, 24, 5133, 1983.
34. **House, H. O. and Fischer, W. F., Jr.**, The chemistry of carbanions. XVII. The addition of methyl organometallic reagents to cyclohexenone derivatives, *J. Org. Chem.*, 33, 949, 1968.

35. **Posner, G. H.,** Conjugate addition reactions of organocopper reagents, *Org. Reactions,* 19, 1, 1972.
36. **House, H. O. and Fischer, W. F.,** The chemistry of carbanions. The addition of methyl organometallic reagents to cyclohexenone derivatives, *J. Org. Chem.,* 33, 949, 1968.
37. **Agami, G., Fadlallah, M., and Levisalles, J.,** Influence du degre de substitution de la double liaison èthylènique sur la stèrèochimie de l'hydrocyanation 1,4 cètones conjugées, *Tetrahedron,* 37, 909, 1981.
38. **Posner, G. H., Mallamo, J. P., Hulce, M., and Frye, L. L.,** Asymmetric induction during organometallic conjugate addition to enantiomerically pure 2-(arylsulfinyl)-2-cyclopentenones, *J. Am. Chem. Soc.,* 104, 4180, 1982.
39. **Sallay, S. I.,** The total synthesis of *dl*-ibogamine, *J. Am. Chem. Soc.,* 89, 6762, 1967.
40. **Alexander, R. P. and Paterson, I.,** Alkene carbosulphenylation and carboselenylation. The use of allyl-trimethyl silane and O-silylated enolates, *Tetrahed. Lett.,* 24, 5911, 1983.
41. **Snider, B. B., Cordova, R., and Price, R. T.,** Reactions of the formaldehyde-trimethylaluminium complex with alkenes, *J. Org. Chem.,* 47, 3643, 1982.
42. **Snider, B. B., Rodini, D. J., Karras, M., Kirk, T. C., Deutsch, E. A., Cordova, R., and Price, R. T.,** Alkylaluminium halides. Lewis acid catalysts which are Bronsted bases, *Tetrahedron,* 37, 3927, 1981.
43. **Brown, H. C. and Krishnamurthy, S.,** Forty years of hydride reductions, *Tetrahedron,* 35, 567, 1979.
44. **Wigfield, D. C.,** Stereochemistry and mechanism of ketone reduction by hydride reagents, *Tetrahedron,* 35, 449, 1979.
45. **Hickmott, P. W.,** Enamines. Recent advances in synthetic, spectroscopic, mechanistic and stereochemical aspects — I, *Tetrahedron,* 38, 1975, 1982.
46. **Hickmott, P. W.,** Enamines. Recent advances in synthetic, spectroscopic, mechanistic, and stereochemical aspects — II, *Tetrahedron,* 38, 3363, 1982.
47. **Whitesell, J. K. and Felman, S. W.,** Asymmetric induction. 2. Enantioselective alkylation of cyclohexanone *via* a chiral enamine, *J. Org. Chem.,* 42, 1663, 1977.
48. **Valentine, D. and Scott, J. W.,** Asymmetric synthesis, *Synthesis,* 329, 1978.
49. **ApSimon, J. and Seguin, R. P.,** Recent advances in asymmetric synthesis, *Tetrahedron,* 35, 2797, 1979.
50. **Izumi, Y. and Tai, A.,** *Stereo-Differentiating Reactions,* Academic Press, New York, 1977.
51. **Ahlbrecht, H.,** 3-Metallated enamines, *Chimia,* 31, 391, 1977.
52. **Enders, D. and Eichenauer, H.,** Asymmetrische Synthesen via metallierte chirale hydrazone. Enantioselective alkylierung von cyclischen ketonen und aldehyden, *Ber. Deut. Chem. Ges.,* 112, 2933, 1979.
53. **Meyers, A. I., Williams, D. R., Erickson, G. W., White, S., and Druelinger, M.,** Enantioselective alkylation of ketones via chiral nonracemic lithioenamines. An asymmetric synthesis of α-alkyl- and α,α′-dialkyl cyclic ketones, *J. Am. Chem. Soc.,* 103, 3081, 1981.
54. **Meyers, A. I., Williams, D. R., White, S., and Ericson, G. W.,** An asymmetric synthesis of acyclic and macrocyclic α-alkyl ketones. The role of (E)- and (Z)-lithioenamines, *J. Am. Chem. Soc.,* 103, 3088, 1981.
55. **Blarer, S. J., Schweizer, W. B., and Seebach, D.,** Asymmetrische Michael-Additionen. Praktisch vollständig diastereo- und enantioselective alkylierungen des enamins aus cyclohexanon und prolinyl methyläther durch ω-nitrostyrole zu μ-2-(1′-aryl-2-′-nitroäthyl) cyclohexanonen, *Helv. Chim. Acta,* 65, 1637, 1982.
56. **Stork, G., Nelson, G. L., Rouessac, F., and Gringore, O.,** A versatile synthesis of cyclopentenones, *J. Am. Chem. Soc.,* 93, 3091, 1971.
57. **Bernasconi, S., Gariboldi, P., Jommi, G., Montanari, S., and Sisti, M.,** Total synthesis of pinguisone, *J. Chem. Soc. Perkin Trans.,* 1, 2394, 1981.
58. **Bernasconi, S., Ferrari, M., Gariboldi, P., Jommi, G., and Sisti, M.,** Synthetic study of pinguisane terpenoids, *J. Chem. Soc. Perkin Trans.,* 1, 1994, 1981.
59. **Corey, E. J. and Ensley, H. E.,** Highly stereoselective conversion of prostaglandin A_2 to the 10,11α-oxido derivative using a remotely placed exogenous directing group, *J. Org. Chem.,* 38, 3187, 1973.
60. **Fried, J., Sih, J. C., Lin, C. H., and Dalven, P.,** Regiospecific epoxide opening with acetylenic alanes. An improved total synthesis of E and F prostaglandins, *J. Am. Chem. Soc.,* 94, 4343, 1972.
61. **Newton, R. F. and Roberts, S. M.,** Steric control in prostaglandin synthesis involving bicyclic and tricyclic intermediates, *Tetrahedron,* 36, 2163, 1980.
62. **Caton, M. P. L.,** A survey of novel and useful reactions discovered through research on prostaglandins, *Tetrahedron,* 35, 2705, 1979.
63. **Mitra, A.,** *The Synthesis of Prostaglandins,* J. Wiley & Sons, New York, 1977.
64. **Davies, S. G.,** *Organotransition Metal Chemistry: Application to Organic Synthesis,* Pergamon Press, Oxford, 1982.
65. **Arzeno, H. B., Barton, D. H. R., Davies, S. G., Lusinchi, X., Meunier, B., and Pascard, C.,** Synthèse de la 10(S)-méthyl-codéine et de la 10(S)-méthyl-morphine, *Nouveau J. Chimie,* 4, 369, 1980.
66. **Pearson, A. J., Kole, S. L., and Chen, B.,** Approach to stereochemically defined cycloheptadiene derivatives using organo-iron chemistry, *J. Am. Chem. Soc.,* 105, 4483, 1983.
67. **Ikota, N. and Ganem, B.,** Shikimate-derived metabolites. 2. Synthesis of a bacterial natural product illustrating a concerted S_N2' reaction, *J. Am. Chem. Soc.,* 100, 351, 1978.

68. **Coblens, K. E., Muralidharan, V. B., and Ganem, B.,** Shikimate derived metabolites. 12. Stereocontrolled total synthesis of shikimic acid and 6β-deuterioshikimate, *J. Org. Chem.*, 47, 5041, 1982.
69. **Hoare, J., Policastro, P. P., and Berchtold, G. A.,** Improved synthesis of racemic chorismic acid. Claisen rearrangement of 4-epi-chorismic acid and dimethyl 4-epi-chorismate, *J. Am. Chem. Soc.*, 105, 6264, 1983.
70. **Herranz, E. and Sharpless, K. B.,** Improvements in the osmium-catalyzed oxyamination of olefins by chloramine-T, *J. Org. Chem.*, 43, 2544, 1978.
71. **Patrick, D. W., Truesdale, L. K., Biller, S. A., and Sharpless, K. B.,** Stereospecific vicinal oxyamination of olefins by alkylimidoosmium compounds, *J. Org. Chem.*, 43, 2628, 1978.
72. **Overman, L. E. and Flippin, L. A.,** Facile aminolysis of epoxides with diethylaluminium amides, *Tetrahed. Lett.*, 22, 195, 1981.
73. **Pauls, H. W. and Fraser-Reid, B.,** A short, efficient route to a protected daunosamine from L-rhamnose, *J. Chem. Soc. Chem. Commun.*, 1031, 1983.
74. **Knapp, S., Sabastian, M. J., and Ramanathan, H.,** Total synthesis of (±)-fortamine and (±)-2-deoxyfortamine, *J. Org. Chem.*, 48, 4788, 1983.
75. **Overman, L. E., Petty, C. B., Ban, T., and Huang, G. T.,** Preparation and Diels-Alder reaction of 1,3-dienes containing both sulfur and nitrogen substituents. Complete orientational control by acylamino group, *J. Am. Chem. Soc.*, 105, 6335, 1983.
76. **Bennet, G. B.,** The Claisen rearrangement in organic synthesis, 1967 to January 1977, *Synthesis,* 589, 1977.
77. **Wehrli, R., Belluš, D., Hausen, H. J., and Schmid, H.,** The Cope rearrangement — a reaction with a manifold mechanism ?, *Chimia,* 30, 416, 1976.
78. **Evans, D. A. and Nelson, J. V.,** Stereochemical study of the [3,3] sigmatropic rearrangement of 1,5-diene-3-alkoxides. Application to the stereoselective synthesis of (±)-juvabione, *J. Am. Chem. Soc.*, 102, 774, 1980.
79. **Ziegler, F. E. and Piwinski, J. J.,** Interception of the Cope chair-like transition-state product during the tandem Cope-Claisen rearrangement. A route to an ambrosanolide synthon, *J. Am. Chem. Soc.*, 102, 6576, 1980.
80. **Ziegler, F. E. and Piwinski, J. J.,** Tandem Cope-Claisen rearrangement: Scope and stereochemistry, *J. Am. Chem. Soc.*, 104, 7181, 1982.
81. **Ireland, R. E., Thaisrivongs, J., Wilcox, C. S.,** Total synthesis of lasalocid A (X537A), *J. Am. Chem. Soc.*, 102, 1155, 1980.
82. **Greene, A. E., Teixeira, M. A., Barreiro, E., Cruz, A., and Crabbé, P.,** The total synthesis of prostaglandins by the tropolone route, *J. Org. Chem.*, 47, 2553, 1982.
83. **Paquette, L. A. and Crouse, G. D.,** Stereocontrolled preparation of precursors to all primary prostaglandins from butadiene, *Tetrahedron,* 37, 281, 1981.
84. **Dowle, M. D. and Davies, D. I.,** Synthesis and synthetic utility of halolactones, *Chem. Soc. Rev.*, 8, 171, 1979.
85. **Kočovský, P. and Tureček, F.,** Mechanism and structural effects in bromolactonization, *Tetrahedron,* 39, 3621, 1983.
86. **Holbert, G. W. and Ganem, B.,** Shikimate derived metabolites. 3. Total synthesis of senepoxide and seneol according to a biogenetic proposal, *J. Am. Chem. Soc.*, 100, 352, 1978.
87. **Nicolaou, K. C. and Lysenko, Z.,** Phenylsulphenyl-lactonization. An easy and synthetically useful lactonization procedure, *J. Chem. Soc. Chem. Commun.*, 293, 1977.
88. **Clive, D. L. J. and Chittattu, G.,** New route to bicyclic lactones. Use of benzeneselenyl chloride, *J. Chem. Soc. Chem. Commun.*, 484, 1977.
89. **Tureček, F.,** Preparation of 7-oxabicyclo [4.3.0] nonanes and 2-oxabicyclo [4.4.0] decanes specifically labeled with deuterium, *Collect. Czech. Chem. Commun.*, 47, 858, 1982.
90. **Corey, E. J., Ravindranathan, T., and Terashima, S.,** A new method for the 1,4-addition of the methylenecarbonyl unit (-CH_2-CO-) to dienes, *J. Am. Chem. Soc.*, 93, 4326, 1971.
91. **Ranganathan, S., Ranganathan, D., and Mehrotra, A. K.,** Ketene equivalents, *Synthesis,* 289, 1977.
92. **Kočovský, P.,** Synthesis of 14-deoxy-14α-strophanthidin, *Collect. Czech. Chem. Commun.*, 45, 2998, 1980.
93. **Holton, R. A.,** Prostaglandin synthesis via carbopalladation, *J. Am. Chem. Soc.*, 99, 8083, 1977.
94. **Trost, B. M., Taber, D. F., and Alper, J. B.,** An approach to the stereocontrolled creation of an acyclic side chain of some natural products, *Tetrahed. Lett.*, 3857, 1976.
95. **Burke, S. D. and Grieco, P. A.,** Intramolecular reactions of diazocarbonyl compounds, *Org. Reactions,* 26, 361, 1979.
96. **Corey, E. J., Arnold, Z., and Hutton, J.,** Total synthesis of prostaglandins E_2 and $F_{2\alpha}$ *(dl) via* a tricarbocyclic intermediate, *Tetrahed. Lett.*, 307, 1970.
97. **Lukina, R. Y. and Gadshtein, M.,** Reaction of cyclopropyl hydrocarbons with mercury (II) salts, *Dokl. Akad. Nauk SSSR,* 71, 65, 1950.

98. **Collum, D. B., Mohamad, F., and Hallock, J. S.,** Mercury (II) mediated opening of cyclopropanes. Effects of proximate internal nucleophiles on stereo- and regioselectivity, *J. Am. Chem. Soc.*, 105, 6882, 1983.
99. **Fringuelli, F. and Taticchi, J.,** Acid-catalysed cleavage of the cyclopropane ring in some ketones of the carane series, *J. Chem. Soc. (C)*, 297, 1971.
100. **Roberts, R. A., Schüll, V., and Paquette, L. A.,** Electrophile initiated ring-opening reactions of 2-methylene-6,6-dimethylbicyclo [3.1.0] hexanes. New methodology for the synthesis of highly functionalized 1,2,3,-trisubstituted cyclopentenes, *J. Org. Chem.*, 48, 2076, 1983.
101. **King, J. F. and de Mayo, P.,** in *Molecular Rearrangements*, Vol. 2, de Mayo, P., Ed., Interscience, New York, 1964, 807.
102. **Černý, V.,** Reaction of steroidal β-oxo cyclopropanes with Jacques' reagent, *Collect. Czech. Chem. Commun.*, 38, 1563, 1973.
103. **Ficini, J.,** Ynamine. A versatile tool in organic synthesis, *Tetrahedron*, 32, 1449, 1976.
104. **Ficini, J. and Krief, A.,** Stereochemical control in the hydrolysis of an ynamine-cyclopentenone adduct. A stereoselective route to diastereoisomeric 2-(1-cyclopentyl 3-oxo) propionic acids, *Tetrahed. Lett.*, 1397, 1970.
105. **Trost, B. M., Taber, D. F., and Alper, J. B.,** An approach to the stereocontrolled creation of an acyclic side chain of some natural products, *Tetrahed. Lett.*, 3857, 1976.
106. **D'Silva, T. D. J. and Peck, D. W.,** Convenient synthesis of frontalin - 1,5-dimethyl-6,8-dioxabicyclo [3.2.1] octane, *J. Org. Chem.*, 37, 1828, 1972.
107. **Woodward, R. B., Bader, F. E., Bickel, H., Frey, A. J., and Kierstead, R. W.,** The total synthesis of reserpine, *Tetrahedron*, 2, 1, 1958.
108. **Takano, S., Sasaki, M., Kanno, H., Shishido, K., and Ogasawara, K.,** New synthesis of (±)-emetine from tetrahydroprotoberberine precursors *via* α-diketone mono thioketal intermediate, *J. Org. Chem.*, 43, 4169, 1978.
109. **Brewster, D., Myers, M., Ormerod, J., Otter, P., Smith, A. C. B., Spinner, M. E., and Turner, S.,** Prostaglandin synthesis. Design and execution, *J. Chem. Soc. Perkin Trans.*, 1, 2796, 1973.
110. **Bridgeman, J. E., Butchers, C. E., Jones, E. R. H., Kasal, A., Meakins, G. D., and Woodgate, P.,** Studies in the steroid group. Part LXXX. Preparation of 2- and 16-oxo and 3,16- and 2,16-dioxo-5α-androstane, and 2-oxo-5α-cholestane, *J. Chem. Soc. (C)*, 2440, 1970.
111. **Marshall, J. A. and Roebke, H.,** A nonoxidative method for ketone transposition, *J. Org. Chem.*, 34, 4188, 1969.
112. **Trost, B. M., Hiroi, K., and Kurozumi, S.,** New synthetic methods. 1,2-(Alkylative) carbonyl transpositions, *J. Am. Chem. Soc.*, 97, 438, 1975.
113. **Montury, M. and Goré, J.,** Obtention en oxydation de Baeyer-Villiger de quelques acetylcyclenes. Application a la transformation ceto-3 → ceto-2 steroides, *Tetrahedron*, 33, 2819, 1977.
114. **Kane, V. V., Singh, V., Martin, A., and Doyle, D. L.,** The chemistry of 1,2-carbonyl transposition, *Tetrahedron*, 39, 345, 1983.
115. **Larson, G. and Fuentes, L. M.,** A reductive 1,2-transposition of acyclic ketones, *Synth. Commun.*, 9, 841, 1979.
115a. **Danishefsky, S., Kitahara, T., Schuda, P. T., and Etheredge, S. J.,** A remarkable epoxide opening. An expeditious synthesis of vernolepin and vernomenin, *J. Am. Chem. Soc.*, 98, 3028, 1976.
116. **Kočovský, P.,** Synthesis of 3,4-, 4,5-, and 5,6-unsaturated 19-substituted cholestane derivatives and related epoxides, *Collect. Czech. Chem. Commun.*, 45, 3008, 1980.
117. **Magid, R. M. and Fruchey, O. S.,** Stereochemistry of the S_N2' reaction of an allylic chloride with a secondary amine, *J. Am. Chem. Soc.*, 99, 8368, 1977.
118. **Gallina, C. and Ciattini, P. G.,** Conversion of allylic carbamates into olefins with lithium dimethylcuprate. A new formal S_N2' reaction, *J. Am. Chem. Soc.*, 101, 1035, 1979.
119. **Trost, B. M. and Klun, T. P.,** Chirality transfer in acyclic systems via organocopper chemistry, *J. Org. Chem.*, 45, 4256, 1980.
120. **Stork, G. and Schoofs, A. R.,** Concerted intramolecular displacement with rearrangement in allylic systems. Displacement of an allylic ester with a carbanion, *J. Am. Chem. Soc.*, 101, 5081, 1979.
121. **Stohrer, W.-D.,** Zur Stereochemie der S_N2'-Reaktion, *Angew. Chem.*, 95, 642, 1983.
122. **Bordwell, F. G.,** Are nucleophilic bimolecular concerted reactions involving four or more bonds a myth?, *Acc. Chem. Res.*, 3, 281, 1970.
123. **Magid, R. M.,** Nucleophilic and organometallic displacement reactions of allylic compounds: Stereo- and regiochemistry, *Tetrahedron*, 36, 1901, 1980.
124. **Overton, K. H.,** Concerning stereochemical choice in enzymic reactions, *Chem. Soc. Rev.*, 8, 447, 1979.
125. **Cane, D. E.,** The stereochemistry of allylic pyrophosphate metabolism, *Tetrahedron*, 36, 1109, 1980.
126. **Overman, L. E., Campbell, C. B., and Knoll, F. M.,** Mild procedures for interconverting allylic oxygen functionality. Cyclization-induced [3,3] sigmatropic rearrangement of allylic carbamates, *J. Am. Chem. Soc.*, 100, 4822, 1978.

127. **Overman, L. E. and Knoll, F. M.,** Palladium(II) — catalyzed rearrangement of allylic acetate, *Tetrahed. Lett.,* 321, 1979.
128. **Keinan, E. and Roth, Z.,** Regioselectivity in organo-transition-metal chemistry. A new indicator substrate to classification of nucleophiles, *J. Org. Chem.,* 48, 1769, 1983.
129. **Grieco, P. A., Takigawa, T., Bongers, S. L., and Tanaka, H.,** Complete transfer of chirality in the [3,3]-sigmatropic rearrangement of allylic acetates catalyzed by palladium (II). Application to stereocontrolled syntheses of prostaglandins possessing either the C-15(S) or C-15(R) configuration, *J. Am. Chem. Soc.,* 102, 7587, 1980.
130. **Trost, B. M.,** Organopalladium intermediates in organic synthesis, *Tetrahedron,* 33, 2615, 1977.
131. **Overman, L. E.,** Allylic and propargylic imidic esters in organic synthesis, *Acc. Chem. Res.,* 13, 218, 1980.
132. **Fleet, G. W. J. and Spensley, C. R. C.,** Model studies on the synthesis of pseudomonic acid, *Tetrahed. Lett.,* 23, 109, 1982.
133. **Ueno, Y., Sano, H., and Okawara, M.,** A new method for the 1,3-hydroxy-transposition in allylic alcohols via allylic stananes, *Synthesis,* 1011, 1980.
134. **Matsubara, S., Takai, K., Nozaki, H.,** Isomerization of primary allylic alcohols to tertiary ones by means of $Me_3SiOOSiMe_3$-$VO(acac)_2$ catalyst, *Tetrahed. Lett.,* 24, 3741, 1983.
135. **Hrubiec, R. T. and Smith, M. B.,** Homoallylic substitution reaction of piperidine with 1-bromo-1-cyclopropylalkanes, *Tetrahed. Lett.,* 24, 5031, 1983.
136. **Fajkoš, J. and Joska, J.,** 5,7-Cyclosteroids with an oxygen function in position 1, *Collect. Czech. Chem. Commun.,* 39, 1773, 1973.
137. **Wharton, P. S. and Bohlen, D. H.,** Hydrazine reduction of α,β-epoxy ketones to allylic alcohols, *J. Org. Chem.,* 26, 3615, 1961.
138. **Stein, A. R.,** Racemization and bromide exchange studies on 1-phenylbromoethane and the question of the ion pair mechanism for bimolecular nucleophilic substitutions at saturated carbon, *J. Org. Chem.,* 41, 519, 1976.
139. **San Fillippo, J., Jr., Chern, C., and Valentine, J. S.,** The reaction of superoxide with alkyl halides and tosylates, *J. Org. Chem.,* 40, 1678, 1975.
140. **Kruizinga, W. H., Strijtveen, B., and Kellog, R. M.,** Cesium carboxylates in dimethylformamide. Reagents for introduction of hydroxyl groups by nucleophilic substitution and for inversion of configuration of secondary alcohols, *J. Org. Chem.,* 46, 4321, 1981.
141. **March, J.,** *Advanced Organic Chemistry,* 2nd ed., McGraw-Hill, New York, 1977.
142. **Carey, F. A. and Sundberg, R. J.,** *Advanced Organic Chemistry,* Part B, Plenum Press, New York, 1983.
143. **Sawyer, D. T. and Gibian, M. J.,** The chemistry of superoxide ion, *Tetrahedron,* 35, 1471, 1979.
144. **Mitsunobu, O.,** The use of diethylazodicarboxylate and triphenylphosphine in synthesis and transformation of natural products, *Synthesis,* 1, 1981.
145. **Plattner, P. A., Fürst, A., Koller, F., and Lang, W.,** Zur Herstellung des Epi-cholesterins über das 3α,5-Dioxy-cholestan, *Helv. Chim. Acta,* 31, 1455, 1948.
146. **Contreras, R. and Mendoza, L.,** The reduction of 5α-cholestan-3-one and 5β-cholestan-3-one by some boranes and hydroborates, *Steroids,* 34, 121, 1979.
147. **Rétey, J. and Robinson, J. A.,** *Stereospecificity in Organic Chemistry and Enzymology,* Verlag Chemie, Weinheim, 1982.
148. **Grob, C. A., Seckinger, K., Tamm, S. W., and Traber, R.,** Nucleophilic displacement at tertiary carbon, *Tetrahed. Lett.,* 3051, 1973.
149. **Morera, E. and Ortar, G.,** Preparative and stereochemical feature of the sulfoxide-sulfenate [2,3] sigmatropic rearrangement in 17-vinyl-17-hydroxy steroids, *J. Org. Chem.,* 48, 119, 1983.
150. **Hoffmann, R. W.,** Stereochemie [2,3] sigmatroper Umlagerung, *Angew. Chem.,* 91, 625, 1979.
151. **Nakata, T., Schmid, G., Vranesic, B., Okigawa, M., Smith-Palmer, T., and Kishi, Y.,** A total synthesis of lasalocid A, *J. Am. Chem. Soc.,* 100, 2933, 1978.
152. **Bartlett, P. A. and Jernstedt, K. K.,** "Phosphate extension" A strategem for the stereoselective functionalization of acyclic homoallylic alcohols, *J. Am. Chem. Soc.,* 99, 4829, 1977.
153. **Fukuyama, T., Vranesic, B., Negri, D. P., and Kishi, Y.,** Synthetic studies on polyether antibiotics-II. Stereocontrolled syntheses of epoxides of bishomoallylic alcohols, *Tetrahed. Lett.,* 2741, 1978.
154. **Still, W. C. and Darst, K. P.,** Remote asymmetric induction. A stereoselective approach to acyclic diols via cyclic hydroboration, *J. Am. Chem. Soc.,* 102, 7385, 1980.
155. **Ando, M., Tajima, K., and Takase, K.,** Studies on the synthesis of sesquiterpene lactones. 8. Synthesis of saussurea lactone, 8-deoxymelitensin, and 11,12-dehydro-8-deoxymelitensin via a novel fragmentation reaction, *J. Org. Chem.,* 48, 1211, 1983.
156. **Anet, F. A. L.,** Dynamics of eight-membered ring in the cyclooctane class, *Fortsch. Chemisch. Forsch.,* 45, 169, 1974.

157. **Anet, F. A. L. and Radwah, T. N.,** Cyclodecane, Forced field calculations and ^{1}H-NMR spectra of deuterated isotopomers, *J. Am. Chem. Soc.,* 100, 7166, 1978.
158. **Anet, F. A. L., Cheng, A. K., and Wagner, J. J.,** Determination of conformational energy barriers in medium- and large-ring cycloalkanes by [1]H- and ^{13}C nuclear magnetic resonance, *J. Am. Chem. Soc.,* 94, 9250, 1972.
159. **Anet, F. A. L., Cheng, A. K., and Krane, J.,** Conformations and energy barriers in medium- and large-ring ketones. Evidence from ^{13}C and ^{1}H nuclear magnetic resonance, *J. Am. Chem. Soc.,* 95, 7877, 1973.
160. **Borgen, G. and Dale, J.,** The preferred conformation of 1,1,4,4-tetramethylcyclononane and some derivatives. Low temperature nuclear magnetic resonance study, *Chem. Commun.,* 1105, 1970.
161. **Anet, F. A. L., Degen, P. J., and Yvari, I.,** Conformations of azocane (azocyclooctane), *J. Org. Chem.,* 43, 3021, 1978.
162. **Groth, P.,** Crystal structure of cycloundecanone at − 165°C, *Acta Chem. Scand.,* 28, 294, 1974.
163. **Ermer, O., Dunitz, J. D., and Bernal, I.,** The structure of medium-ring compounds. XVIII. X-ray and neutron diffraction analysis of cyclodecane-1,6-trans-diol, *Acta Crystallogr. Sect. B.,* 29, 2278, 1973.
164. **Wiberg, K. B.,** A scheme for strain energy minimization. Application to the cycloalkanes, *J. Am. Chem. Soc.,* 87, 1070, 1965.
165. **Hendricson, J. B.,** Molecular geometry. V. Evaluation of functions and conformations of medium rings, *J. Am. Chem. Soc.,* 89, 7036, 1967.
166. **Hendricson, J. B.,** Molecular geometry. VI. Methyl-substituted cycloalkanes, *J. Am. Chem. Soc.,* 89, 7043, 1967.
166a. **Hendricson, J. B.,** Molecular geometry. VII. Modes of interconversion in the medium rings, *J. Am. Chem. Soc.,* 89, 7047, 1967.
167. **Bixon, M. and Lifson, S.,** Potential functions and conformations in cycloalkanes, *Tetrahedron,* 23, 769, 1967.
168. **Engler, E. M., Andose, J. D., and von R. Schleyer, P.,** Critical evaluation of molecular mechanics, *J. Am. Chem. Soc.,* 95, 8005, 1973.
169. **Allinger, N. L. and Sprague, J. T.,** Conformational analysis. LXXXIV. A study of the structures and energies of some alkenes and cycloalkenes by the force field method, *J. Am. Chem. Soc.,* 94, 5734, 1972.
170. **Allinger, N. L., Tribble, M. T., and Miller, M. A.,** Conformational analysis - LXXIX. An improved force field for the calculation of the structures and energies of carbonyl compounds, *Tetrahedron,* 28, 1173, 1972.
171. **Anet, F. A. L. and Rawdah, T. N.,** Iterative force-field calculations of cycloundecane, cyclotridecane, and cyclopentadecane, *J. Am. Chem. Soc.,* 100, 7810, 1978.
172. **Eliel, E. L., Allinger, N. L., Angyal, S. J., and Morrison, G. A.,** *Conformational Analysis,* John Wiley & Sons, New York, 1965.
173. **Anet, F. A. L. and Cheng, A. K.,** Conformation of cyclohexadecane, *J. Am. Chem. Soc.,* 97, 2420, 1975.
174. **Allinger, N. L., Gorden, B., and Profeta, S., Jr.,** On the conformational structure of cyclohexadecane, the corresponding ketone, the 1,9-dione and the corresponding ketals, *Tetrahedron,* 36, 859, 1980.
175. **Dale, J.,** Exploratory calculations of medium and large rings. Part 1. Conformational minima of cycloalkanes, *Acta Chem. Scand.,* 27, 1115, 1973.
176. **Dale, J.,** Exploratory calculations of medium and large rings. Part 2. Conformational interconversions in cycloalkanes. *Acta Chem. Scand.,* 27, 1130, 1973.
176a. **Dale, J.,** Exploratory calculations of medium and large rings. Part 3. Mono- and bis(gem-dimethyl)cycloalkanes, *Acta Chem. Scand.,* 27, 1149, 1973.
177. **Anet, F. A. L., St. Jacques, M., Henrichs, P. M., Cheng, A. K., Krane, J., and Wong, L.,** Conformational analysis of medium-ring ketones, *Tetrahedron,* 30, 1629, 1974.
178. **Still, W. C. and Galynker, I.,** Chemical consequences of conformation in macrocyclic compounds. An effective approach to remote asymmetric induction, *Tetrahedron,* 37, 3981, 1981.
179. **Still, W. C.,** ($\pm$)-Periplanone B. Total synthesis and structure of the sex excitant pheromone of the American cockroach, *J. Am. Chem. Soc.,* 101, 2493, 1979.
180. **Allinger, N. L.,** Conformational analysis. 130. MM2. A hydrocarbon force-field utilizing V_1 and V_2 torsional terms, *J. Am. Chem. Soc.,* 99, 8127, 1977.
181. **House, H. O., Phillips, W. V., and Van Derveer, D.,** Chemistry of carbanions. 34. Alkylation of a 1-decalone enolate with abnormal geometry, *J. Org. Chem.,* 44, 2400, 1979.
182. **Vedejs, E. and Gapinski, D. M.,** Local conformer control in medium-ring olefin epoxidation and osmylation, *J. Am. Chem. Soc.,* 105, 5058, 1983.
183. **Still, W. C. and Novack, V. J.,** Macrocyclic stereocontrol. Total synthesis of ($\pm$)-3-deoxyrosaranolide, *J. Am. Chem. Soc.,* 106, 1148, 1984.
184. **Still, W. C. and Mobilio, D.,** Synthesis of asperdiol, *J. Org. Chem.,* 48, 4785, 1983.
185. **Aoki, M., Tooyama, Y., Uyehara, T., and Kato, T.,** Synthesis of ($\pm$)-asperdiol, a marine anticancer cembrenoid, *Tetrahed. Lett.,* 24, 2267, 1983.

186. **Corey, E. J., Hopkins, P. B., Kim, S., Yoo, S., Nambiar, K. P., and Falck, J. R.,** Total synthesis of erythromycins. 5. Total synthesis of erythronolide A, *J. Am. Chem. Soc.*, 101, 7131, 1979.
187. **Misumi, A., Iwanaga, K., Furuta, K., and Yamamoto, H.,** Simple, asymmetric construction of carbocyclic framework. Direct coupling of dimenthyl succinate with 1,ω-dihalides, *J. Am. Chem. Soc.*, 107, 3343, 1985.
188. **Dorsch, D., Kunz, E., and Helmchen, G.,** Syntheses of dictyopterene B (hormosirene) and its enantiomer via asymmetric S_CN' reactions, *Tetrahed. Lett.*, 26, 3319, 1985.
189. **Colobert, F. and Genet, J.-P.,** Synthesis of (+)-dictyopterene, a constituent of marine brown algae and (+)-dictyopterene C' by chirality transfer of optically active allylic benzoate with palladium(O) catalyst, *Tetrahed. Lett.*, 26, 2779, 1985.
190. **Maruoka, K., Fukutani, Y., Yamamoto, H.,** Trialkylaluminum-alkylidene iodide. A powerful cyclopropanation agent with unique selectivity, *J. Org. Chem.*, 50, 4412, 1985.
191. **Kozikowski, A. P. and Stein, P. D.,** Intramolecular nitrile oxide cycloaddition route to carbocyclics. A formal total synthesis of $PGF_{2\alpha}$, *J. Org. Chem.*, 49, 2302, 1984.
192. **Trost, B. M. and Burgess, K.,** Addition-cyclization catalysed by palladium, *J. Chem. Soc. Chem. Commun.*, 1084, 1985.
193. **Trost, B. M. and Lautens, M.,** An Unusual dichotomy in the regioselectivity of a metal catalyzed versus thermal ene reaction, *Tetrahed. Lett.*, 26, 4887, 1985.
194. **Yamaguchi, M., Tsukamoto, M., and Hirao, I.,** A highly stereoselective synthesis of carbocyclic compounds by the Michael induced intramolecular alkylation. A stereocontrol of extracyclic chiral centers, *Tetrahed. Lett.*, 26, 1723, 1985.
195. **Danheiser, R. L., Bronson, J. J., and Okano, K.,** Carbanion-accelerated vinylcyclopropane rearrangement. Application in a general, stereocontrolled annulation approach to cyclopentene derivatives, *J. Am. Chem. Soc.*, 107, 4579, 1985.
196. **Sakane, S., Maruoka, K., and Yamamoto, H.,** Asymmetric cyclization of unsaturated aldehydes catalyzed by a chiral Lewis acid, *Tetrahed. Lett.*, 26, 5535, 1985.
197. **Dolling, U. H., Davis, P., and Grabowski, E. J. J.,** Efficient catalytic asymmetric alkylations. 1. Enantioselective synthesis of (+)-indacrinone via chiral phase-transfer catalysis, *J. Am. Chem. Soc.*, 106, 446, 1984.
198. **Collum, D. B., Kahne, D., Gut, S. A., DePue, R. T., Mohamadi, F., Wanat, R. A., Clardy, J., and Van Duyne, G.,** Substituent effects on the stereochemistry of substituted cyclohexanone dimethylhydrazone alkylations. An X-ray crystal structure of lithiated cyclohexanone dimethylhydrazone, *J. Am. Chem. Soc.*, 106, 4865, 1984.
199. **Wanat, R. A. and Collum, D. B.,** On the origin of the stereoselectivity of hydrazone alkylations. Investigation of aggregation effects and solution kinetics, *J. Am. Chem. Soc.*, 107, 2078, 1985.
200. **Godleski, S. A. and Villhauer, E.,** Metal catalyzed stereospecific Michael reaction equivalent, *J. Org. Chem.*, 49, 2246, 1984.
201. **Posner, G. H. and Asirvatham, E.,** A short, asymmetric synthesis of natural (−)-methyl jasmonate, *J. Org. Chem.*, 50, 2589, 1985.
202. **Luo, F. T. and Negishi, E.,** Palladium-catalyzed allylation of lithium 3-alkenyl-1-cyclopentenolates-triethylborane and its application to a selective synthesis of methyl (Z) jasmonate, *Tetrahed. Lett.*, 26, 2177, 1985.
203. **Taylor, R. J. K.,** Organocopper conjugate addition — enolate trapping reactions, *Synthesis*, 364, 1985.
204. **Larock, R. C. and Leach, D. R.,** Organopalladium approaches to prostaglandins. 3. Synthesis of bicyclic and tricyclic 7-oxaprostaglandin endodperoxide analogues via oxypalladation of norbornadiene, *J. Org. Chem.*, 49, 2144, 1984.
205. **Bruno, G., Siv, C., Pfeifer, G., Triantaphylides, C., Denis, P., Mortreus, A., and Pettit, F.,** Threophos: A new chiral aminophosphine phosphonite (AMPP) ligand highly efficient in asymmetric hydrovinylation of cyclohexa-1,3-diene catalyzed by nickel complexes, *J. Org. Chem.*, 50, 1781, 1985.
206. **Heumann, A., Reglier, M., and Waegel, B.,** Oxidation mit Palladiumsalzen: stereo- und regiospezifische Acetoxylierung von 4-Vinylcyclohexenderivaten, *Angew. Chem. Suppl.*, 922, 1982.
207. **Bäckvall, J.-E., Vågberg, J., and Nordberg, R. E.,** Palladium catalyzed 1,4-acetoxy-trifluoroacetoxylation of 1,3-dienes, *Tetrahed. Lett.*, 25, 2717, 1984.
208. **Bäckvall, J.-E., Nyström, J.-E., and Nordberg, R. E.,** Stereo- and regioselective palladium-catalyzed 1,4-acetoxy-chlorination of 1,3-dienes. 1-Acetoxy-4-chloro-2-alkenes as versatile synthons in organic transformations, *J. Am. Chem. Soc.*, 107, 3676, 1985.
209. **Bäckvall, J.-E., Byström, S., and Nordberg, R. E.,** Stereo- and regioselective palladium-catalyzed 1,4-diacetoxylation of 1,3-dienes, *J. Org. Chem.*, 49, 4619, 1984.
210. **Bäckvall, J.-E.,** Palladium in some selective oxidation reactions, *Accounts Chem. Res.*, 16, 335, 1983.
211. **Byström, S. E., Aslanian, R., and Bäckvall, J.-E.,** Synthesis of protected allylamines via palladium-catalyzed amide addition to allylic substrates, *Tetrahed. Lett.*, 26, 1749, 1985.

212. **Deardorff, D. R., Myles, D. C., and MacFerrin, K. D.,** A palladium-catalyzed route to mono- and diprotected cis-2-cyclopentene-1,4-diols, *Tetrahed. Lett.*, 26, 5615, 1985.
213. **Trost, B. M. and Angle, S. R.,** Palladium-mediated vicinal cleavage of allyl epoxides with retention of stereochemistry: a cis hydroxylation equivalent, *J. Am. Chem. Soc.*, 107, 6123, 1985.
214. **Danishefsky, S. and Bednarski, M.,** On the acetoxylation of 2,3-dihydro-4-pyrones: a concise, fully synthetic route to the glucal stereochemical series, *Tetrahed. Lett.*, 26, 3411, 1985.
215. **Fristad, W. E. and Peterson, J. R.,** Manganese(III)-mediated γ-lactone annulation, *J. Org. Chem.*, 50, 10, 1985.
216. **Corey, E. J. and Gross, A. W.,** Carbolactonization of olefins under mild conditions by cyanoacetic and malonic acids promoted by manganese(III) acetate, *Tetrahed. Lett.*, 26, 4291, 1985.
217. **Brown, H. C., Jadhav, P. K., and Bhat, K. S.,** Asymmetric synthesis of the diastereomeric 1-(2-cyclohexenyl)-1-alkanols in high optical purity via a stereochemically stable allylic borane, B-2-cyclohexen-1-yldiisopinocamphenylborane, *J. Am. Chem. Soc.*, 107, 2564, 1985.
218. **Brown, H. C., Imai, T., Desai, M. C., and Singaram, B.,** Chiral synthesis via organoboranes. 3. Conversion of boronic esters of essentially 100% optical purity to aldehydes, acids, and homologated alcohols of very high enantiomeric purity, *J. Am. Chem. Soc.*, 107, 4980, 1985.
219. **Zoretic, P. A., Chambers, R. J., Marbury, G. D., and Riebiro, A. A.,** Stereospecific synthesis of secondary allylic alcohols: Selenoxide chemistry, *J. Org. Chem.*, 50, 2981, 1985.
220. **Saddler, J. C. and Fuchs, P. L.,** Enantiospecific syntheses of γ-substituted enones: Organometallic S_N2' conjugate-addition reactions of epoxy vinyl sulfones, *J. Am. Chem. Soc.*, 103, 2112, 1981.
221. **Hutchinson, D. K. and Fuchs, P. L.,** Stereoselective, chemodirected formal S_N2' addition of organometallic reagents to β'-amino cyclopentenyl sulfone derivatives, *J. Am. Chem. Soc.*, 107, 6137, 1985.
222. **Itoh, T., Kaneda, K., and Teranishi, S.,** Unusual stereochemical course of vanadium-catalysed epoxidation of medium-ring allylic alcohols, *J. Chem. Soc. Chem. Commun.*, 421, 1976.
223. **Dehnel, R. B. and Whitham, G. H.,** Stereochemical aspects of the vanadium-catalysed epoxidation of conformationally biased cyclohex-2-en-1-ols by alkyl hydroperoxides, *J. Chem. Soc. Perkin Trans.*, 1, 953, 1979.
224. **Chamberlain, P., Roberts, M. L., and Whithman, G. H.,** Epoxidation of allylic alcohols with peroxyacids. Attempts to define transition state geometry, *J. Chem. Soc. (B)*, 1374, 1970.
225. **Ponsold, K., Schubert, G., Wunderwald, M., and Tresselt, D.,** Stereoselective epoxidation of 14,15-unsaturated estratriene 3-methyl ethers. The *syn*-directive effect of urethanes, *J. Prakt. Chem.*, 323, 819, 1981.
226. **Černý, V. and Kočovský, P.,** Neighboring group participation in hypobromous acid addition to 19-substituted 5α-cholest-1-enes and acid cleavage of their epoxy analogs, *Collect. Czech. Chem. Commun.*, 47, 3062, 1982.
227. **Dryuk, V. G.,** Progress in olefin epoxidation, *Usp. Khim.*, 54, 1674, 1985 (in Russian).
228. **McKittrick, B. A. and Ganem. B.,** *syn*-Stereoselective epoxidation allylic ethers using CF_3CO_3H, *Tetrahed. Lett.*, 26, 4895, 1985.
229. **Corey, E. J., Kang, J., and Kyler, K.,** Activation of methylenetriphenylphosphorane by reaction with t-butyl- or sec-butyllithium, *Tetrahed. Lett.*, 26, 555, 1985.
230. **Overman, L. E. and Sugai, S.,** A convenient method for obtaining trans-2-aminocyclohexanol and trans-2-aminocyclopentanol in enantiomerically pure form, *J. Org. Chem.*, 50, 4154, 1985.
231. **Tömösközi, I., Gruber, L., and Gulácsi, E.,** A simple enantiocomplementary route to prostanoids; inversion of chirality of 2,5-di-hydroxycyclopent-3-enyl-acetic acid lactone derivatives, *Tetrahed. Lett.*, 26, 3141, 1985.
232. **Ogawa, S. and Takagaki, T.,** Total synthesis of (+)-pipoxide and (+)-β-senepoxide and their diene precursors, *J. Org. Chem.*, 50, 2356, 1985.
233. **Curran, D. P. and Rakiewicz, D. M.,** Tandem radical approach to linear condensed cyclopentanoids. Total synthesis of (±)-hirsutene, *J. Am. Chem. Soc.*, 107, 1448, 1985.
234. **Whitten, J. P., McCarthy, J. R., and Whalon, M. R.,** Facile synthesis of the four 3-aminocyclopentane-1,2-diol stereoisomers, *J. Org. Chem.*, 50, 4399, 1985.
235. **Faller, J. W., Murray, H. H., White, D. L., and Chao, K. H.,** Stereoselective syntheses of some cyclohexene derivatives using complexes of molybdenum, *Organometallics*, 2, 400, 1983.
236. **Pearson, A. J. and Khan, M. N. L.,** Stereocontrolled functinalization of cyclohexene using organomolybdenum chemistry, *Tetrahed. Lett.*, 25, 3507, 1984.
237. **Pearson, A. J. and Khan, M. N. I.,** Stereocontrolled lactone synthesis using diene-molybdenum chemistry, *J. Am. Chem. Soc.*, 106, 1872, 1984.
238. **Pearson, A. J., Knole, S. L., and Ray, T.,** Regio- and stereocontrolled functionalization of cycloheptadiene using organoiron and organoselenium chemistry, *J. Am. Chem. Soc.*, 106, 6060, 1984.
239. **Pearson, A. J. and Khan, M. N. I.,** Stereo- and regio-controlled functionalization of cycloheptene using organomolybdenum chemistry, *Tetrahed. Lett.*, 26, 1407, 1985.

240. **Pearson, A. J. and Yoon, J.,** Effect of ligand environment on nucleophile addition to cyclohexadienyliron cations, *Tetrahed. Lett.*, 26, 2399, 1985.
241. **Pearson, A. J. and Ray, T.,** A convergent organoiron approach to steroid synthesis, *Tetrahed. Lett.*, 26, 2981, 1985.
242. **Pearson, A. J. and Chen, B.,** A convenient preparation of dicarbonyl (γ^5-cycloheptadienyl) (triphenylphosphite) iron tetrafluoroborate, a potential macrolide antibiotic precursor, *J. Org. Chem.*, 50, 2587, 1985.
243. **Pearson, A. J., Khan, M. N. I., Clardy, J., and Cun-Heng, H.,** Stereocontrolled functionalization in the cyclohexane ring using organomolybdenum chemistry, *J. Am. Chem. Soc.*, 107, 2748, 1985.
244. **Wilhelm, D., Bäckvall, J.-E., Nordberg, R. E., and Norin, T.,** Stereochemistry and mechanism of palladium(II)-induced ring opening of the cyclopropane in a vinylcyclopropane. Chloro- and oxypalladation of (+)-2-carene, *Organometallics*, 4, 1296, 1985.
245. **Bandra, B. M. R., Birch, A. J., and Kelly, L. F.,** Superimposed lateral control of structure and reactivity exemplified by enantiospecific synthesis of (+) and (−)-gabaculine, *J. Org. Chem.*, 49, 2496, 1984.
246. **Birch, A. J., Raverty, W. D., and Stephenson, G. R.,** Asymmetric synthesis of optically active tricarbonyliron complexes of 1,3-dienes, *Tetrahed. Lett.*, 21, 197, 1980.
247. **Deprés, J.-P., Coelho, F., and Greene, A. E.,** A simple procedure for stereospecific vicinal dicarboxylation of olefins, *J. Org. Chem.*, 50, 1972, 1985.
248. **Cainelli, G., Manescalchi, F., Martelli, G., Panunzio, M., and Plessi, L.,** Inversion of configuration of alcohols through nucleophilic displacement promoted by nitrate ions, *Tetrahed. Lett.*, 26, 3369, 1985.
249. **Maruoka, K., Itoh, T., and Yamamoto, H.,** Methylaluminum bis(2,6-di-tert-butyl-4-alkylphenoxide). A new reagent for obtaining unusual equatorial and anti-Cram selectivity in carbonyl alkylation, *J. Am. Chem. Soc.*, 107, 4573, 1985.
250. **Baldwin, J. R., Jones, R. H., Najera, C., and Yus, M.,** Functionalisation of unactivated methyl groups through cyclopalladation reactions, *Tetrahedron*, 41, 699, 1985.

Chapter 8

CHIRAL SYNTHONS

I. INTRODUCTION

In the preceding Chapters we have dealt with the problems of stereoselective synthesis mostly without paying regard to whether it led to a racemic or an optically active product. As mentioned in the introduction to this book, pure enantiomers of the desired compound could be obtained in two ways: either we perform the whole synthesis with racemic intermediates and eventually resolve the enantiomers in one of the last steps, or we start with an optically active compound. This Chapter is devoted to discussion of the latter method; we will focus on the choice of the starting chiral material (synthon) from the viewpoint of its availability and synthetic utility (for recent reviews see References 1 to 6).

When designing a synthesis of an optically active compound, it is necessary to consider *a priori* in which step it would be most economical to start working with an optically active intermediate. As it will be documented with the following examples, the syntheses starting from chiral synthons may turn out to be more complicated than those working with racemates. In the case of chiral synthons the synthetic possibilities are limited by the supply of starting compounds (the chiral pool), which are often to be transformed by neck-breaking procedures. From a different point of view, it is then difficult to trace the structural features of the starting chiral synthon in the completed molecule! However, recently developed enantioselective Sharpless epoxidation of allylic alcohols[8-10] and other methods opened new avenues in this area. The simplicity of the Sharpless procedure makes it possible to synthesize a variety of optically active compounds to suit precisely a particular synthetic target (for review see References 10 and 11).

In molecules composed of periodically repeating building units, e.g. in peptides, the choice of the chiral synthons is a trivial matter and, conversely, it would be hardly feasible to synthesize a racemate and then attempt to resolve the mixture to enantiomers, even if the compound was a simple tripeptide. The search for suitable chiral synthons is not trivial, however, when one plans a synthesis of a chemically more complex compound (especially a polycyclic one), and then it may be more advantageous to conduct the synthesis with a racemate and sacrifice large parts of the product as the undesired enantiomer after final separation. Nevertheless, an easily available optically active material can be found even for very complex molecules. In steroid chemistry, for instance, an important starting compound has been 3β-acetoxy-5,16-pregnandien-20-one (1), which is available in bulk quantities by a several-step degradation of the natural sapogenine diosgenin. The acetoxyketone 1 has served for preparation of almost all important steroid compounds (Figure 1), namely, progesterone (2), testosterone (3), estrone (4), cortisole (5),[12] aldosterone (6),[13] strophantidin (7)[14] and its 14-deoxyanalog (8),[15] ecdysone (9),[16] batrachotoxine (10),[17] and many others.

The chemist's task is therefore to consider all facts and decide either for a "racemic" synthesis or for a chiral synthon. Obviously, if there is a starting chiral material at hand, the second way will be preferred. In the other case the synthetic economy would favor the "racemic" way. The homochiral way becomes mandatory if one of the goals is to prove by enantioselective synthesis the absolute configuration of the target compound. Then, regardless of synthetic difficulties, it is necessary to start with a compound of a known absolute configuration. In a convergent synthesis, where several building blocks bearing chirality elements are to be connected, the use of chiral synthons is unavoidable as well.

In this Chapter we will present some possibilities of synthesis of optically active compounds from chiral precursors (synthons) mostly of natural origin.

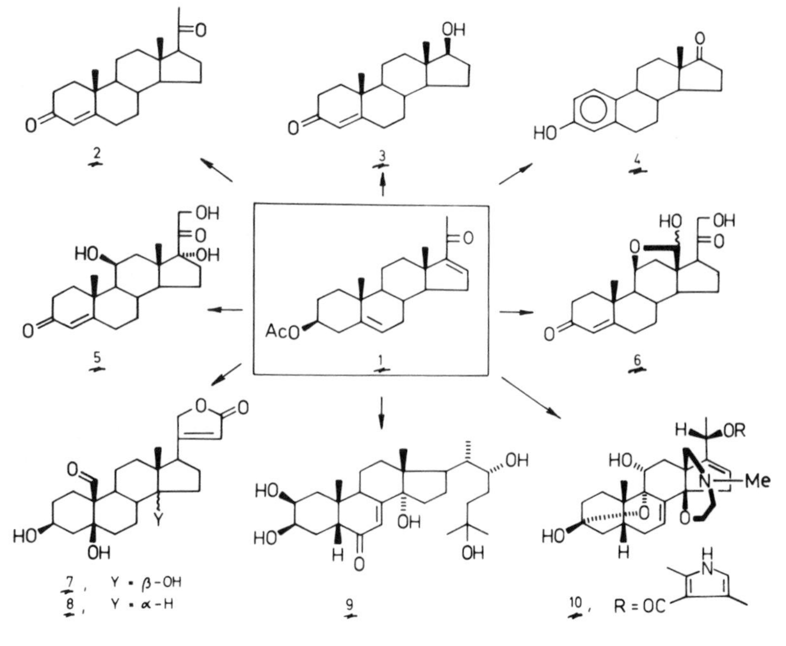

FIGURE 1.

The chiral pool consists mainly of low-molecular weight compounds, such as monoterpenes, amino acids, hydroxy acids, monosaccharides and others, which can, rather universally, be used in total syntheses of various optically active products. We shall focus our attention on the strategy of incorporating the chiral elements into the skeleton and on its fate during the synthesis. More detailed information can be found in the original papers or in recent review articles. For reviews on chiral synthons see References 1 to 5, for use of carbohydrates as chiral synthons see Reference 6, and for synthesis of optically active pheromones see Reference 18.

II. TERPENES

In this Section there will be shown some simple ways of using chiral synthons, namely, ring cleavage in cyclic terpenes, cyclization, ring expansion and contraction, and modifications of the side-chain.

A. Limonene (11)

Limonene occurs in various ethereal oils, particularly in oils of lemon, orange, caraway, dill, and bergamot. The lemon, orange, bergamot, and tangerine peel oil contain (R)-(+)-limonene (11), while South-European turpentine of *Pinus sabininiana, P. pinea, P. canariensis,* and others contains its enantiomer. The optical rotation was found to be $\alpha_D = +123.8°$ (neat) for the (R)-(+)isomer and $\alpha_D = -101.3°$ (neat) for the (S)-enantiomer.[19]

(R)-(+)-Limonene was utilized in the synthesis of (+)-juvabione (19) (Figure 2),[20] a compound inhibiting metamorphosis in insects.[16] Hydroboration of the disubstituted double bond in 11 with the bulky disiamylborane proceeded with excellent regiospecificity yielding the separable diastereoisomers 12 and 13. The alcohol 12 was converted to nitrile 14 which was treated with an organolithium reagent to give the ketone 15. Three-step oxidation of the ring methyl group (15→16→17→18) yielded the acid 18 which was esterified to natural (+)-juvabione (19) with defined absolute configuration on both chirality centers. The same

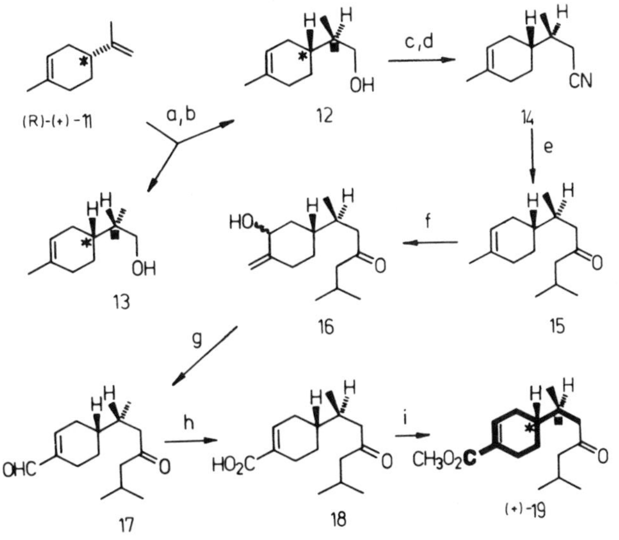

FIGURE 2. (a) $Me_2CHCH(Me)–BH_2$; (b) H_2O_2, OH^-; (c) TsCl; (d) NaCN; (e) Me_2CHCH_2Li; (f) O_2, $h\nu$, Hematoporphyrine; (g) H_2CrO_4; (h) Ag_2O; (i) CH_2N_2.

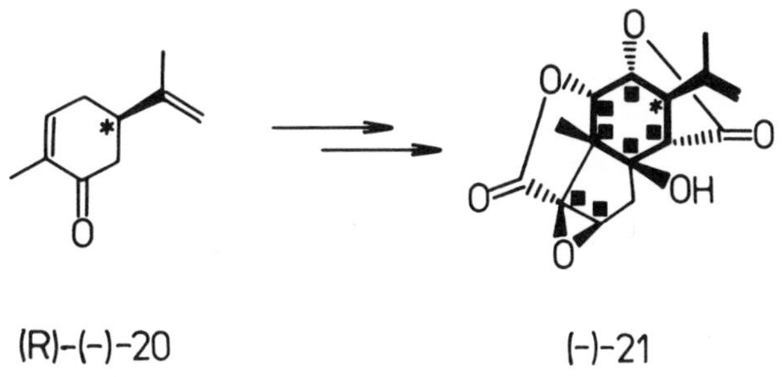

FIGURE 3.

authors[20] prepared for comparison the juvabione enantiomer from (S)-(−)-limonene and thus corrected the previously suggested[21,22] absolute configuration of the natural compound.

(R)-(+)-Limonene was also employed in the synthesis and determination of absolute configuration of (−)-acorenone,[23,24] and in the synthesis of a cyclic analog of a juvenile hormone.[25]

Of compounds derived from limonene, for instance (S)-(−)-perilla aldehyde has been used as a starting material in the synthesis of (+)-phyllantocin, which determined the absolute configuration of the latter.[26]

B. Carvone (20)

Natural (R)-(−)-carvone (20) is obtained from spearmint and kuromoji oil, its enantiomer is contained in caraway and dill seed oils,[19] and in mandarin peel (*Citrus reticulata*).[27] The racemate was found in gingergrass oil.[28] The highest value of optical rotation for (R)-carvone was measured as $\alpha_D = -62.5°$ (neat) and for its enantiomer as $\alpha_D = +61.2°$ (neat).[19] Optically pure carvone is available synthetically from (+)-limonene[26] or from (−)-α-pinene.[30,31]

R-(−)-Carvone was utilized by Corey and Pearce[32] as a chiral synthon in the synthesis of (−)-picrotoxinine (21) (Figure 3) and picrotine.[33] The mixture of these substances, known

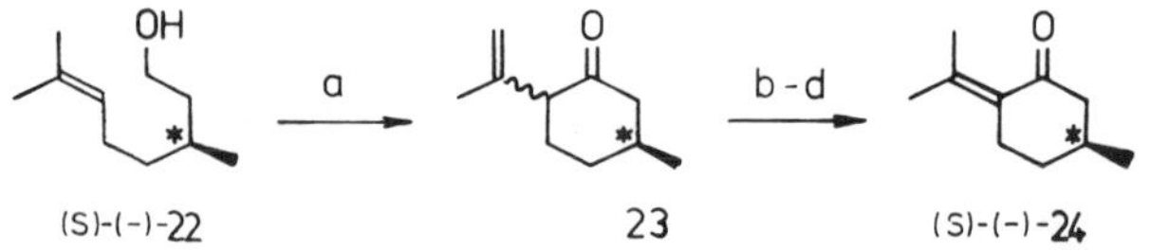

FIGURE 4. (a) $CrO_3 \cdot C_5H_5N \cdot HCl$; (b) NaOH, EtOH; (c) H_2NNH–CO–NH_2, crystallization; (d) CH_3COCH_2OH.

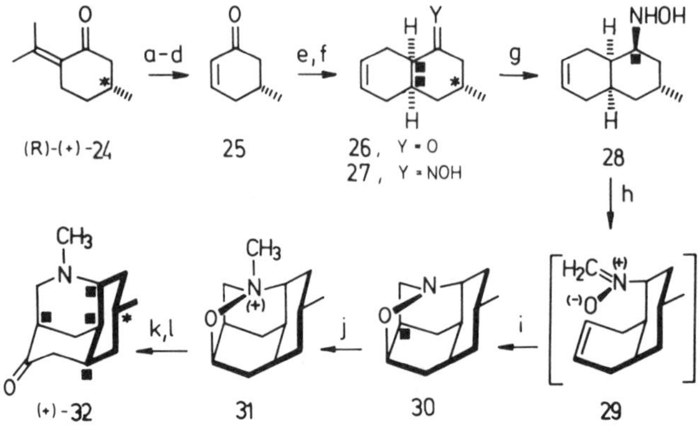

FIGURE 5. (a) OH^-; (b) $(i\text{-}Pr)_2NLi$, $(PhS)_2$; (c) m–Cl–$C_6H_4CO_3H$; (d) 65°C, CCl_4; (e) CH_2=CH–CH=CH_2, $SnCl_4$; (f) NH_2OH; (g) $NaBH_3CN$; (h) CH_2=O; (i) 115°C; (j) CH_3OSO_2F; (k) $LiAlH_4$; (l) CrO_3, H_2SO_4, Me_2CO.

as ''picrotoxin'', was isolated from berries of *Menispermum coculus*. Both compounds are antagonists of γ-amino butyric acid in nerve synapsis. Figure 3 shows the fate of the carbon atoms originating from the carvone molecule, and simultaneously illustrates how the single chirality center in the starting compound (*) induced the stereoselective formation of the other seven centers (■).

(R)-(−)-Carvone (20) was further used for the synthesis of optically active 4,11-epoxy-*cis*-eudesmane,[34] the defense secretion from termites *Amitermes evuncifer*.

C. Pulegone (24)

Pulegone is found in oils derived from plants of the *Labiatae* family. In nature it occurs in both (+) and (−) forms, but its optical purity is not high. Dextrorotatory pulegone is obtained from pennyroyal oils from *Mentha pulegium, M. longifolia,* and others.[19] Laevorotatory pulegone is the major constituent of *Agastocha formosanum* oil.[35]

Corey worked out a method for preparation of optically pure S-(−)-pulegone (24) from (S)-(−)-citronellol (22), the optical purity of which is only 85% (Figure 4).[35] This method can also afford R-(+)-pulegone, if the starting compound is (+)-citronellol. A different method was reported by Japanese authors.[36]

Oppolzer and Petrzilka[37] employed R-(+)-pulegone as a chiral synthon in their synthesis of luciduline. This alkaloid was isolated from *Lycopodium lucidulum* and assigned[38] structure 32. The synthesis by the Swiss authors[37] unambiguously proved the absolute configuration of 32 (Figure 5): (R)-(+)-Pulegone was first cleaved by retroaldolization and then converted to the enone 25. The second carbocyclic ring was constructed by Lewis acid catalyzed Diels-Alder reaction of 25 with butadiene. The third ring was closed by the following procedure: the ketone 26 was converted to nitrone 29 which thermally cyclized to compound 30; after N-methylation the nitrogen-oxygen bond was cleaved with lithium aluminum hydride reduction and the resulting alcohol was oxidized to (+)-luciduline (32), identical with the

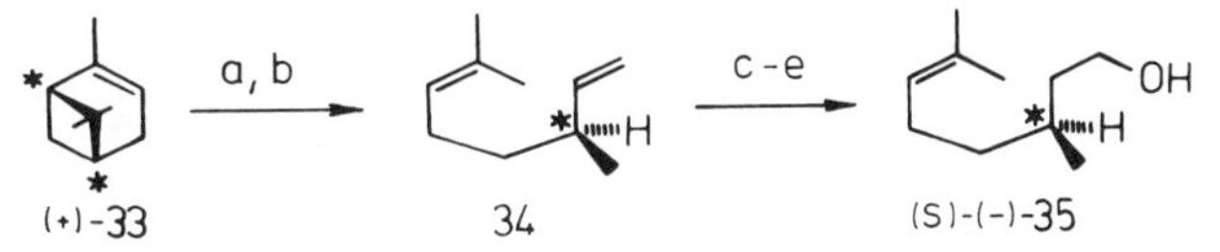

FIGURE 6. (a) H_2, Raney-Ni, 80°C, 50 atm; (b) 385—450°C; (c) $(i\text{-}Bu)_2AlH$ 100°C; (d) O_2; (e) H_2O.

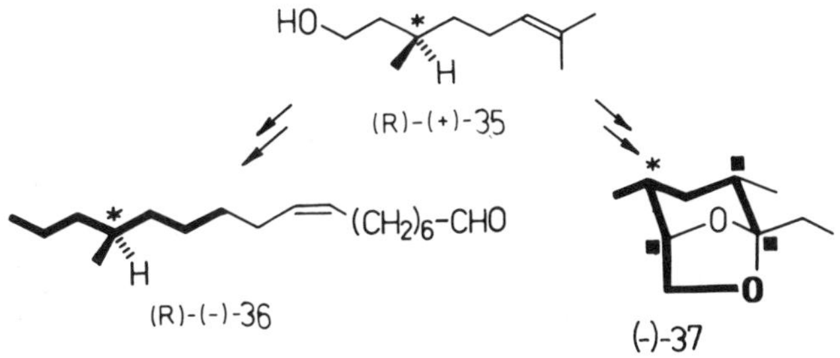

FIGURE 7.

natural compound. The chiral center of pulegone (*) was preserved during the whole synthesis and, in turn, it determined the configuration of the other four centers (■).

Pulegone also served for the preparation of chiral intermediates in the synthesis of a series of pheromones,[39-41] (R)-(−)-sarcomycine,[42] (−)-prezizaene,[43,44] Corey lactone,[45] actinidine,[46,47] retigeranic acid,[48] and citronellol and citronellic acid.[49]

D. β-Citronellol (35)

(R)-(+)-β-Citronellol (35) is obtained from essential oils of lemon and orange coming from Java and Ceylon or from other sources.[19] Its enantiomer occurs in the rose and geranium oils of German and Soviet origin.[19] The optical purity of either form is not too high and therefore a synthetic procedure was developed starting from α-pinene (33), which furnishes citronellol in 60% yield and 90% optical purity. (S)-(−)-Citronellol (35) is prepared from (+)-α-pinene (33) (Figure 6) and, similarly, the R-(+)-enantiomer is obtained synthetically from (−)-α-pinene[50,51] and (+)-pulegone,[49] For other syntheses see References 36 and 52 to 56. The optical activity of the (R)-isomer was determined as $\alpha_D = +5.22°$, that of the (S)-enantiomer is $\alpha_D = -4.76°$.[19]

(R)-(+)-Citronellol (35) was utilized as a chiral synthon in the synthesis of acyclic pheromones possessing one chirality center.[18] For instance, the pheromone of the female Dermestid beetle (*Trogoderma inclusum*) 36 was prepared in this way (Figure 7).[57] Its enantiomer was also prepared from (S)-(−)-citronellol and turned out to be 250 times less active then the (R)-form.[57]

(−)-Multistriatin (37), one of the components of the aggregation pheromone of the European elm bark beetle (*Scolytus multistriatus*), was prepared from (R)-(+)-citronellol, too, (Figure 7)[58] but the optical purity achieved was only 40%. Another component of the same pheromone, *threo*-4-methylheptan-3-ol, was synthesized from methyl (R)-citronellate and in this case the synthesis succeeded in establishing the absolute configuration of the product as (3S,4S).[59,60]

(−)-Citronellene, prepared from citronellol, was employed by Ireland et al. in the synthesis of antibiotic lasalocide A,[61,62] and by Anderson for the synthesis of Red scale pheromone.[63]

E. β-Pinene (38)

Pinenes (α- and β-) are the main components of turpentine. They differ in the position

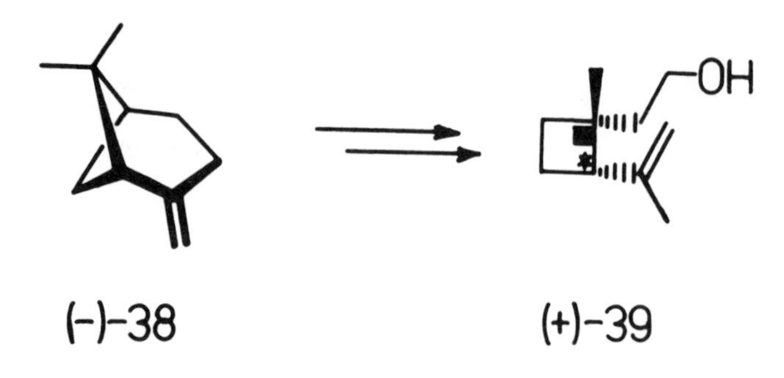

FIGURE 8.

of the double bond and can be separated by distillation. In nature they occur in both (+) and (−) form. European turpentine contains mostly the (−)-enantiomer, while American turpentine contains the other form. The optical activity of the dextrorotatory enantiomer is $\alpha_D = +51.4°$, the lavorotatory form has $\alpha_D = -51.28°$. For preparation of optically pure (+) and (−)-α-pinene see Reference 63a.

Both pinene isomers can be employed as a source of the four-membered ring, as illustrated with the synthesis of (+)-grandisol (39) from (−)-β-pinene (Figure 8).[64] Grandisol represents the main constituent of the four-component synergist substance of the boll weevil (*Anthronomus grandis*) pheromone.[65] The boll weevil is one of the most important insect pests in the U.S. in terms of crop loss economics; more than 75% of all cotton losses caused by insects have been attributed to it, and therefore the synthesis of grandisol, which at the same time led to elucidation of its absolute configuration, was highly desirable. Of the two original chirality centers (*) in the starting pinene, one is retained in the final product, while an additional center had to be created (■) during the synthesis.

Pinene and its derivatives can also be conceived as precursors of compounds containing three-membered rings. The cyclohexane ring in pinene can be contracted by the Favorskii rearrangement of a proper bromo ketone, as it was demonstrated by the synthesis of methyl-(+)-*trans*-chrysanthemate, starting from (+)-α-pinene.[66] Pinene and its derivatives have also been utilized in the synthesis of tetrahydrocannabinol;[67] other terpenes have served for this purpose as well.[68-71]

F. β-Carene (40)

The hydrocarbon (+)-β-carene (40) is a constituent of turpentine from *Pinus longifolia*, where its content amounts to 30 to 40%, of *P. silvestris* and others. The optical activity was determined as $\alpha_D = +7.68°$.[19] In the synthesis of natural products, carene is an important precursor for compounds containing a three-membered ring with geminal methyl groups. The following example of a synthesis of *trans*-chrysanthemic acid (45, Figure 9)[72] nicely illustrates the case in which both enantiomers of the same compound can be prepared from a single enantiomer of the chiral precursor! Ozonolysis of (+)-β-carene afforded the keto aldehyde 41 which was converted, by two different synthetic procedures, to enantiomers of *trans*-chrysanthemic acid (45). Note that the intermediate lactones 44 and 49 in Figure 9 are actually enantiomers of the same compound. Moreover, two original chirality centers in carene remained intact in the course of the synthesis, save for the epimerization of the carboxyl group in the last step.

Of carene derivatives, (+)-*trans*-2-caren-oxide was used in the synthesis of tetrahydrocannabinol.[71]

G. Camphor (51)

The source of natural camphor is the camphora tree (*Cinnamorum camphora*). The absolute

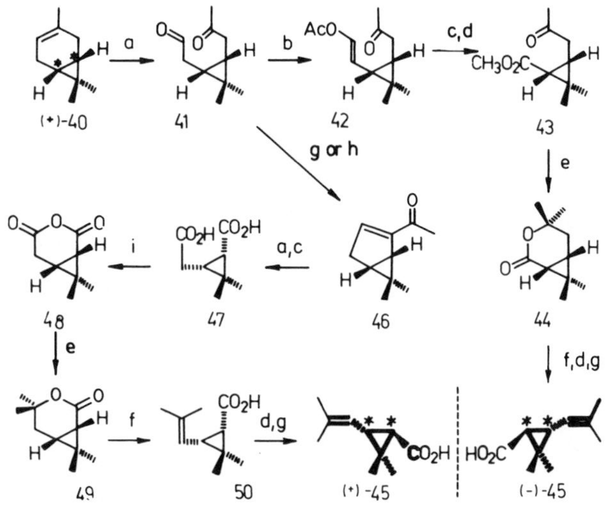

FIGURE 9. (a) O_3; (b) Ac_2O, AcONa; (c) O_3, H_2O_2; (d) CH_2N_2; (e) CH_3MgI; (f) H_2SO_4; (g) Base; (h) H^+; (i) Ac_2O, 140°C.

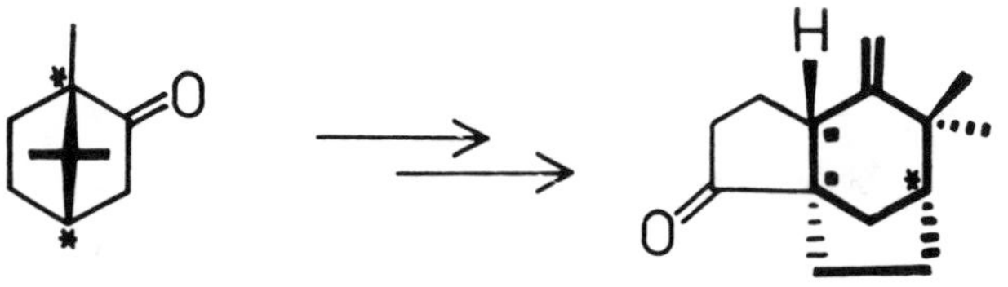

FIGURE 10.

configuration was determined by German authors.[73] Camphor coming from Java and Sumatra, as well as that from Brazil, Japan, and China, is dextrorotatory. On the other hand, some sorts of Soviet camphor are lavarotatory. The racemate is available synthetically. (−)-Camphor is now easily available in bulk by oxidation of (−)-borneol.[74]

(−)-Camphor (51) was employed as a chiral synthon in the synthesis of (−)-khusimonine (52)[75] and for the preparation of an optically active intermediate in the total synthesis of steroids.[74,76] In the khusimonine synthesis (Figure 10), one of the camphor chirality centers (*) vanishes, while two new centers are formed (■). Optically active camphor was also utilized in the synthesis of patchouli alcohol[77] and vitamine B_{12}, as will be elaborated in Section VI.G.

III. AMINO ACIDS

The prime importance of amino acids as chiral synthons undoubtedly lies in the field of peptide synthesis. Here we wish to outline their use in synthesis of different compounds such as alkaloids, pheromones, antibiotics, etc.

A. Norvaline (53)

Swiss authors[78] employed (R)-(−)-norvaline (53) in the total synthesis of pumiliotoxin C, a toxic compound isolated from the skin of the poison dart frog, *Dendrobates pumilio*

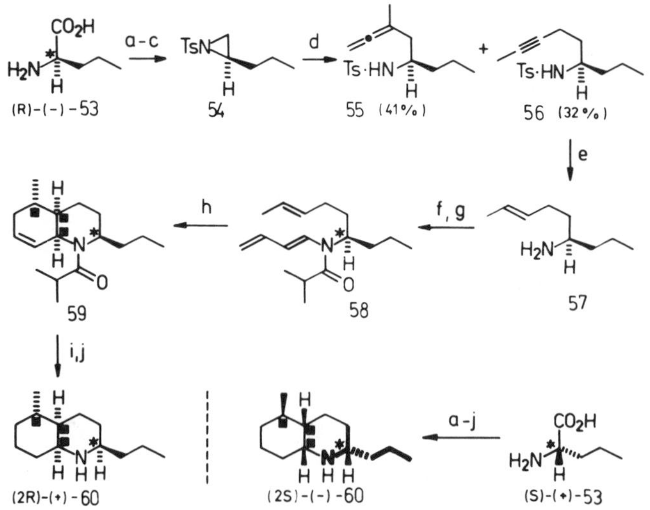

FIGURE 11. (a) $LiAlH_4$; (b) TsCl, C_5H_5N; (c) KOH, CH_3OH; (d) CH_3–C≡C–CH_2MgBr; (e) Na, NH_3; (f) CH_3CH=CH–CHO; (g) $Me_2CHCOCl$; (h) 230°C, toluene, 2% $CH_3CON(SiMe_3)_2$; (i) H_2, Pd; (j) $(i–Bu)_2AlH$.

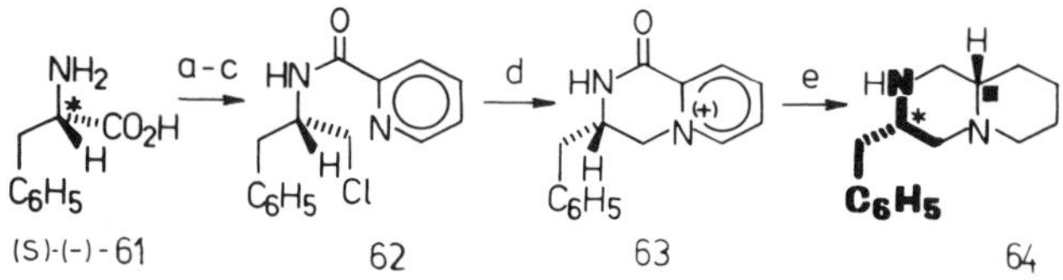

FIGURE 12. (a) Reduction; (b) α-C_5H_4N–CO_2CH_3; (c) $SOCl_2$; (d) heat; (e) H_2, Pt.

and *D. arratus,* living in Costa Rica.[79,80] The compound 60, obtained synthetically, was identical with the natural product except for the opposite optical rotation (Figure 11). The synthesis starting from enantiomeric (S)-(+)-norvaline (+)-(53) then furnished the natural (−)-pumiliotoxin C, which corrected the previously (and incorrectly) assigned[79-82] absolute configuration of the toxin.

B. Phenylalanine (61)

(S)-(−)-Phenylalanine (61) served as a chiral synthon in the synthesis of verruculotoxin (64), produced by the strain *Pennicilium verruculosum* (Figure 12).[83]

C. Glutamic Acid (65)

Natural (S)-(+)-glutamic acid (65) and its enantiomer have found wide-spread use in syntheses of optically active insect pheromones and related compounds. Beside being easily available, glutamic acid can be converted to lactone 68 with almost complete retention of configuration (Figure 13).[84] Decomposition of the diazonium salt 66 proceeds with participation by the neighboring carboxyl group to form the transient α-lactone 67 which is cleaved by the second carboxyl giving rise to lactone 68. In this way the double inversion of configuration at the α-carbon atom results in 95% retention of the original configuration. A similar course of diazotation with subsequent decomposition of the diazonium salt has been observed for leucine,[85,86] α-aminobutyric acid,[87] and methionine,[88] wherein the corresponding α-lactone intermediates are cleaved by water with the same stereochemical outcome.

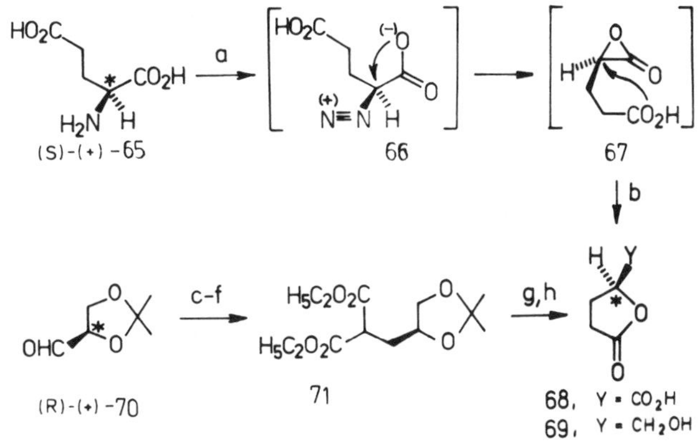

FIGURE 13. (a) HNO_2; (b) $BH_3 \cdot SMe_2$; (c) $LiAlH_4$; (d) TsCl, C_5H_5N; (e) NaI, Me_2CO; (f) $Na \cdot CH(CO_2Et)_2$; (g) NaOH, CH_3OH, H_2O; (h) H_2SO_4.

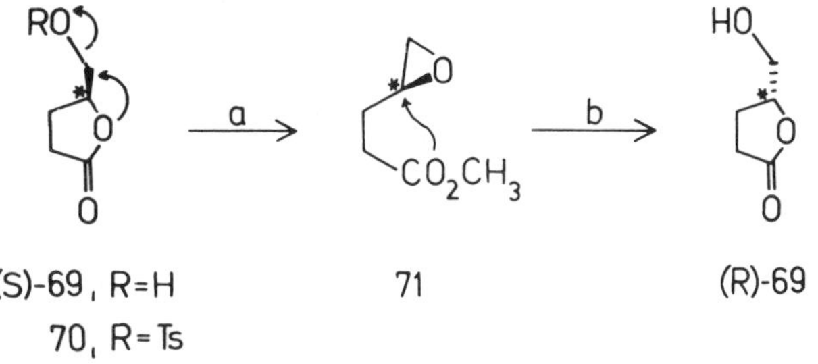

FIGURE 14. (a) CH_3ONa, CH_3OH; (b) CF_3CO_2H, $-10°C$.

The free carboxylic group in 68 can be selectively reduced to produce the (S)-hydroxylactone 69 (Figure 13), which is also frequently used a chiral synthon (*vide infra*). The lactone 69 may be prepared from (R)-(+)-glyceraldehyde (70) as well.[89] The latter procedure is even more advantageous, for the chirality center is not involved in the synthetic transformations and the product is practically optically pure. The enantiomeric (R)-69 can be obtained from the corresponding enantiomers of glutamic acid or glyceraldehyde,[89] or by a three-step epimerization of 69, via tosylate 70 and epoxide 71 (Figure 14).[90]

Examples of the synthetic use of glutamic acid in the chemistry of insect products are depicted in Figure 15. Hydride reduction of 68 affords the triol 72 whose vicinal diol moiety can be protected as an isopropylidene derivative, which is suitable for further syntheses.[91,94]

Hydrogenation of the acid chloride 73 gives rise to the aldehyde 74 which on Wittig coupling affords the unsaturated lactone 75,[91] the sexual pheromone of the Japanese beetle (*Popilia japonica*). Another synthetic sequence, i.e., coupling of the chloride 73 with decylcadmium (e), reduction of the ketone function followed by protection of the hydroxyl group as a tetrahydropyranyl ether, and selective reduction of the lactone to ketal followed by Wittig reaction (h) and hydrogenation, gave rise to derivative 77 which was converted to (+)-dispalure (78),[92] the sexual attractant of the female gypsy moth (*Portheria dispar*). The synthetic product 78 contained only 5.8% of the other enantiomer.[93]

The hydroxylactone 69, prepared from 68 by borane reduction, was converted to tosylate 70 which served as a starting compound in numerous syntheses of compounds biologically active in insects.[91,94] Thus, coupling 70 with lithium dimethylcuprate afforded (R)-(+)-lactone 79, a component of the sexual pheromone of the dermestid beetle (*Trogoderma glubrum*).[95] The corresponding enantiomer was obtained by enzymatic synthesis.[96] Lactone 79 was further reduced to the corresponding acetal which was coupled with the organometallic

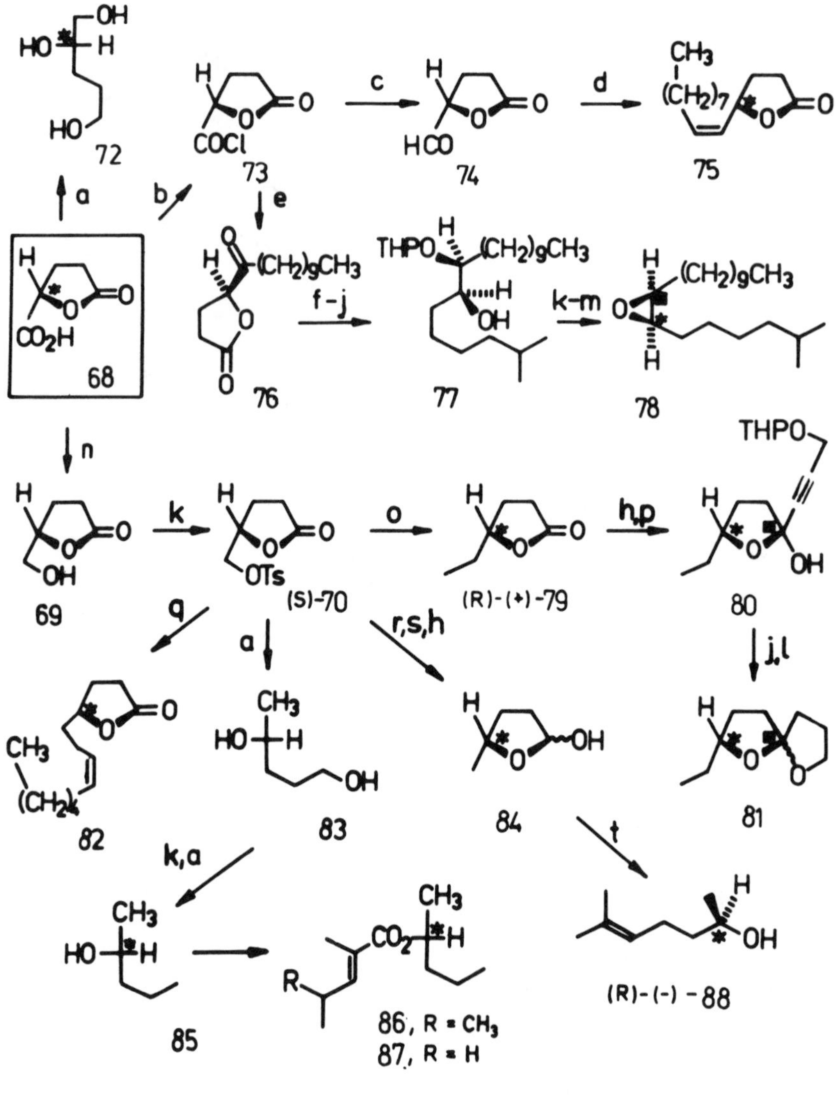

FIGURE 15. (a) $LiAlH_4$; (b) $(COCl)_2$; (c) H_2, $Pd/BaSO_4$; (d) $Ph_3P{=}CH(CH_2)_7CH_3$; (e) $(C_{10}H_{21})Cd$; (f) $NaBH_4$; (g) Dihydropyran; (h) $(i\text{-}Bu)_2AlH$; (i) $Ph_3P{=}CH{-}CHMe_2$; (j) H_2, Pt; (k) TsCl; (l) H_3O^+; (m) KOH; (n) $BH_3{\cdot}SMe_2$; (o) Me_2CuLi; (p) $LiC{\equiv}C{-}CH_2OTHP$; (q) $(Me_2CH{=}CH{-}CH_2)_2CuLi$; (r) LiI, Me_2CO; (s) Raney –Ni, EtOH, $CaCO_3$; (t) $Ph_3P{=}CHMe_2$.

reagent (p) to give 80. Hydrogenation of the latter, followed by acidic cyclization, gave rise to a mixture of diastereoisomeric ketals 81.[97] This mixture (in a 1:1 ratio) is the active component of chalcogran, the aggregation pheromone of the beetle *Pityogeneses chalcographus.*[98]

The reaction of the tosylate 70 with lithium-bis-(octenyl)-cuprate afforded another pheromone 82.[91,99] Hydride reduction of 70 led to diol 83 which was converted in two steps to alcohol 85 and the latter was esterified to 86 and 87.[91] Both esters are components of the pheromone of the beetle *Rhyzopertha dominica.*

In another synthesis, (S)-tosyloxylactone 70 was converted to acetal 84 which yielded on Wittig reaction the R-(−)-alcohol 88.[100] An analogous procedure starting from R-(−)-glutamic acid led to the (S)-(+)-enantiomer of 88. The mixture of the enantiomers (88) in a ratio (R):(S) = 35:65 is known as sulcatol, an aggregation pheromone excreted by the

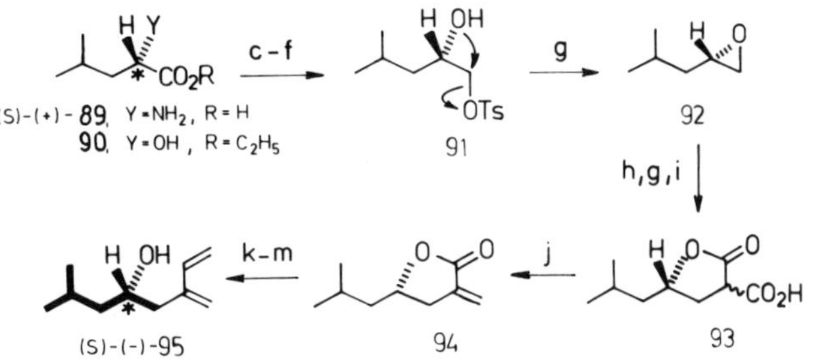

FIGURE 16. (a) HNO_2; (b) EtOH, H^+(89→90); (c) Dihydropyran, H^+; (d) $LiAlH_4$; (e) TsCl, C_5H_5N; (f) AcOH, THF, H_2O; (g) KOH; (h) $CH_2(CO_2Et)_2$, EtONa; (i) H_3O^+; (j) $CH_2{=}O$, Et_2NH; (k) $(PhSe)_2$, $NaBH_4$; (l) $(i\text{-}Bu)_2AlH$; (m) $Ph_3P{\cdot}CH_3Br$, NaH, Me_2SO.

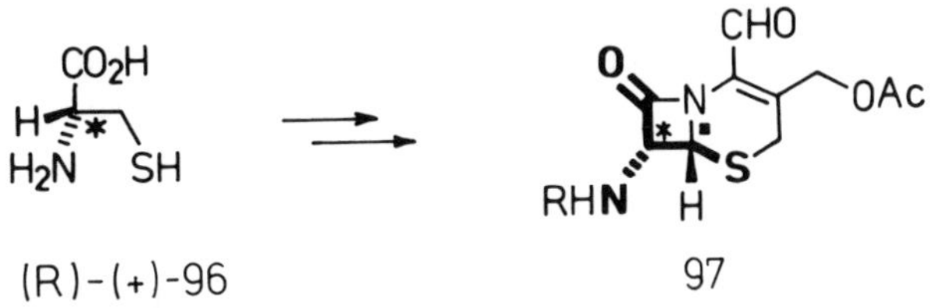

FIGURE 17.

male ambrosia beetle (*Gnathoricus sulcatus*).[101] Laboratory and field experiments proved that sulcatol is active only when containing both enantiomers. Suprisingly, the racemate is even more active than the natural mixture![102]

Glutamic acid has been utilized in synthesis and determination of absolute configuration of various other natural products, e.g. (−)-steganone,[103] (−)-isosteganone,[104] (+)-*trans*-burserane,[104] (−)-steganacin,[105] (+)-quebrachamine,[106] (−)-velbamine,[107] and others.

D. Leucine (89)

Mori[85,86] has used natural (S)-(+)-leucine (89) for the synthesis of (S)-(−)-ipsenol (95), (Figure 16) in a similar way as has been done for glutamic acid. (S)-(+)-Leucine (89) of 89% optical purity was converted to the ethyl ester of the corresponding hydroxy acid (reactions a and b in Figure 16) with retained configuration at C-2. The monotosylate 91, prepared therefrom, was treated with base affording the epoxide 92 which was condensed with diethyl malonate. The resulting lactone 93 was converted in several steps to (S)-(−)-ipsenol 95, a component of the sexual attractant of the bark beetle *Ips paraconfusus*. This pheromone consists of both ipsenol enantiomers and (+)-*cis*-verbenol. Neither individual component affects the beetle's behavior; they are active only in a mixture.[108] By contrast, the related spined engraver beetle (*Ips grandicolis*) aggregates only in response to (S)-(−)-ipsenol.[109]

Of leucine derivatives, (S)-leucinamide was used by Bláha et al. as a chiral synthon in the synthesis and determination of the absolute configuration of 5,8-diazatricyclo [6.3.0.0^{1,5}]undecan-4,9-dione.[110]

E. Cysteine (96)

Natural (R)-(+)-cysteine (96) was employed by Woodward et al.[111] in their synthesis of cephalosporin C (97). Figure 17 illustrates the fate of the cysteine atoms and the formation of a new chirality center (■).

Uskoković et al.[112] started with (R)-(+)-cysteine (96) in their synthesis of D-(+)-biotin.

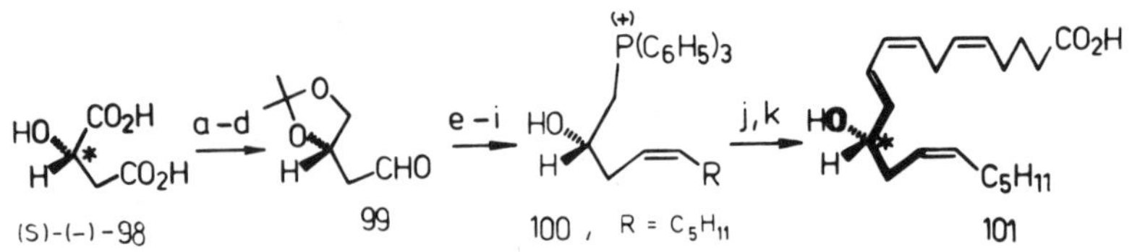

FIGURE 18. (a) $CH_2{=}C(CH_3)(OCH_3)$; (b) $LiAlH_4$, THF; (c) BF_3, 23°C; (d) $CrO_3{\cdot}2C_5H_5N$; (e) $Ph_3P{=}CH{-}C_6H_{11}$, THF; (f) 1 M-HCl, THF, 47°C; (g) $Me_3C_6H_2{-}SO_2Cl$, C_5H_5N, 0°C 1 hr; (h) NaI, Me_2CO, 25°C; (i) Ph_3P, C_6H_6, 40°C; (j) CH_3Li, HMPA, then THF $O{=}CH{-}CH{=}CH{-}CH_2{-}CH{=}CH{-}(CH_2)_3{-}CO_2CH_3$; (k) OH^-.

Of other amino acids, we note the use of (R)-(−)-α-aminobutyric acid in the synthesis of chalcogran, the principal pheromone of Kupferstecher (*Pityogeneses chalcographus*) pesting on Norway spruce.[87] L-Hydroxyproline played a role in the synthesis of anthramycin methylether.[113] (−)-Deoxyprosophylline[114] and dihydrosphingosine[115] were prepared from (S)-serine; (S)-lysine has served as a source of optically active α-aminoadipic acid of the same configuration.[116] Tyrosine[117] and (−)-DOPA[118] have also found some use as chiral synthons.

IV. HYDROXY ACIDS

As we have seen in the preceding section, the use of amino acids as chiral synthons often necessitated converting the amino group to a hydroxyl with a defined stereochemistry. Nature offers an alternative solution by providing us with chiral hydroxy acids, namely, malic, lactic, and tartaric acid which are easily accessible in an optically active form. While the first two acids contain by one chirality center, tartaric acid represents a potential synthon of two vicinal centers. The versatility of these hydroxy acids is due to the presence of the carboxylic groups which can be modified, often selectively, and this fact forms the strategic basis for most syntheses.

A. Malic Acid (98)

Malic acid occurs in nature as an L-enantiomer [(S)-(−)-configuration]; it is prepared by fermentation. The D-enantiomer can be obtained by resolving the synthetic racemate.[119] The natural (S)-(−)form has found use as a chiral synthon in the synthesis of prostaglandins and related compounds;[1,121,122] the absolute configuration of alkaloid (+)-sesbanine[123] was determined by means of the chiral center provided by (S)-(−)-malic acid. Other examples can be found in Seebach's papers.[124-126]

For illustration we present here the Corey synthesis of (S)-12-hydroxy-5,8,14-*cis*-10-*trans*-eikosatetraenic acid (101), starting from (S)-(−)-malic acid (Figure 18).[122] The acid 101 is a human metabolite of arachidonic acid. The synthesis enabled the authors to confirm the structure of 101 and establish its absolute configuration. Practically, the synthesis provided a sample of the acid in an amount necessary for biological studies.[1] The synthesis is based on a sequential elongation of the carboxyl ends in malic acid by Wittig couplings as strategic steps, without affecting the original chiral center. Besides the selective formation of Z-double bonds it was necessary to differentiate the reactivity of carboxyl groups in the starting malic acid by a proper combination of protective groups and chemical transformations.

B. Lactic Acid (102)

Lactic acid obtained by fermentation is laevorotatory. In contrast, the enantiomeric form occurs in muscles. Both enantiomers are available by resolving the synthetic racemate.[127,128] Recently, Brown devised a procedure for the preparation of optically pure t-butyl ester of (S)-(−)-lactic acid by asymmetric reduction of t-butyl pyruvate.[129]

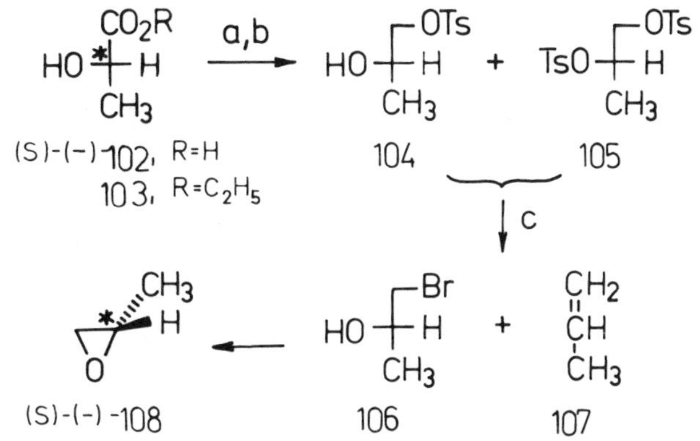

FIGURE 19. (a) $LiAlH_4$; (b) TsCl, C_5H_5N, THF, −30°C; (c) LiBr, Me_2CO; (d) 50%–KOH, H_2O.

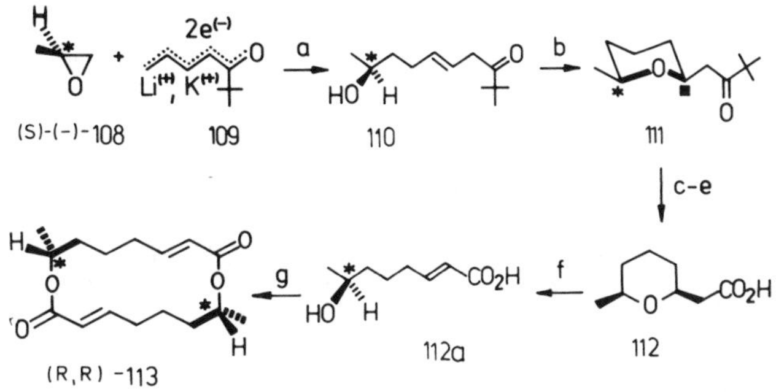

FIGURE 20. (a) THF, −78 → 0°C; (b) CH_3ONa, CH_3OH, THF, 20°C; (c) $NH_2OH{\cdot}HCl$, AcONa, CH_3OH, reflux; (d) PCl_5, Et_2O; (e) 50%–KOH, Ethyleneglycol, 150°C; (f) $(i\text{–}Pr)_2NLi$, −78 → 0°C; (g) $EtO_2C\text{–}N{=}N\text{–}CO_2Et$, Ph_3P, THF, toluene.

(S)-(−)-Lactic acid can be rather easily converted to (S)-(−)-1,2-epoxypropane (108), a valuable three-carbon chiral synthon (Figure 19).[130] Ethyl lactate (103) is reduced to a diol which affords on tosylation the monotosylate 104, together with a small amount of ditosylate 105. The undesired ditosylate need not be separated, for in the next step it is converted to propylene, while the monotosylate 104 gives the bromohydrin 106. The latter is then cyclized in alkaline medium to (S)-(−)-epoxide 108 (α_D = −6.9°).[130,131]

(S)-(−)-Propylene oxide 108 was employed by Seebach in his ingenious synthesis of (R,R)-(−)-dideoxypyrenophorine (113) (Figure 20) a compound exerting antibacterial activity.[132] Reaction of 108 with the dianion 109 yielded the keto alcohol 110 which was converted to the tetrahydropyran 111. The tert-butyl group was split off by the Beckmann cleavage followed by hydrolysis to give the acid 112. The ring was opened by treating 112 with base, yielding the α,β-unsaturated acid 112a. All these operations left the chiral center intact. The last step which forms the large ring is not a simple lactonization, but a two-fold nucleophilic substitution at the chirality centers which proceeds with inversion of the original configuration, so that the final product 113 acquires the desired (R,R)-configuration! This simple and elegant strategy was further employed by Seebach in his synthesis of natural pyrenophorine.[133]

Nonactine[134] is another antibiotic prepared from (S)-(−)-lactic acid. Ethyl-(S)-(−)-lactate served as a chiral synthon in the synthesis of the wasp workers pheromone (*Paravespula vulgaris*).[135]

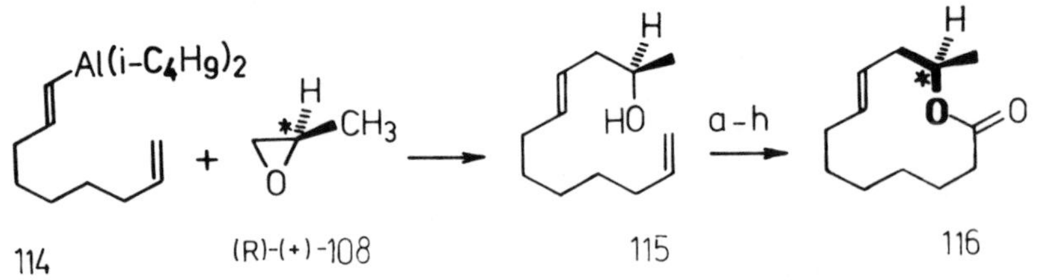

FIGURE 21. (a) Dihydropyran; (b) R_2BH; (c) H_2O_2, NaOH; (d) NCS, Me_2S, Et_3N; (e) AcOH, H_2O, THF; (f) Ag_2O; (g) $(\alpha\text{-}C_5H_4N\text{-}S)_2$, Ph_3P; (h) $AgBF_4$.

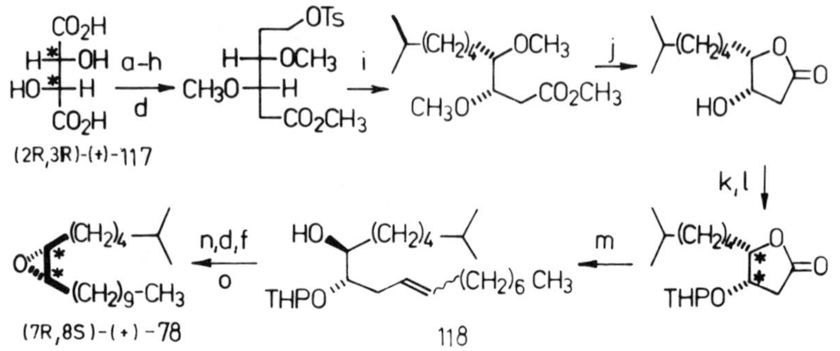

FIGURE 22. (a) EtOH, H^+; (b) CH_3I, Ag_2O; (c) $LiAlH_4$; (d) TsCl; (e)NaCN, Me_2SO; (f) H^+, CH_3OH; (g) 1 eq. KOH; (h) B_2H_6; (i) $[(CH_3)_2\ CHCH_2CH_2]_2CuLi$; (j) BCl_3; (k) dihydropyran; (l) $(i\text{-}Bu)_2AlH$; (m) $Ph_3P^+\text{-}(CH_2)_7CH_3Br(-)BuLi$; (n) H_2, Pd – C; (o)KOH, CH_3OH.

1,2-Epoxypropane (108) of an opposite configuration (R) was used by Japanese authors[136] in the synthesis of (R)-recifeiolide (116, Figure 21). The final lactonization was achieved by treating the corresponding hydroxy acid with α,α-bipyridyl sulfide and sodium fluoroborate.

Other examples of the use of optically active epoxides can be found in papers on the synthesis of erythronolide B[137,138] and in References 8 to 11 and 139 to 142.

C. Tartaric Acid (117)

Tartaric acid occurs in nature in the L-(+)-form, i.e. with (2R,3R)-configuration (for notation see Reference 143). The D-(−)-form (or 2S,3S) can be prepared from the natural enantiomer by racemization in alkaline medium followed by necessary resolution by means of L-asparagine or (+)-methylamphetamine.[144]

Mori has made use of the two vicinal chirality centers in tartaric acid to synthesize a chiral oxirane. Starting from L-(+)-tartaric acid he was able to prepare the sexual attractant of the gypsy moth (*Portheria dispar*), (7R,8S)-(−)-dispalure (78), in enantiomeric purity exceeding 98% (Figure 22).[93,145] As with the acid 101 (Figure 18), the Mori synthesis of 78 is based on a gradual construction of the hydrocarbon chains from carboxyl groups. The vicinal diol grouping of tartaric acid forms the synthon for the oxirane ring, whereby both chirality centers are preserved. Of necessity, the ring closure (118 → 78) is accompanied by the inversion of configuration at one center.

L-(+)-Tartaric acid played the pivotal role in Seebach's synthesis of antibiotic LLP-880,[146] in the Mori synthesis of the sexual pheromone of pine sawfly,[147] Rapoport synthesis of anthopleurine[148] and in the synthesis of N-benzoyl-L-ristosamine by Italian authors.[149] Starting from D-(−)-tartaric acid, Mori prepared (1R,5S,7R)-(+)-exobrevicomin,[145] the attractant of *Dendroctomus brevicomis,* and thus proved that only the (+)-enantiomer is biologically active. The synthesis simultaneously provided proof of the absolute configuration of the natural pheromone.

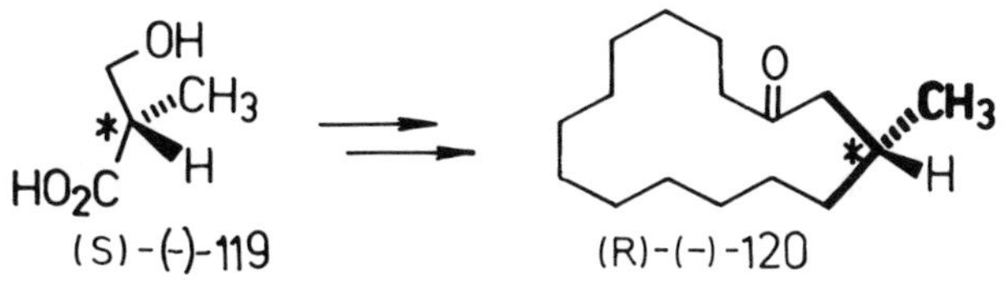

(S)-(-)-119 → (R)-(-)-120

FIGURE 23.

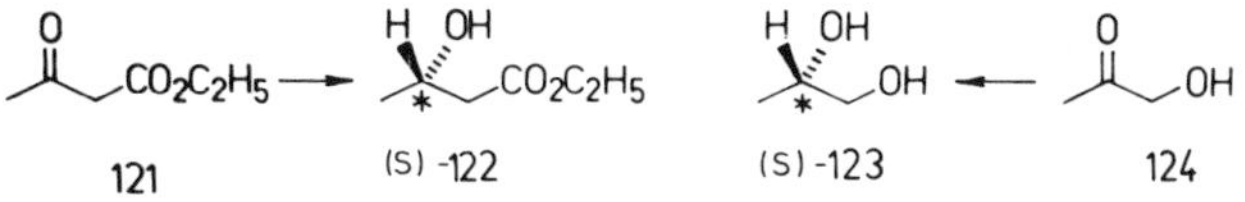

121 → (S)-122 (S)-123 ← 124

FIGURE 24.

Japanese authors[150] have utilized D-(−)-tartaric acid for the preparation of prostaglandin intermediates. The reader will find more examples in References 140 to 142 and 151 to 155.

D. Other Hydroxy Acids and Related Compounds

While the hydroxy acids occurring in nature undoubtedly constitute one of the most important groups of the chiral pool, we should also mention several synthetic chiral hydroxy acids which can be rather easily prepared in an optically active form.

(S)-2-Methyl-3-hydroxypropionic acid (119), prepared by microbial hydroxylation of isobutyric acid, provided the chirality center in the synthesis of R-(−)-muscone (120) by Swiss authors (Figure 23).[156] The same acid was employed by Evans et al.[157] in the synthesis of optically active kaleimycin (ionophore A-23187) and by Meyers et al. in the synthesis of (−)-maysine.[158] The corresponding chiral aldehyde supplied one of the chirality centers in the synthesis of monensin.[159]

Ethyl acetoacetate (121) can be reduced with baker's yeast (*Saccharomyces cerevisiae*) to give optically active ethyl (S)-3-hydroxybutyrate (122).[2,160] In the same way, hydroxyacetone (124) is reduced to (S)-diol 123 (Figure 24),[161] which is also accessible by reducing ethyl (S)-(−)-lactate with lithium aluminum hydride. Other α-hydroxy acids may be prepared by asymmetric reduction.[129]

Another chiral synthon was prepared by hydroxylation of dimethyl acrylic acid with the strain *Pseudomonas putida*.[162] Finally we note that the synthon prepared from levulinic acid[163-165] was used in the synthesis of (1R,5S)-(+)-frontalin,[166,167] an aggregation pheromone isolated from hindgut extracts of the male western pine beetle (*D. frontalis*). (S)-(+)-Citramalic acid was used in a prostaglandin synthesis.[168]

V. CARBOHYDRATES

So far we have dealt with simple organic compounds, possessing one or two chirality centers, that come into consideration as chiral synthons for synthesis of natural products. Carbohydrates, on the other hand, have an irreplacable position as synthons providing many chiral centers. One of the great advantages of carbohydrates is their optical purity which is not endangered by easy racemization, a common problem encountered with other compounds possessing one chirality center. Due to the variety of relative configurations, carbohydrates make possible the introduction of several chiral centers at the very beginning of the synthesis, which otherwise would have to be built from several building blocks in a more complicated procedure. This considerably reduces the difficulties associated with formation of undesirable stereoisomers in the course of a gradual build-up of the skeleton (for a review see References 3 and 6).

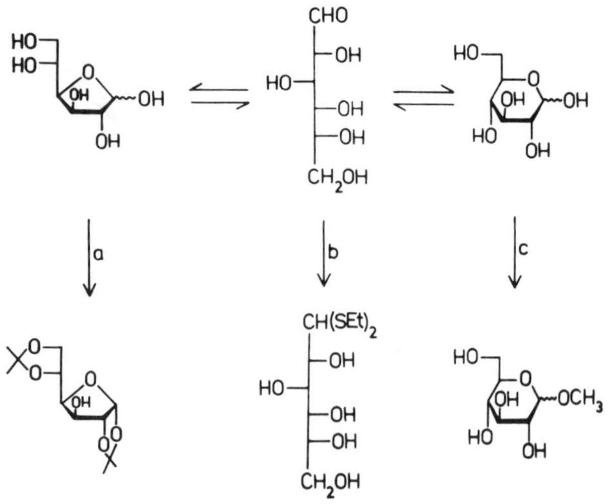

FIGURE 25. (a) Me_2CO, H^+; (b) EtSH, H^+; (c) CH_3OH, H^+.

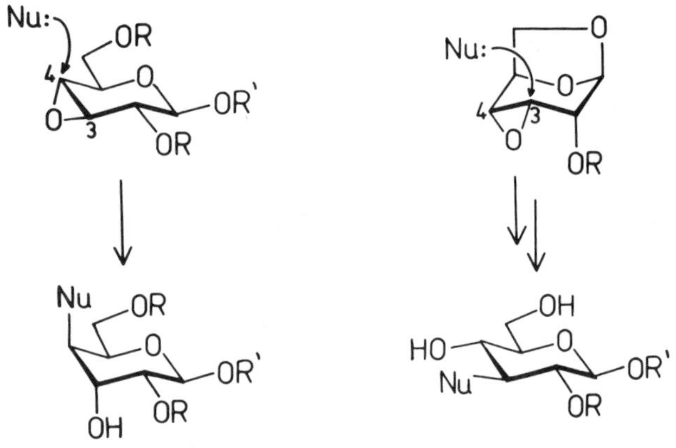

FIGURE 26.

A typical monosaccharide, e.g. glucose (Figure 25),[6] exists in solution as an equilibrium mixture of one acyclic and two cyclic forms (furanose and pyranose). The individual forms can be trapped by simple chemical transformations. For instance, the acyclic form of glucose can be trapped as a thioacetal or cyanohydrine, the furanose form as a diisopropylidene derivative, and the pyranose form as a glucoside. This, together with other selective transformations known from carbohydrate chemistry, enables us to distinguish selectively the individual positions in the synthon, which is essential for the synthetic implementation.

Furthermore, the pyranose form offers another means of influencing the regio- and stereoselectivity of synthetic transformations, as the most stable chair-form can be inverted to another chair-form with all substituents axial. This is achieved by closing a five-membered ring in a 1,6-anhydro derivative (Figure 26)[169,170] which locks the system in an energetically unfavorable conformation. For instance, cleavage of the 3,4-oxirane ring in a 1,6-anhydro derivative provides a diaxial diol[169] which formally corresponds to a diequatorial cleavage of an analogous epoxide in a glucopyranose form (compare Reference 171).

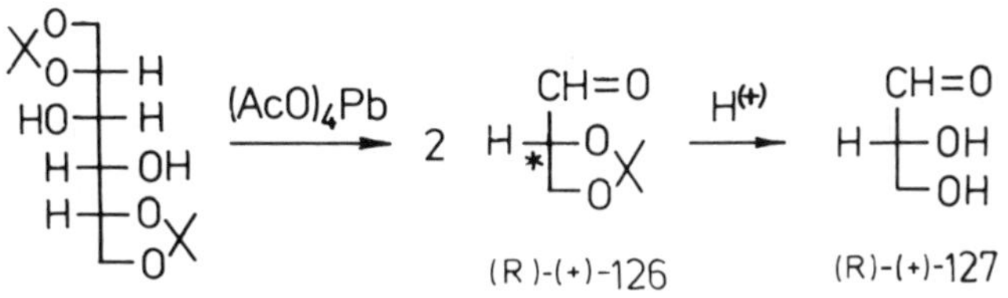

FIGURE 27.

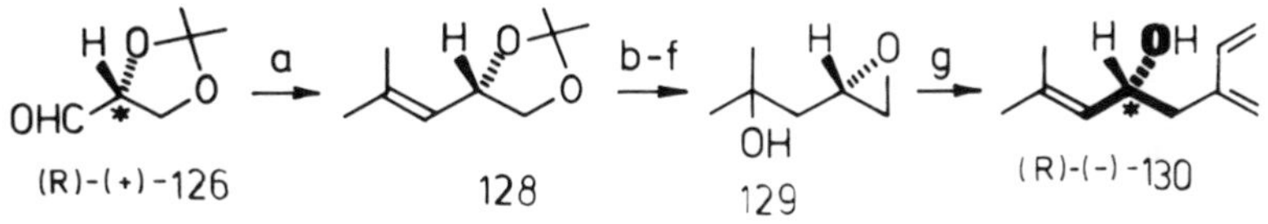

FIGURE 28. (a) Ph_3P^+–$CHMe_2$ I^-, NaH, Me_2SO; (b) $(AcO)_2Hg$; (c) $NaBH_4$; (d) HCl; (e) TsCl; (f) KOH; (g) cf. Figure 16.

A. Glyceraldehyde (127)

D-Glyceraldehyde, i.e. (R)-(+)-127, and its isopropylidene derivative 126 are the simplest and easily accessible chiral synthons derived from carbohydrates. Compound 126 is prepared from 1,2:5,6-di-O-isopropylidene-D-mannitol (125) by oxidation with lead tetraacetate (Figure 27).[6,172]

(R)-(+)-Glyceraldehyde serves as a starting point for preparation of various chiral C_3-synthons.[169-179] Moreover, (R)-(+)-glyceraldehyde can be converted to its (S)-(−)-enantiomer by switching the positions of the hydroxymethylene and the aldehyde group. This is carried out by an oxidation-reduction sequence combined with selective protection, while the chiral center at C-2 remains untouched.[180,181] Unnatural L-glyceraldehyde can otherwise be prepared from L-arabinose,[182] L-ascorbic acid,[183] L-mannitol,[184,185] or L-serine.[186]

The 2,3-isopropylidene derivative of (R)-(+)-glyceraldehyde has found wide application in syntheses of various natural products by Mori et al.[187-189] For instance, ipsdienol (130), one of the components of the sexual attractant of the bark beetle *Ips paraconfusus,* was synthesized from protected glyceraldehyde 126 via the epoxide 129 (Figure 28), as in the synthesis of ipsenol.[188,189] The enantiomeric epoxide 129 was obtained from (R)-(+)-malic acid and converted to natural (S)-(+)-ipsdienol (130). Beside the preparative goals the synthesis unambiguously determined the absolute configuration of this natural product.[190]

R-(+)-Glyceraldehyde and its derivatives have also served as starting compounds in the synthesis of multistriatin (an aggregation pheromone of the European elm bark beetle),[191] prostaglandin E_1,[192,193] (−)-carnitine (vitamine B_3),[183] in the construction of the side chain of (24R)-24,25-dihydroxy-vitamine D_3,[194] and in the synthesis of leukotrienes.[195,196] Glyceraldehyde has also been employed for induction of chirality in the Diels-Alder reaction.[197]

B. Glucose (131)

The molecule of D-glucose offers the synthetic chemist a rich variety of vicinal chirality centers, while the availability of glucose in optically pure form makes it a fundamental chiral synthon.[6,169-171,198] The implementation of D-glucose in syntheses of several structurally differing, optically active compounds is demonstrated below. Other examples can be found in References 6 and 169 to 171.

(+)-Cerulenin (139) is an antibiotic isolated from *Cephalosporium caerulens.* The structure and absolute configuration of 139 was proved by synthesis starting from D-glucose (Figure 29).[199] The C-2 and C-3 positions in D-glucose make up the carbon part of the oxirane ring in (+)-cerulenin (139). While the configuration at C-2 was preserved during

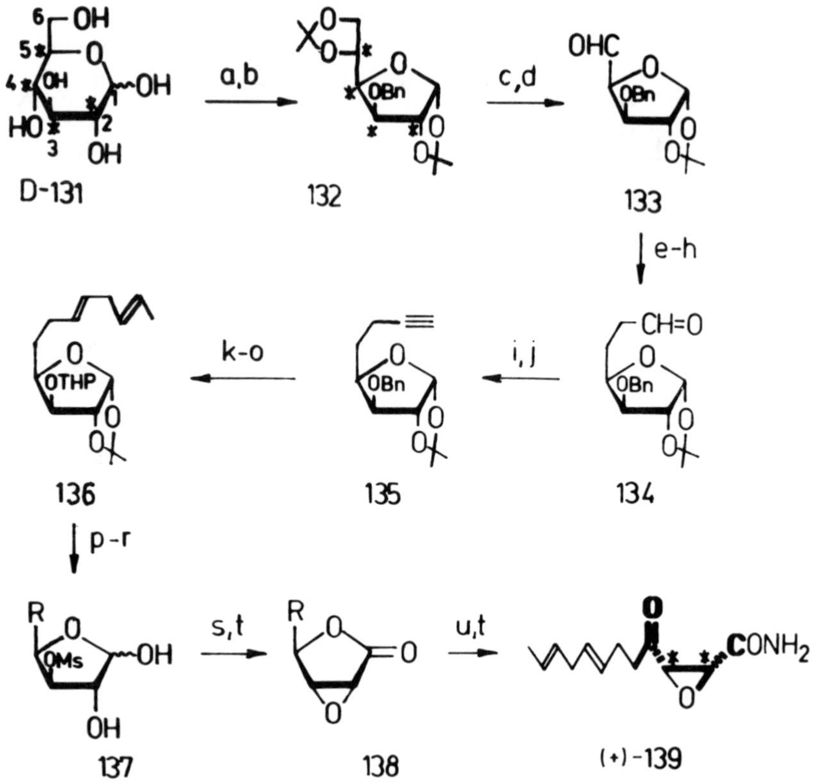

FIGURE 29. (a) Me_2CO, H^+; (b) $PhCH_2Cl$; (c) H^+; (d) HIO_4; (e) $(EtO)_2P(O){-}CH_2CO_2Et$, NaH, Et_2O; (f) H_2, Pd/C; (g) $LiAlH_4$; (h) $(COCl)_2$, Me_2SO, CH_2Cl_2, $-60°C$; (i) Ph_3P, CBr_4, Zn, CH_2Cl_2; (j) BuLi, THF; (k) EtMgBr, THF, 60°C 1 hr; (l) $CH_3CH{=}CH{\cdot}CH_2Cl$; (m) 1.9 eq Li, NH_3, THF, 0.5 eq t-BuOH, $(NH_4)_2SO_4$; (n) Dihydropyran, $C_5H_5N{\cdot}TsOH$; (o) 3 eq. Li, NH_3, THF; (p) $C_5H_5N{\cdot}TsOH$, EtOH; (q) CH_3SO_2Cl, C_5H_5N, CH_2Cl_2; (r) CF_3CO_2H, H_2O, 0°C, 2.5 hr; (s) CH_3ONa, THF, 0°C, 5 hr; (t) $CrO_3{\cdot}C_5H_5N{\cdot}HCl$; (u) NH_3, CH_3OH.

the synthesis, the configuration at C-3 was reversed in closing the oxirane ring (137 → 138). (+)-Cerulenin was also prepared by several other routes.[200-202]

The series of chirality centers provided by D-glucose was employed by Stork and his group in their elegant syntheses of prostaglandin $F_{2\alpha}$[203] and related compounds. The fate of the glucose carbon atoms in the synthesis of $PGF_{2\alpha}$ is shown in Figure 30 (bold marks). All the six glucose atoms were built into the target molecule. Of the original chirality centers (*), C-5 was retained and formed the C-11 position in the prostaglandin. Other two original centers (C-3 and C-4) vanished, but the latter was recreated again by Claisen rearrangement which simultaneously destroyed the chirality at C-2; this step gave rise to the double bond of the lower side chain which was also formed in a stereoselective manner.

Glucose derivatives 141 and 142 have been employed in the synthesis of thromboxane B_2 (143). In both cases the synthesis was designed to preserve the pyranose ring which constituted the terahydropyran ring in 143 (Figure 31).[204-206] For other approaches to prostaglandins see Reference 207.

In addition to the above-mentioned examples, D-glucose has been involved in a number of other stereoselective syntheses. These include 11-oxa-prostaglandin,[208] both enantiomers of carboxychrysanthemic acid,[209] (−)-*cis*-roseoxide,[210] frontalin,[211] multistriatin,[212] chalcogran,[213] biotin,[214] (+)-azimic and (+)-carpamic acid,[215] antibiotic A26711B,[216] (+)-furanomycin,[217,218] carbomycin,[219] one part of the molecule of leukomycin A_3,[220] (−)-

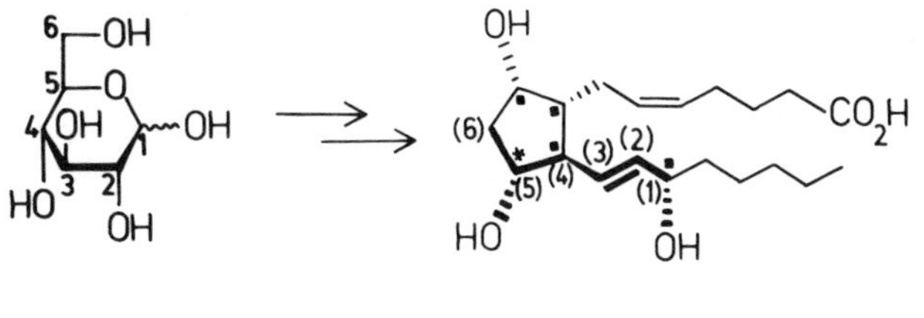

D-131

140

FIGURE 30.

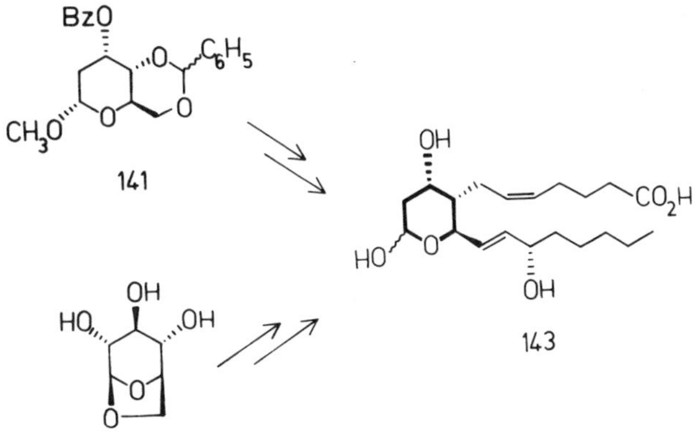

141

142

143

FIGURE 31.

avenaciolide,[221,222] (−)-isoavenaciolide,[222,223] chiral cryptates,[224] chiral Wilkinson catalysts,[225-227] and other compounds.[6,169,228-230] Japanese authors[231] have revised the absolute configuration of detoxinine D_1 by using D-glucose as a chiral synthon. In order to mention some glucose derivatives as well, we note that 2-amino-2-deoxy-D-glucose served as a starting compound for the preparation of 3-hydroxyhistidine.[228a]

C. Mannose (144)

The importance of D-mannose for the synthesis of optically active compounds is exemplified by the stereoselective preparation of (+)-biotin (152, Figure 32).[232] The biotin side chain carrying the carboxylic group was constructed from the aldehyde 146 by the Wittig reaction followed by hydrogenation of the double bond (146 → 147). The dimesylate 148 that contains three original chiral centers of mannose was reacted with sodium sulfide, affording thiolan 149; the reaction proceeds with configurational inversion at C-4 (mannose numbering). The thiolan was converted to dimesylate 150 which gave a diazide (again with inversion of configuration at both asymmetric centers). Reduction of the azide groups followed by acetylation afforded the *bis*-acetamido derivative 151. After hydrolyzing the acetamide groups in 151, the synthesis was completed by closing the 2-oxo-imidazolidine ring to give (+)-biotin (152). It should be noted that of the six mannose carbon atoms, five were retained, and of the four chirality centers (disregarding C-1), three (*) were involved in the target molecule 152. All these centers, however, underwent inversion of configuration in the course of nucleophilic substitution.

D-Mannose has also been implemented in the synthesis of nonactinic acid[233,234] and ciliarine.[235]

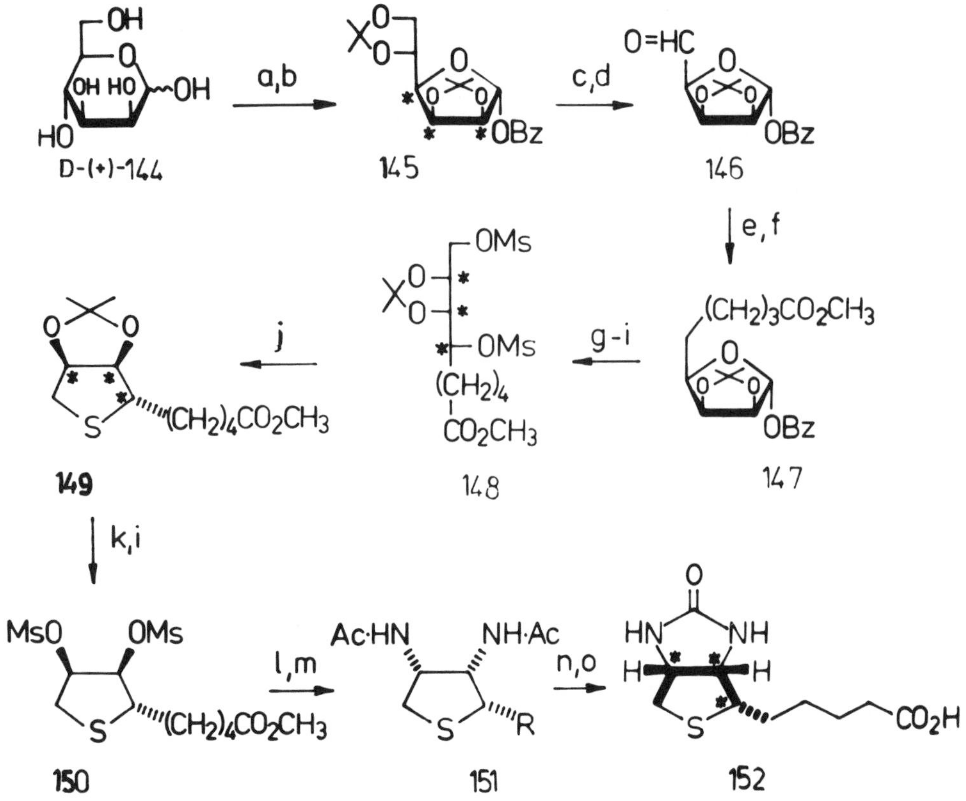

FIGURE 32. (a) Me_2CO, H^+; (b) PhCOCl, C_5H_5N; (c) 70%-AcOH, 20°C, 48 hr; (d) HIO_4, Me_2CO, H_2O; (e) $Ph_3P{=}CH(CH_2)_2CO_2CH_3$; (f) H_2, Pd/C, CH_3OH; (g) CH_3ONa; (h) $NaBH_4$; (i) CH_3SO_2Cl, C_5H_5N; (j) Na_2S, HMPA; (k) 90%-HCO_2H; (l) NaN_3, HMPA, 80°C, 7 hr; (m) H_2, Pt, CH_3OH, Ac_2O, 20°C, 4 hr; (n) $Ba(OH)_2$, H_2O, 140°C, 14 hr; (o) $COCl_2$.

D. Other Carbohydrates

Carbohydrates other than D-glucose and D-mannose have appeared much less frequently as chiral synthons and thus we present a rather brief account here.

D-Ribose was a starting compound in the Corey synthesis of leukotriene C-1,[236] L-erythrose was employed by Stork and Raucher[237] in their synthesis of prostaglandin A_2, and D-erythrose found application in the preparation of one of the intermediates in the synthesis of leukotriene C-1.[238]

The synthesis of (+)-muscarine,[239,240] its derivatives[241] and thermozymocidine[245] made use of L-arabinose as a chiral synthon. (+)-Muscarine has also been prepared from D-mannitol.[242] Prelog-Djerassi lactone was synthesized from D-allal.[243] A number of chiral synthons and derivatives of 2,4-di-C-methyl-D-hexopyranoses have been prepared to be implemented in syntheses of macrolide antibiotics.[244] Other examples can be found in References 6, 169, 170, 171, and 245 to 247.

VI. MISCELLANEOUS CHIRAL SYNTHONS

In the last few years the literature has abounded in applications of various chiral synthons, either prepared by ingenious enantioselective synthetic methods, or obtained by resolving racemic material. Since quoting all applications of this kind is far beyond the scope of this book, we have selected several examples to highlight the topic (Figure 33). Others can be found in Chapter 2 in Volume II.

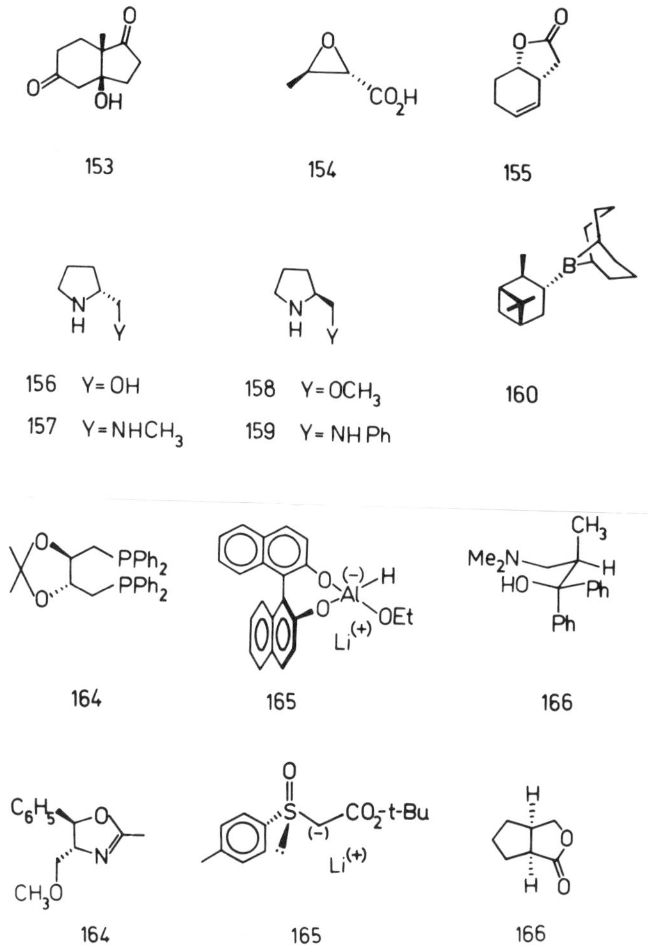

153 154 155

156 Y = OH
157 Y = $NHCH_3$

158 Y = OCH_3
159 Y = NHPh

160

164 165 166

164 165 166

FIGURE 33.

The chiral bicyclic ketone 153 (Figure 33) has played the key role in the total synthesis of steroids and polycyclic isoprenoids (see also Chapter 3, Section I.A.2).

The (+)-epoxy acid 154 was built into the molecule of erythronolide B in a synthesis by Corey.[137,138] (−)-Lactone 155 represents another chiral synthon serving for the synthesis of optically active prostaglandins. The optically active 155 was obtained by resolving the corresponding racemic acid with (+)-1-(1-naphthyl)-ethylamine.[248,249] For other chiral cyclohexane and cyclopentane derivatives see References 250 to 254.

Lactone 166 also appears to be a useful chiral synthon;[255] it is prepared from the corresponding achiral diol by stereoselective oxidation of one hydroxy group with horse liver dehydrogenase, followed by lactonization of the acid formed[255] (see also References 256 and 257).

Another important group of chiral synthons consists of compounds that serve to induce chirality in the course of synthesis. Some are temporality incorporated into the synthetic molecule, others simply function as chiral reagents. For instance, pyrrolidine derivatives 156 and 157 (useful in asymmetric α-alkylation of ketones or asymmetric reductions) were obtained in an optically active form from L-(+)-glutamic acid,[258] while compounds 158 and 159 with opposite absolute configuration were prepared from L-(−)-proline[259-263] (see also References 250, 251, and 264 to 266).

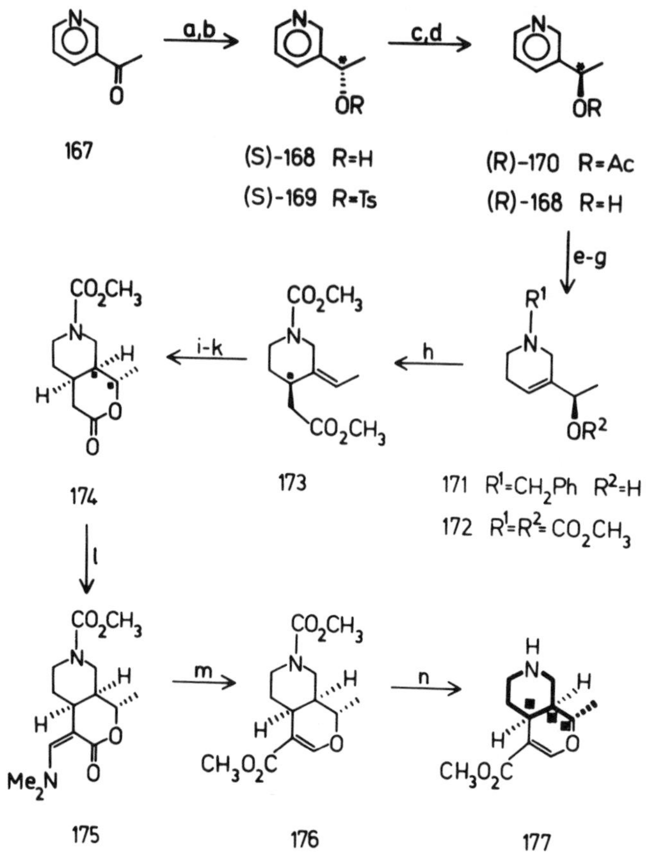

FIGURE 34. (a) Spirotrichum exile (QM-1250); (b) NaH, TsCl; (c) $AcONEt_4$, Me_2CO, 60°C; (d) NaOH, CH_3OH; (e) $PhCH_2Cl$; (f) $NaBH_4$; (g) $ClCO_2CH_3$, CH_2Cl_2; (h) $CH_3C(OMe)_3$; (i) R_2BH; (j) H_2O_2, NaOH; (k) HCl; (l) t-BuO–CH$(NMe_2)_2$, 20°C, 60 hr; (m) HCl, CH_3OH, 120°C, 24 hr (sealed tube); (n) HBr, AcOH.

The chiral borane 160, prepared from α-pinene and 9-borabicyclo[3.3.1]nonane, was transformed to a chiral hydride exerting high enantioselectivity.[129,267-269] Chiral phosphines, such as DIOP (164), and related compounds, have been used extensively as ligands in chiral Wilkinson catalysts (for review see Reference 270).

Chiral aluminum hydrides can be prepared by partial solvolysis of lithium aluminum hydride with optically active alcohols, such as quinine,[271,272] (−)-2,2′-dihydroxy-1,1′-binaphtyl (162),[273-276] or (+)-4-dimethylamino-3-methyl-1,2-diphenyl-2-butanol (163).[277] For other syntheses of optically active alcohols see References 278, 279.

Optically active oxazolines 164[280-284] have proved to be useful for preparation of carboxylic acids having a chirality center at the α-position. On the other hand, carboxylic acids with a chirality center at the β-position can be prepared by using chirality transfer from the optically active sulfoxide 165.[285-290] For other methods of stereoselective synthesis of carboxylic acids see References 291 to 296.

The last example of this section shows the formation of a chirality center by enzymatic reaction. The chiral center works as a control element in subsequent stereoselective steps in the synthesis of compound 177 (Figure 34), an intermediate in the synthesis of indole alkaloids.[297] The starting chiral synthon (S)-168 was prepared by microbial reduction of β-acetylpyridine (167). Since the configuration of the chirality center in 168 is opposite to

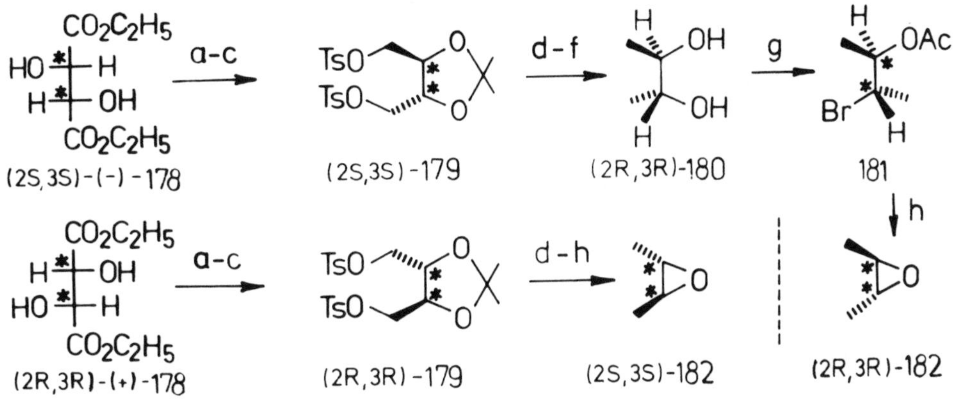

FIGURE 35. (a) Me_2CO, H^+; (b) $LiAlH_4$; (c) TsCl, C_5H_5N; (d) NaI, Me_2CO; (e) Zn; (f) H^+; (g) HBr, AcOH; (h) NaOH.

what was needed, the alcohol (S)-168 was converted to its enantiomer (R)-168 via a tosylate 169. In the next steps the pyridine ring was reduced (171) and the chirality of the side chain was transcribed into the ring through a Claisen rearrangement, giving rise to the chiral ester 173. Stereoselective hydroboration of the double bond in 173 with 9-BBN was controlled by the chiral center (■) and led to formation of a new chirality center at the bridgehead position (■), recreating simultaneously the original chirality center that had vanished earlier by Claisen rearrangement. The lactone 174 was then converted to the target compound 177 in several steps.

Other examples of the utilization of chiral synthons in synthesis of natural products can be found in References 1 and 298 to 331. Enantioselective syntheses based on Sharpless oxidation have been recently reviewed (see Chapter 2 in Volume II).[10,11]

VII. SYNTHESES INVOLVING SEVERAL CHIRAL SYNTHONS

In the preceding sections we have shown selected examples, where the optically active target compound was synthesized from a single chiral synthon. The formation of new chirality centers was controlled by those already present in the molecule; however, such an intramolecular chirality induction is limited to forming new centers in close vicinity of the parent one(s). If one plans a convergent synthesis of a molecule with remote chiral elements, the building blocks carrying these elements must be optically active, i.e. their absolute configuration must be unequivocally defined in order to obtain a product with required relative, as well as absolute configuration (see Figure 10 in Chapter 1). This section is devoted to discussing various strategies that combine two or more chirality synthons differing in structure or absolute configuration.

A. The Synthesis of the Sexual Pheromone of Pine Sawflies

The sexual pheromone of pine sawflies, (2R,3R,7R)-3,7-dimethyl-2-pentadecanol (189), contains two groups of remote chirality centers. Mori and Tamada[332] have synthesized four stereoisomers of 189 by coupling enantiomeric 2,3-epoxybutanes 182 with cuprates 188 (Figure 37) prepared from bromo derivatives 186. The enantiomeric pair of epoxides was prepared from enantiomeric tartaric acids as shown in Figure 35. On the other hand, the enantiomeric bromides 186 were obtained from a single chiral precursor, (R)-(+)-citronellyl tosylate (183) (Figure 36)! The natural pheromone (2R,3R,7R)-189 was obtained by coupling the (R)-cuprate 188 with the (2R,3R)-epoxide 182 (Figure 37).

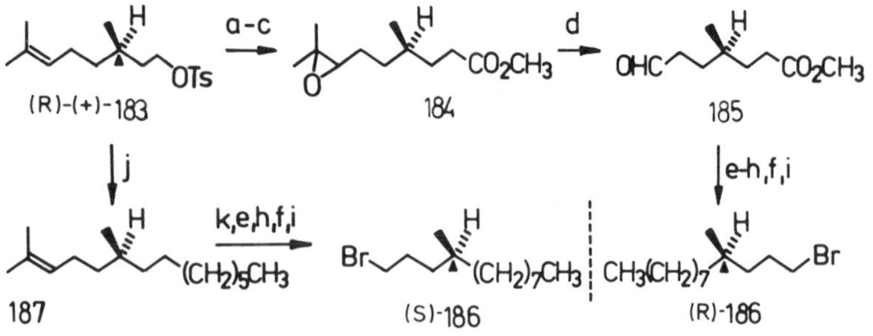

FIGURE 36. (a) NaCN, Me_2SO; (b) CH_3OH, H^+; (c) RCO_3H; (d) $NaIO_4$; (e) $NaBH_4$; (f) TsCl, C_5H_5N; (g) $(C_5H_{11})_2CuLi$; (h) $LiAlH_4$; (i) LiBr, Me_2CO; (j) $(C_6H_{13})_2CuLi$; (k) O_3.

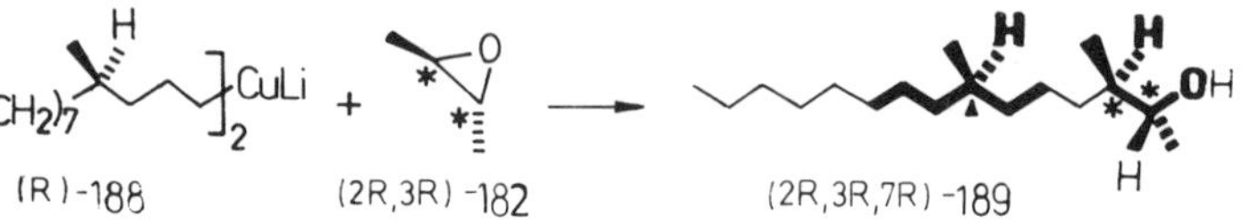

FIGURE 37.

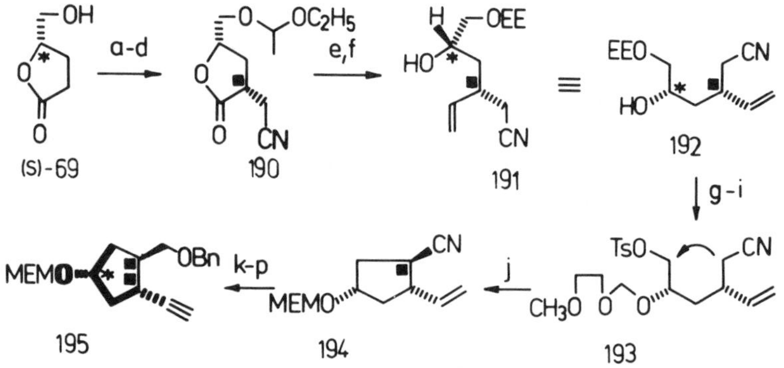

FIGURE 38. (a) $CH_2{=}O$, Et_2NH; (b) EtI, THF; (c) NaCN, DMF; (d) $CH_2{=}CH{-}OEt$; (e) $(i{-}Bu)_2AlH$; (f) $Ph_3P{=}CH_2$; (g) $MeO{-}CH_2CH_2{-}OCH_2Cl$, NaH; (h) 75%-AcOH, 35°C, 30 min; (i) TsCl, C_5H_5N; (j) $NaN(SiMe_3)_2$, C_6H_6, reflux 20 min; (k) NaOH, CH_3OH, H_2O; (l) CH_2N_2; (m) $LiAlH_4$; (n) $PhCH_2Cl$, NaH; (o) $C_5H_5N{\cdot}HBr_3$, $CHCl_3$; (p) $NaNH_2$.

B. The Synthesis of Brefeldin A

(+)-Brefeldin A (203) is an interesting compound showing antibiotic and antiviral activity. In addition to several synthetic routes to racemic brefeldin A, a synthesis was published leading to the natural (+)-enantiomer (Figures 38 to 40).[333] The target molecule was assembled from two chiral synthons 195 and 198 which were prepared from enantiomeric (S)- and (R)-hydroxylactones 69. The (S)-enantiomer 69 was obtained from D-glyceraldehyde, (R)-69 was prepared from (R)-(−)-glutamic acid. The target molecule 203 contains by one chirality center from each (S)- and (R)-lactone 69 (denoted * and ▲), and three additional centers (■) formed in the course of the synthesis, the configuration of which is determined by the parent centers.

C. The Synthesis of the Molecular Fragment of Griseoviridine (205)

Two structurally different synthons have been used in the synthesis of the nine-membered macrolide ring of a synthetic precursor of griseoviridin (205), (Figure 41).[334] The starting optically active compounds were the protected (S)-(+)-cystine 204 and (S)-(+)-3-hydroxybutyric acid (119).

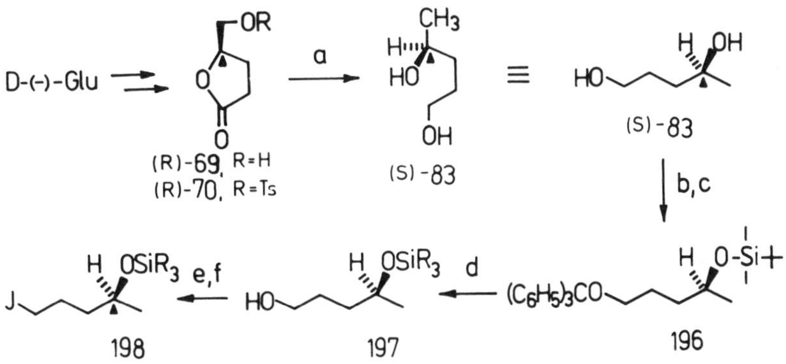

FIGURE 39. (a) $LiAlH_4$, THF; (b) Ph_3CCl, C_5H_5N; (c) t–$BuMe_2SiCl$; (d) Na, NH_3; (e) TsCl, C_5H_5N; (f) NaI, Me_2CO, reflux.

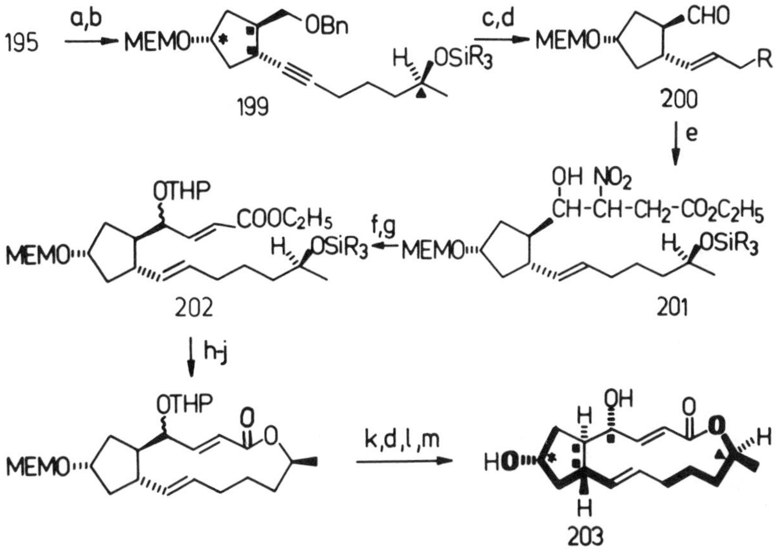

FIGURE 40. (a) BuLi, THF, HMPA; (b) 198; (c) Na, NH_3; (d) $CrO_3{\cdot}C_5H_5N{\cdot}HCl$, AcONa; (e) $O_2N–CH_2CH_2CO_2Et$, $(i–Pr)_2NLi$, Me_2SO; (f) Piperidine, HMPA; (g) Dihydropyran; (h) OH^-; (i) Bu_4NF; (j) $(\alpha$-$C_5H_4N–S)_2$; (k) H^+; (l) $NaBH_4$; (m) $TiCl_4$.

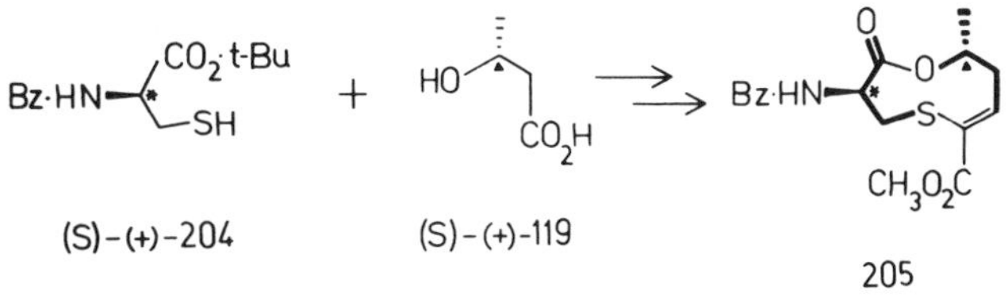

FIGURE 41.

D. The Synthesis of Cytochalasin B (210)

The antibiotic cytochalasin B (210) contains eight asymmetric centers which, together with other intriguing structural features, represent a challenge to the synthetic skills of organic chemists. Stork et al.[335] (Figure 42) have used three chiral synthons, namely, (R)-(+)-citronellyl acetate (206, center *), (R)-(+)-acetylmalic acid (207, center ▲), and a derivative

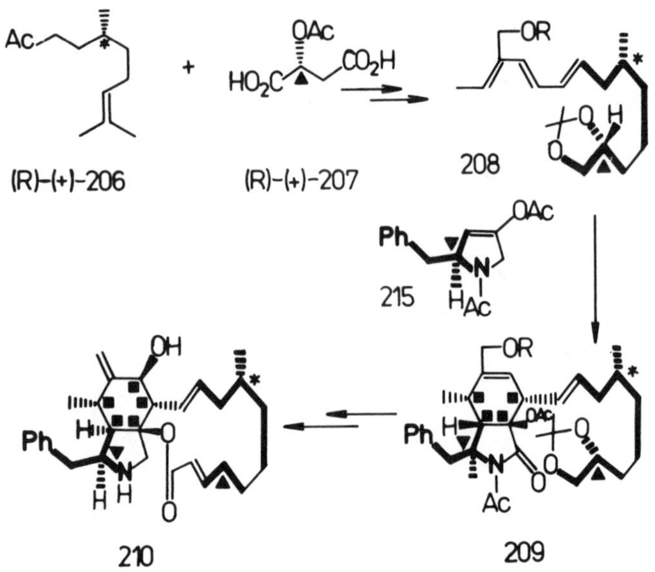

FIGURE 42.

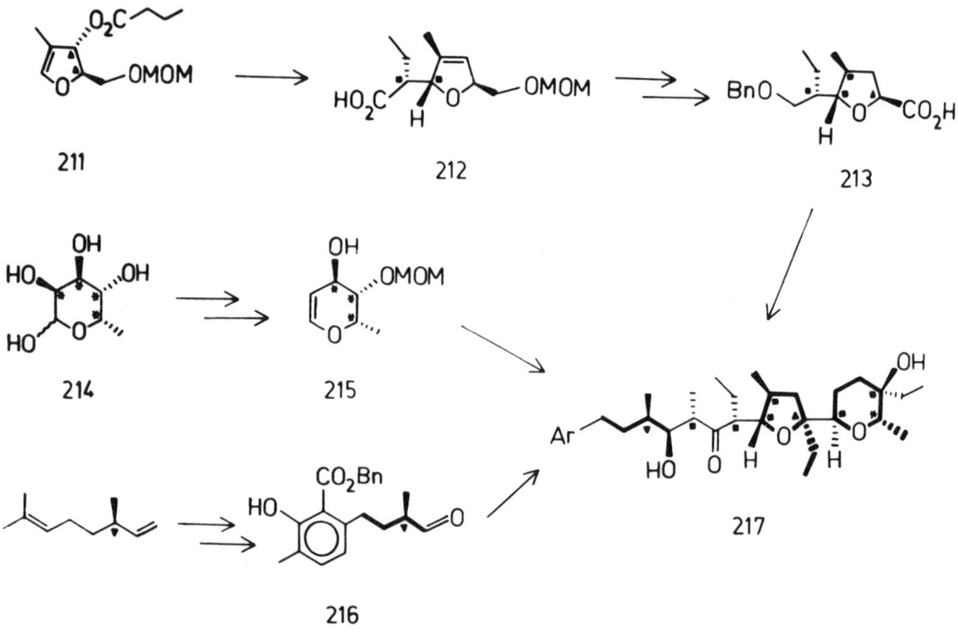

FIGURE 43.

of (S)-(−)-phenylalanine (215, center ▼), to construct the target molecule. The fate of the individual parts including the original chiral centers is visualized in Figure 42. All three parent centers were retained (*, ▲, ▼), while five new centers (■) were formed during the synthesis; four of them were formed in a single step (Diels-Alder reaction 208 → 209).

E. The Synthesis of Lasalocid A (217)

Three chiral synthons have also been used in the synthesis of another antibiotic, lasalocid A (Figure 43).[336-340] The part of the molecule that contains the tetrahydrofuran ring was prepared from derivative 211. The strategic step involved Claisen rearrangement (211 → 212) in which one chirality center vanished (▲), while two new ones were formed (■). The tetrahydropyran part comes from 6-deoxy-gulose (214), whereas the side chain was constructed by attaching the third synthon 216, prepared from (R)-citronellene. A different synthesis of lasalocid A was reported by Kishi et al.[341,342]

F. The Synthesis of Monensin (228)

The antibiotic monensin (228), produced by the strain *Streptomyces cinnamonensis,* is the longest-known member of a group containing over forty natural polyether antibiotics isolated to date. The stereoselective formation of the 17 chirality centers (giving over 130,000 stereoisomeric combinations) on the 26 carbon skeleton could have been hardly conceived without chiral synthons. The first synthesis of this complex natural product was reported by Kishi.[320-322]

Second synthesis, accomplished by Still et al.,[159] started from four chiral fragments (221, 222, 223, and 224, Figure 44), which were assembled (Figure 45) to form monensin. The individual chiral synthons (221 to 224) were prepared from optically active compounds (218, 98, 219, and 220 respectively), each containing one chiral center. Note that 218 and 220 are derived from the same hydroxyaldehyde. The four parent chirality centers (twice *, ▲, and ▼) were implanted into the target molecule and simultaneously served as control elements in forming the remaining 13 centers.

G. The Synthesis of Vitamin B_{12}

We are closing this Chapter with the synthesis of vitamin B_{12}, in which the authors, R. B. Woodward and A. Eschenmoser, have employed four chiral synthons, (−)-camphor (51), indole derivative 229, heterocycle 239, and (+)-camphor. The synthetic strategy was based on forming and connecting four pyrrole segments according to the scheme: A + D → AD, B + C → BC, AD + BC → ABCD.[343-347] Because the great complexity of the synthesis exceeds the scope of this book, we present here only an outline.

The AD block was formed (Figure 46) by linking the acyl chloride 230 to the indole 229 (229 + 230 → 231), followed by extensive transformations which eventually resulted in the substituted bipyrrole 240. Note that only the indole five-membered ring of 231 was retained in 240, while the other parts of the original synthon underwent profound reorganization. The block C (238) was synthesized from (+)-camphor in several steps (Figure 47). The block B (239) was connected with 238 to form the BC part 241. The two moieties (AD + BC, Figure 48) were assembled by a multistep sequence and, after many difficulties, the synthesis of cobyric acid (242) has been successfully completed. This feat is considered as a formal synthesis of vitamin B_{12}, since its preparation from cobyric acid has been described earlier.[348] To summarize the origin of the nine chirality centers in 242, let us note that one (denoted * in ring D) comes from (−)-camphor, one (* in C ring) from (+)-camphor, two (▲ in ring A) from 229, and two (▼ in ring B) come from 239. The remaining three centers (■) were formed in the course of the synthesis.

VIII. NOTES ADDED IN PROOF

(+)-Carvone has been prepared from (+)-car-3-ene[349] and a new synthesis of (S)-β-citronellol appeared.[350] Acyclic stereoselection leading to replicating 1,5-, 1,4-, and 1,3-disubstitution patterns has been achieved starting from L-glutamic acid.[351-353]

L-Glutamic acid also served as a chiral synthon in a synthesis of (S)-(−)-3-piperidinol.[354] This route turned out to be superior to that starting from (S)-(−)-malic acid.[354] Another use of (S)-(−)-malic acid led to the synthesis of (+)-heliotridene[355] and to the enantiomers of 3-hydroxy-1,7-dioxaspiro[5.5]undecane, a minor component of the olive fly pheromone.[356]

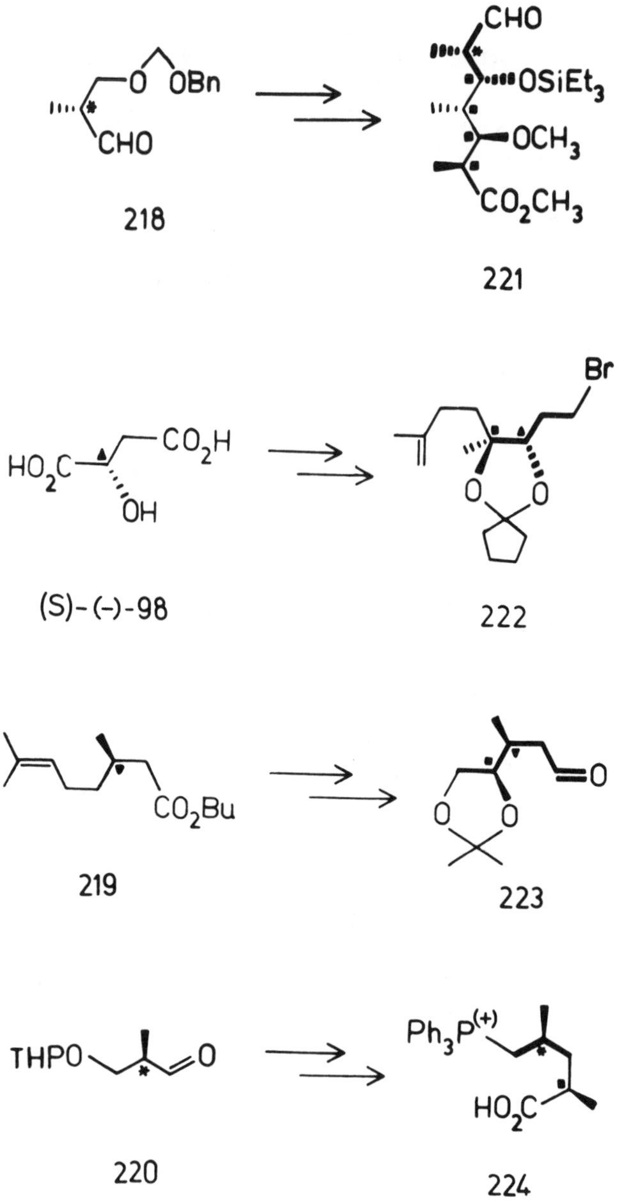

FIGURE 44.

Two molecules of (R)-(+)-1,2-epoxypropane were built into the molecule of antibiotic (−)-grahamimycin A_1 as the only source of chirality.[357] L-Tartaric acid was implemented as a chiral starting material in the syntheses of anantiomerically pure 4-O-benzyl-2,3-dideoxy-L-threo-hex-2-eno-1,5-lactone[358] and ionophore antibiotic X-14547A.[359]

The reaction of chiral epichlorohydrin (derived from glyceraldehyde) with soft carbon nucleophiles has been shown to split the oxirane ring leaving the C–Cl bond intact.[360] This finding has been utilized in enantioselective syntheses of (−)- and (+)-vincadiformine and (−)-tabersonine.[360] For reactions of chiral benzyl glycidol with C-nucleophiles see Reference 361. Annulation of a levoglucosane derivative to aromatics has been described.[362] A conversion of halo-furanoses to substituted chiral cyclopentanes by one carbon insertion in-

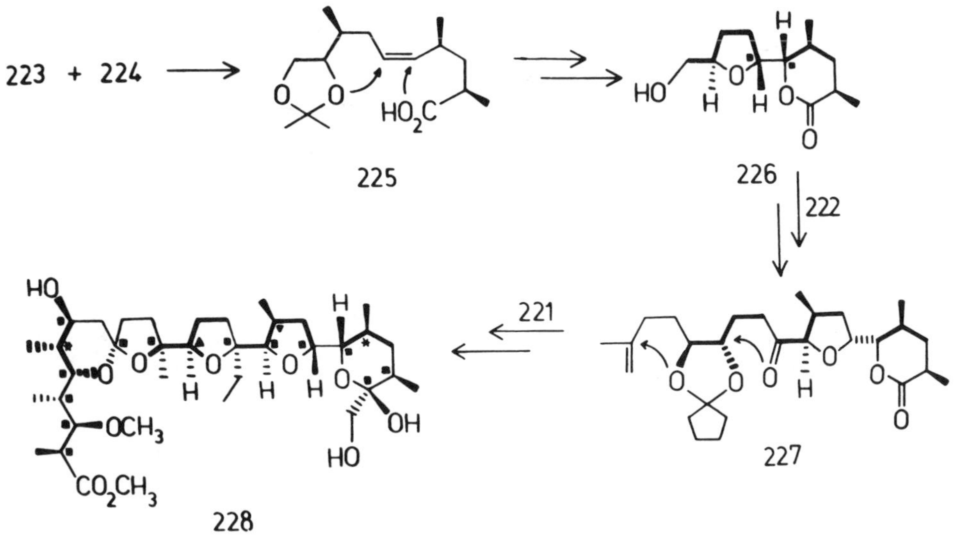

223 + 224 → 225 → 226 → (222) → 227 → (221) → 228

FIGURE 45.

volving a radical cyclization has been developed.[363] β-Ketoesters may be reduced to the corresponding β-hydroxy acid esters on the Raney nickel modified with tartaric acid and sodium bromide. This method gives up to 89% ee.[364,365]

Another approach to monensin has been reported starting from fructose, xylose, and mannose as sources of chirality.[366-368]

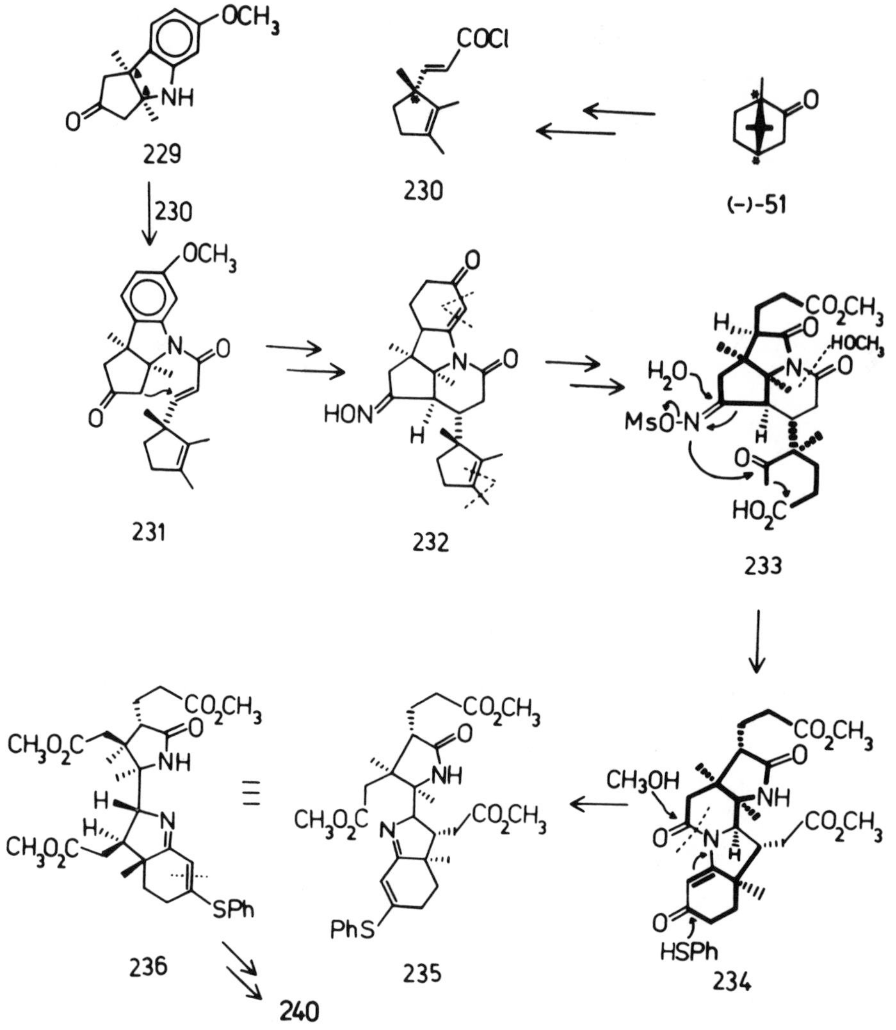

FIGURE 46.

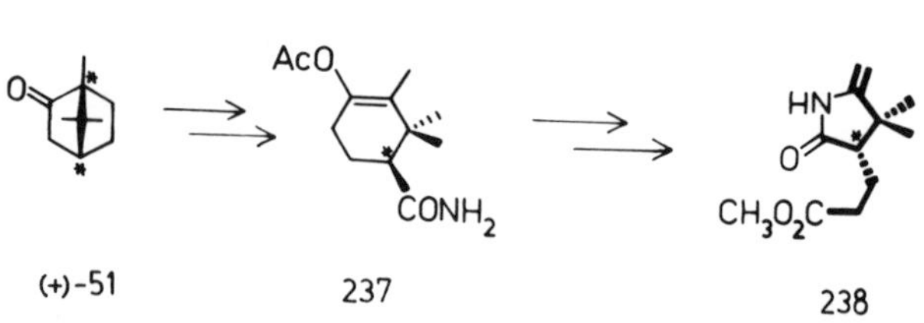

FIGURE 47.

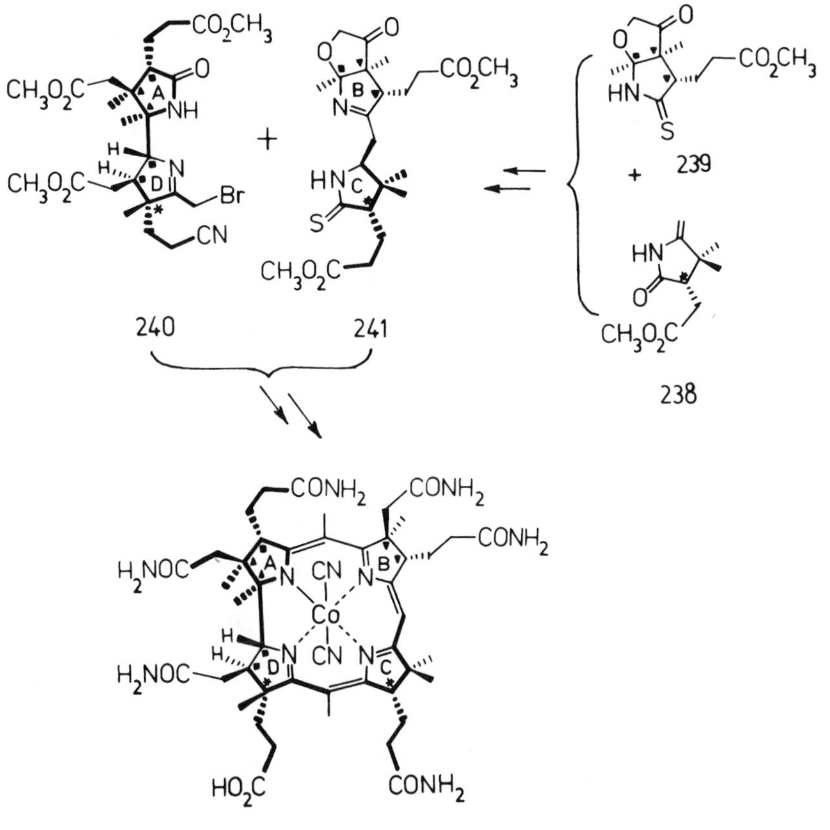

FIGURE 48.

REFERENCES

1. **Kočovský, P.,** Chiral synthons and intermediates in organic synthesis I, *Chem. Listy,* 76, 1147, 1982 (in Czech).
2. **Kočovský, P. and Černý, M.,** Chiral synthons and intermediates in organic synthesis II, *Chem. Listy,* 77, 373, 1983 (in Czech).
3. **Hanessian, S.,** *Synthetic Design with Chiral Templates,* Pergamon Press, Oxford, 1983.
4. **Fischli, A.,** Chirale Ökonomie, *Chimia,* 30, 4, 1976.
5. **Seebach, D. and Hungerbühler, E.,** in *Modern Synthetic Methods,* Vol. 2, Scheffold, R., Ed., Salle and Sauerländer, Frankfurt, 1980.
6. **Fraser-Reid, B. and Anderson, R. C.,** Carbohydrate Derivatives in the Asymmetric Synthesis of Natural Products, in *Fortschritte der Chemie Organischer Naturstoffe,* Vol. 39, Hertz, W., Griesbach, H., and Kirby, G. W., Eds., Springer Verlag, Wien, 1980, 1.
7. **Seebach, D. and Kalinowski, H. O.,** Enantiomerenreine Naturstoffe und Pharmaka aus billigen Vorlaüfern (Chiral pool), *Nachr. Chem. Techn.,* 24, 415, 1976.
8. **Katsuki, T. and Sharpless, K. B.,** The first practical method for asymmetric epoxidation, *J. Am. Chem. Soc.,* 102, 5974, 1980.
9. **Rossiter, B. E., Katsuki, T., and Sharpless, K. B.,** Asymmetric epoxidation provides shortest routes to four chiral epoxy alcohols which are key intermediates in syntheses of methylmycin, leukotriene C-1 and dispalure, *J. Am. Chem. Soc.,* 103, 464, 1981.

10. **Sharpless, K. B., Behrens, C. H., Katsuki, T., Lee, A. W. M., Martin, V. S., Takatani, M., Viti, S. M., Walker, F. J., and Woodard, S. S.,** Stereo and regioselective opening of chiral 2,3-epoxy alcohols. Versatile routes to optically pure natural products and drugs. Unusual kinetic resolution, *Pure Appl. Chem.*, 55, 589, 1983.
11. **Behrens, C. H. and Sharpless, K. B.,** New transformations of 2,3-epoxy alcohols and related derivatives. Easy route to homochiral substances, *Aldrichim. Acta,* 16, 67, 1983.
12. **Fieser, L. F. and Fieser, M.,** *Steroids,* Reinhold, New York, 1959.
13. **Kirk, D. N. and Hartshorn, M. P.,** *Steroid Reaction Mechanisms,* Elsevier, Amsterdam, 1968.
14. **Yoshii, E., Oribe, T., Tumura, K., and Koizumi, T.,** Studies on the synthesis of cardiotonic steroids. 4. Synthesis of strophanthidin, *J. Org. Chem.*, 43, 3946, 1978.
15. **Kočovský, P.,** Synthesis of 14-deoxy-14α-strophanthidin, *Tetrahed. Lett.*, 21, 555, 1980.
16. **Sláma, K., Romaňuk, M., and Šorm, F.,** *Insect Hormones and Bioanalogues,* Springer, Wien, 1974.
17. **Imhof, R., Gösinger, E., Graf, W., Berger-Fenz, L., Berner, H., Schaufelberger, P., and Wherli, H.,** Steroide und Sexualhormone. 246. Die Partialsynthese von Batrachtoxinin A, *Helv. Chim. Acta,* 56, 139, 1973.
18. **Mori, K.,** The synthesis of insect pheromones, in *The Total Synthesis of Natural Products,* Vol. 4, ApSimon, J., Ed., J. Wiley & Sons, New York, 1981, 1.
19. The Merck Index, 9th edition, Merck and Co., Rahway, N.J., 1976.
20. **Pawson, B. A., Cheung, H. C., Gubraxani, S., and Saucy, G.,** Syntheses of natural (+)-juvabione, its enantiomer (−)-juvabione, and their diastereoisomer (+)- and (−)- epijuvabione, *J. Am. Chem. Soc.*, 92, 336, 1970.
21. **Nakazaki, M. and Isobe, S.,** The structure of todomatuic acid. Synthesis of (+)-dihydrodesoxo-todomatuic acid, *Bull. Chem. Soc. Jpn.*, 34, 741, 1961.
22. **Nakazaki, M. and Isobe, S.,** The structure of todomatuic acid. The synthesis of (±)-dihydrodesoxo-todomatuic acid, *Bull. Chem. Soc. Jpn.*, 36, 1198, 1963.
23. **Lange, G. L., Orom, W. J., and Wallace, D. J.,** A stereoselective synthesis of the spirosesquiterpene (−)-acorenone, *Tetrahed. Lett.*, 4479, 1977.
24. **Ruppert, J. F., Avery, M. A., and White, J. D.,** Synthesis of (−)-acorenone B, *J. Chem. Soc. Chem. Commun.*, 978, 1976.
25. **Wawrenczyk, C. and Zabża, A.,** Insect growth regulants. VIII. Synthesis of a cyclic analog of juvenile hormone — II, *Tetrahedron,* 36, 3091, 1980.
26. **McGuirk, P. R. and Collum, D. B.,** Total synthesis of (+)-phyllatocin, *J. Am. Chem. Soc.*, 104, 4496, 1982.
27. **Kugler, E. and Kováts, E.,** Zur Kentnnis des Mandarinenschalen-Öls, *Helv. Chim. Acta,* 46, 1480, 1963.
28. **Walbaum, H. and Hüthig, O.,** Über das Gingergrasöl, *J. Prakt. Chem.*, 71, 459, 1905.
29. **Royals, E. E. and Horne, S. E., Jr.,** Conversion of *d*-limonene to *l*-carvone, *J. Am. Chem. Soc.*, 73, 5856, 1951.
30. **Shono, T., Nishigushi, I., Yokoyama, T., and Nitta, M.,** A novel synthesis of l-carvone from l-α-pinene, *Chem. Lett.*, 433, 1975.
31. U.S. Pat. 2,796,428 (1957 to Gilden), Booth and Klein.
32. **Corey, E. J. and Pearce, H. L.,** Total synthesis of picrotoxinin, *J. Am. Chem. Soc.*, 101, 5841, 1979.
33. **Corey, E. J. and Pearce, H. L.,** Total synthesis of Picrotin, *Tetrahed. Lett.*, 21, 1823, 1980.
34. **Baker, L., Evans, D. A., and McDowell, P. G.,** Stereospecific synthesis of 4,11-epoxy-cis-eudesmane, a tricyclic sesquiterpene secretion from the termite *Amitermes evuncifer, J. Chem. Soc. Chem. Commun.*, 111, 1977.
35. **Corey, E. J., Ensley, H. E., and Suggs, J. W.,** *J. Org. Chem.*, 41, 380, 1976.
36. **Sato, T., Kawara, T., Nishizawa, A., and Fujisawa, T.,** A novel synthetic method for optically active terpenes by the ring-opening reaction of (R)-(+)-β-propiolactone, *Tetrahed. Lett.*, 21, 3377, 1980.
37. **Oppolzer, W. and Petrzilka, M.,** An enantioselective total synthesis of natural (+)-luciduline, *Helv. Chim. Acta,* 61, 2755, 1978.
38. **Ayer, W. A., Masaki, N., and Nkunika, D. S.,** Luciduline: a unique type of *lycopodium* alkaloid, *Can. J. Chem.*, 46, 3631, 1968.
39. **Sepulveda, J. and Polo, M. C.,** Acyclic aldehydes from cyclic terpenes. Preparation of (3R)-3,7-dimethyl-7-methoxy-6-oxooctanal from (+)-pulegone, *Ann. Quim,* 75, 398, 1979; *Chem. Abstr.*, 91, 193432c, 1979.
40. **Johnson, F. and Whitehead, A.,** The stereochemistry of 2-substituted cyclohexanone amines and the corresponding Schiff's bases, *Tetrahed. Lett.*, 3825, 1964.
41. **Schaffer, H. J. and Jain, V. K.,** The synthesis of dehydrocycloheximide and the conversion of *cis*-2,4-dimethylcyclohexanone to its *trans* isomer, *J. Org. Chem.*, 29, 2595, 1964.
42. **Boeckman, R. K.,Naegely, P. C., and Arthur, S. D.,** Efficient enantioselective synthesis of the antitumor agent sarkomycin, *J. Org. Chem.*, 45, 752, 1980.
43. **Vettel, P. R. and Coates, R. M.,** Total synthesis of (−)-prezizaene and (−)-prezizanol, *J. Org. Chem.*, 45, 5432, 1980.

44. **Marx, N. and Norman, L. R.,** Synthesis of (−)-acorone and realted spirocyclic sesquiterpenes, *J. Org. Chem.*, 40, 1602, 1975.
45. **Corey, E. J. and Ensley, H. E.,** Preparation of an optically active prostaglandin intermediate via asymmetric induction, *J. Am. Chem. Soc.*, 97, 6908, 1975.
46. **Wuest, J. D., Madonik, A. M., and Gordon, D. C.,** Vinylketenes. Synthesis of (+)-actinidine, *J. Org. Chem.*, 42, 2111, 1977.
47. **Sakan, T., Fujimoto, A., Murai, F., Butsugan, I., and Suzui, A.,** On the structure of actinidine and matatabilactone, the effective components of *Actinidia polygama*, *Bull. Chem. Soc. Jpn.*, 32, 315, 1959.
48. **Paquette, L. A.,** Recent synthetic developments in polyquinane chemistry, *Topics in Current Chemistry*, 119, 1, 1984.
49. **Plešek, J.,** Die Hydrolyse der C-C-Bindung in einigen β-Halogenketonen II. Darstellung der (+)-citronellsäure aus (+)-pulenone, *Collect. Czech. Chem. Commun.*, 22, 644, 1957.
50. **Rienäcker, R. and Ohlhof, G.,** Optisch aktives β-Citronellol aus (+)- oder (−)-Pinan, *Angew. Chem.*, 73, 240, 1961.
51. **Ziegler, K., Krupp, F., and Zozel, K.,** Metalloorganische Verbindungen, XL: Synthese von Alkoholen aus Organoaluminium-Verbindunge, *Justus Liebegs Ann. Chem.*, 629, 241, 1960.
52. **Rienäcker, R. and Ohlof, G.,** Optisch aktives β-Citronellol aus (+)- oder (−)-Pinan, *Angew. Chem.*, 73, 240, 1961.
53. **Naves, Y.-R. and Tullen, P.,** Synthèses des *cis*- et *trans*-(méthyl-2-propène-1)-yl-2-méthyl-4-tétrahydropyranes, *Helv. Chim. Acta*, 44, 1867, 1961.
54. **Eschinazi, H. E.,** Structural studies in the citronellyl and rhodinyl series. The synthesis of rhodinal and rhodinol, *J. Org. Chem.*, 26, 3072, 1961.
55. **Shono, T., Matsumura, Y., Hibino, K., and Miyawaki, S.,** Novel synthesis of *l*-citronellol from *l*-menthone, *Tetrahed. Lett.*, 1295, 1974.
56. **Hidai, M., Ishiwatari, H., Yagi, H., Tanaka, E., Onozawa, K., and Uchida, Y.,** Synthesis of (+)- or (−)-citronellol from isoprene, *J. Chem. Soc. Chem. Commun.*, 170, 1975.
57. **Mori, K., Suguro, T., and Uchida, M.,** Synthesis of optically active forms of (Z)-14-methyl-hexadec-8-enal, the pheromone of female Dermestid beetle, *Tetrahedron*, 34, 3119, 1978.
58. **Cenigliaro, G. J. and Kocienski, P. J.,** Synthesis of (−)- α-multistriatin, *J. Org. Chem.*, 42, 3622, 1977.
59. **Mori, K.,** Absolute configuration of (−)-4-methylheptan-3-ol, a pheromone of the smaller European elm bark beetle, as determined by the synthesis of its (3R,4R)-(+)- and (3S,4R)-(+)-isomers, *Tetrahedron*, 33, 289, 1977.
60. **Gore, W. E., Pearce, G. T., and Silverstein, R. M.,** Relative stereochemistry of multistriatin (2,4-dimethyl-5-ethyl-16,8-dioxabicyclo[3.2.1]octane), *J. Org. Chem.*, 40, 1705, 1975.
61. **Ireland, R. E., McGarvey, G. J., Anderson, R. C., Badoud, R., Fitszimmons, B., and Thaisrivongs, S.,** A chiral synthesis of the left-side aldehyde for lasalocid A synthesis, *J. Am. Chem. Soc.*, 102, 6179, 1980.
62. **Ireland, R. E., Thaisrivongs, S., and Wilcox, C. S.,** Total synthesis of Lasalocid A (X537A), *J. Am. Chem. Soc.*, 102, 1155, 1980.
63. **Anderson, R. J., Adams, K. G., Chinn, H. R., and Henrick, C. A.,** Synthesis of the optical isomers of 3-methyl-6-isopropenyl-9-decen-1-yl acetate, a component of the California Red scale pheromone, *J. Org. Chem.*, 45, 2229, 1980.
63a. **Brown, H. C., Jadhav, P. K., and Desai, M. C.,** A convenient procedure for upgrading commercial (+)- and (−)-α-pinene to material of high optical purity, *J. Org. Chem.*, 47, 4583, 1982.
64. **Hobbs, P. D. and Magnus, P. D.,** Studies on terpenes. 4. Synthesis of optically active grandisol, the Boll weevil pheromone, *J. Am. Chem. Soc.*, 98, 4594, 1976.
65. **Tumlinson, J. H.,** Sex pheromones produced by male Boll weevil: Isolation, identification, and synthesis, *Science*, 166, 1010, 1969.
66. **Mitra, R. B. and Khanra, A. S.,** A stereospecific synthesis of methyl (+)-trans-chrysanthemate from (+)-α-pinene, *Syn. Commun.*, 7, 245, 1977.
67. **Mechoulam, R., Braun, R., and Gaoni, Y.,** A stereospecific synthesis of (−)-Δ^1- and (−)-$\Delta^{1(6)}$-tetrahydrocannabinols, *J. Am. Chem. Soc.*, 89, 4552, 1967.
68. **Petrzilka, T., Haefliger, W., Sikemeier, C., Ohlhoff, G., and Eschenmoser, A.,** Synthese und Chiralität des (−)-cannabidiols, *Helv. Chim. Acta*, 50, 719, 1967.
69. **Petrzilka, T., Haefliger, W., and Sikemeier, C.,** Synthese von Haschisch-Inhaltstoffen, *Helv. Chim. Acta*, 52, 1102, 1969.
70. **Razdan, R. K., Handrick, G. R., and Dalzell, H. C.,** A one-step synthesis of (−)-Δ^1-tetrahydrocannabinol from chrysanthenol, *Experientia*, 31, 16, 1975.
71. **Razdan, R. K. and Handrick, G. R.,** A stereospecific synthesis of (−)-Δ^1- and (−)-$\Delta^{1(6)}$-tetrahydrocannabinols, *J. Am. Chem. Soc.*, 92, 6061, 1970.
72. **ApSimon, J.,** *The Total Synthesis of Natural Products*, Vol. II, John Wiley & Sons, New York, 55, 1973.

73. **Freudenberg, K., Lwowski, W., and Hohmann, W.,** Die Konfiguration des tertiären Kohlenstoffatoms, VII. Konfiguration des Camphers II, *Justus Liebigs Ann. Chem.*, 594, 76, 1955.
74. **Stevens, R. V., Gaeta, F. C. A., and Lawrence, D. S.,** Camphorae: Chiral intermediates for the enantiospecific total synthesis of steroids, 1, *J. Am. Chem. Soc.*, 105, 7713, 1983.
75. **Liu, H. and Chan, W. H.,** A total synthesis of (−)-khusimone, *Can. J. Chem.*, 57, 708, 1979.
76. **Stevens, R. V. and Gaeta, F. C. A.,** An experimental determination of "bonding charge" in carbon-carbon bonds, *J. Am. Chem. Soc.*, 99, 6106, 1977.
77. **Büchi, G., MacLeod, W. D., Jr., and Padilla, J.,** Terpenes. XIX. Synthesis of Patchouli alcohol, *J. Am. Chem. Soc.*, 86, 4438, 1964.
78. **Oppolzer, W. and Flaskamp, E.,** An enatioselective synthesis and the absolute configuration of natural pumiliotoxin C, *Helv. Chim. Acta*, 60, 204, 1977.
79. **Witkop, B.,** New direction in the chemistry of natural products: The organic chemist as a pathfinder for biochemistry and medicine, *Experientia*, 27, 1121, 1971.
80. **Habermehl, G., Andress, H., Miyahara, K., Witkop, B., and Daly, J. V.,** Synthese von Pumiliotoxin C, *Justus Liebigs Ann. Chem.*, 1577, 1976.
81. **Habermehl, G. and Andres, H.,** Synthese von rac. Pumiliotoxin C, *Naturwissenschaften*, 62, 345, 1975.
82. **Oppolzer, W., Fröstl, W., and Weber, H. P.,** The total synthesis of (±)-pumiliotoxin C, *Helv. Chim. Acta*, 58, 593, 1975.
83. **MacMillan, J. G., Springer, J. P., Clardy, J., Cole, R. J., and Kirskey, J. W.,** Structure and synthesis of verrucolotoxin, a new mycotoxin from *Penicillium verruculosum* Peyronel, *J. Am. Chem. Soc.*, 98, 246, 1976.
84. **Červinka, O. and Hub, L.,** Asymmetric reactions, XXVII. Absolute configuration of γ-butyrolactone-γ-carboxylic acid and γ-valerolactone-γ-carboxylic acid, *Collect. Czech. Chem. Commun.*, 33, 2927, 1968.
85. **Mori, K.,** Synthesis and absolute configuration of (−)-ipsenol (2-methyl-6-methylene-7-octen-4-ol), the pheromone of *Ips paraconphusus*, *Tetrahed. Lett.*, 2187, 1975.
86. **Mori, K.,** Synthesis of optically active forms of ipsenol, the pheromone of Ips bark beetle, *Tetrahedron*, 32, 1101, 1976.
87. **Mori, K., Sasaki, M., Tamada, S., Suguro, T., and Masuda, S.,** Synthesis of optically active 2-ethyl-1,6-dioxaspiro [4,4]nonane (chalcogran), the principal aggregation pheromone of *Pityogeneses chalcographus* (L.), *Tetrahedron*, 35, 1601, 1979.
88. **Kleeman, A., Lehmann, B., and Martens, J.,** Enantioselektive Synthese der Hydroxyanaloga von D- und L-Methionin, *Angew. Chem.*, 91, 858, 1979.
89. **Kitahara, T., Mori, K., and Matsui, M.,** Total synthesis of (+)-brefeldin A, *Tetrahed. Lett.*, 3021, 1979.
90. **Ho, P.-T. and Davies, N.,** A practical synthesis of (R)-(−)-γ-hydroxymethyl-γ-butyrolactone from natural glutamic acid, *Synthesis*, 462, 1983.
91. **Smith, L. R. and Williams, H. J.,** Glutamic acid in pheromone synthesis, *J. Chem. Educ.*, 56, 698, 1979.
92. **Iwaki, S., Marumo, S., Saito, T., Yamada, M., and Katagiri, K.,** Synthesis of optically active dispalure, *J. Am. Chem. Soc.*, 96, 7842, 1974.
93. **Mori, K., Takigawa, T., and Matsiu, M.,** Stereoselective synthesis of optically active dispalure, the pheromone of the Gypsy moth (*Portheria dispar* L.), *Tetrahed. Lett.*, 3953, 1976.
94. **Rossi, R.,** Insect pheromones. II. Synthesis of chiral components of insect pheromones, *Synthesis*, 413, 1978.
95. **Ravid, U., Silverstein, R. M., and Smith, L. R.,** Synthesis of enantiomers of 4-substituted γ-lactones with known absolute configuration, *Tetrahedron*, 34, 1449, 1978.
96. **Bernardi, R., Fuganti, C., Grasselli, P., and Marinoni, G.,** Synthesis of the enantiomeric forms of 4-hexanolide (γ-caprolactone) from the optically active 5-phenyl-4-pentene-2,3-diol prepared from cinnamaldehyde and baker's yeast, *Synthesis*, 50, 1980.
97. **Smith, L. R., Williams, H. J., and Silverstein, R. M.,** Facile synthesis of optically active 2-ethyl-1,6-dioxaspiro [4,4]nonane, component of the aggregation pheromone of the *Pityogeneses chalcographus* (L.), *Tetrahed. Lett.*, 3231, 1978.
98. **Francke, W., Heeman, V., Gerken, B., Renwick, J. A. A., and Vité, J. P.,** 2-Ethyl-1,6-dioxaspiro [4,4]nonane, principal aggregation pheromone of *Pityogeneses chalcographus* (L.), *Naturwissenchaften*, 64, 590, 1977.
99. **Müller-Schwarze, D., Ravid, U., Cleasson, A., Singer, A. G., Silverstein, R. M., Müller-Schwarze, C., Volkman, N. J., Zemanek, K. F., and Butler, R. G.,** The "deer lactone". Source, chiral properties, and responses by black-tailed deer, *J. Ecol. Chem.*, 4, 247, 1978; *Chem. Abstr.*, 89, 10441Ou, 1978.
100. **Mori, K.,** Synthesis of optically active forms of sulcatol, the aggregation pheromone in the Scolytid beetle, *Gnathotrichus sulcatus*, *Tetrahedron*, 31, 3011, 1975.

101. **Byrne, J. K., Swigar, A. A., Silverstein, R. M., Borden, J. H., and Stokking, E.,** Sulcatol. Population aggregation pheromone in the Scolytid beetle, *Gnathotrichus sulcatus, J. Insect. Physiol.*, 20, 1895, 1974; *Chem. Abstr.*, 82, 28785b, 1975.
102. **Borden, J. H., Chong, L., McLean, J. A., Slessor, K. N., and Mori, K.,** *Gnathotrichus sulcatus.* Synergistic response to enantiomers of the aggregation pheromone sulcatol, *Science,* 192, 894, 1976.
103. **Robin, J. P., Gringore, O., and Brown, E.,** Asymmetric total synthesis of antileukaemic lignan precursor (−)-steganone and revision of its absolute configuration, *Tetrahed. Lett.*, 21, 2709, 1980.
104. **Tomioka, K., Ishiguro, T., and Koga, K.,** Asymmetric total synthesis of the antileukaemic lignans (+)-trans-burseran and (−)-isostegan, *J. Chem. Soc. Chem. Commun.*, 652, 1979.
105. **Tomioka, K., Ishiguro, T., and Koga, K.,** First total synthesis of (+)-steganacin. Determination of absolute stereochemistry, *Tetrahed. Lett.*, 21, 2973, 1980.
106. **Takano, S., Chiba, K., Yonaga, M., and Osagawara, K.,** Enantioselective synthesis of (+)-quebrachamine using L-glutamic acid as a chiral template, *J. Chem. Soc. Chem. Commun.*, 616, 1980.
107. **Takano, S., Yonaga, M., Chiba, K., and Osagawara, K.,** Enantioselective synthesis of (−)-velbamine and (+)-isovelbamine using L-glutamic acid as chiral template, *Tetrahed. Lett.*, 21, 3697, 1980.
108. **Wood, D. L., Stark, R. W., and Silverstein, R. M.,** Unique synergistic effects produced by the principal sex attractant compounds of *Ips paraconfusus* (le Conte) (*Coleoptera: Scolytidae*), *Nature,* 215, 206, 1967.
109. **Vité, J. P., Hedden, R., and Mori, K.,** *Ips grandicolis.* Field response to the optically pure pheromone, *Naturwissenschaften,* 63, 43, 1976.
110. **Bláha, K., Buděšinský, M., Frič, I., Koblicová, Z., Maloň, P., and Tichý, M.,** Polycyclic dilactams with inherently chiral amid chromophores, *Tetrahed. Lett.*, 3949, 1978.
111. **Woodward, R. B., Heusler, K., Gostelli, J., Naegeli, P., Oppolzer, W., Ramage, R., Ranganathan, S., and Vorbrüggen, H.,** The total synthesis of cephalosporin C., *J. Am. Chem. Soc.*, 88, 852, 1966.
112. **Confalone, P. N., Pizzolato, G., Baggiolini, E. G., Lollar, D., and Uskoković, M. R.,** A stereospecific total synthesis of *d*-biotin, *J. Am. Chem. Soc.*, 97, 5936, 1975.
113. **Leimgruber, W., Batcho, A. D., and Czajkowski, R. C.,** Total synthesis of anthramycin, *J. Am. Chem. Soc.*, 90, 5641, 1968.
114. **Saitoh, Y., Moriyama, Y., Takahashi, A., and Khuong-Huu, Q.,** Chiral and stereoselective total synthesis of (−)-deoxyprosopinine and (−)-deoxyprosophylline, *Tetrahed. Lett.*, 21, 75, 1980.
115. **Saitoh, Y., Moriyama, Y., Hirota, H., and Takahashi, T.,** Synthesis of dihydrosphingosine, *Bull. Chem. Soc. Jpn.*, 53, 1783, 1980.
116. **Scott, A. I. and Wilkinson, T. J.,** Synthesis of L-α-aminoadipic acid from L-lysine, *Synthetic Commun.*, 10, 127, 1980.
117. **Wong, C. M.,** Concerning the mechanism of the addition of ethanesulfenyl chloride to 3,4-dihydro-2H-pyran, *Can. J. Chem.*, 46, 1101, 1968.
118. **Yamada, S., Konda, M., and Shiori, T.,** A biogenetic type synthesis of (S)-(+)-laudanosine from L-(−)-dopa, *Tetrahed. Lett.*, 2215, 1972.
119. **McKenzie, A., Plenderleith, H. J., and Walker, N.,** Optical activation of racemic acid by *d*-malic acid, *J. Chem. Soc.*, 123, 2875, 1923.
120. **Paul, K. G., Johnson, F., and Favara, D.,** Prostaglandins I. Direct synthesis of optically active Corey-intermediate from (S)-(−)-malic acid, *J. Am. Chem. Soc.*, 98, 1285, 1976.
121. **Bindra, J. S. and Bindra, R.,** *Prostaglandin Synthesis,* Academic Press, New York, 1977.
122. **Corey, E. J., Niwa, H., and Knolle, J.,** Total synthesis of (S)-12-hydroxy-5,8,14-cis-10-trans-eicosatetraenoic acid (Samuelsson's HETE), *J. Am. Chem. Soc.*, 100, 1942, 1978.
123. **Tomioka, T. and Koga, K.,** An efficient synthesis of (+)-sesbanine, *Tetrahed. Lett.*, 21, 2321, 1980.
124. **Seebach, D. and Wasmuth, D.,** Herstellung von *erythro*-2-hydroxybernsteinsäure-Derivaten aus Äpfelsäure ester, *Helv. Chim. Acta,* 63, 197, 1980.
125. **Hungerbühler, E., Naef, R., Wasmuth, D., Seebach, D., Loosli, H. R., and Wehrli, A.,** Synthese optisch aktiever 2-Methyl- und 2-Äthyl-1,6-dioxaspiro[4.4]nonan und [4.5]dekan — Pheronone aus einem gemeinsamen chirale Vorläufer, *Helv. Chim. Acta,* 63, 1960, 1980.
126. **Züger, M., Weler, T., and Seebach, D.,** 2,3-Disubstituted γ-butyrolactons from the Michael addition of doubly deprotonated, optically active β-hydroxycarboxylates to Nitroolefins, *Helv. Chim. Acta,* 63, 2005, 1980.
127. **Purdie, T. and Walker, J. W.,** Resolution of lactic acid into its optically active components, *J. Chem. Soc.*, 61, 754, 1982.
128. **Borsook, H., Huffman, H. M., and Liu, Y.-P.,** The preparation of crystalline lactic acid, *J. Biol. Chem.*, 102, 449, 1933.
129. **Brown, H. C., Pai, G. G., and Jadhav, P. K.,** Remarkable optical induction in the reduction of α-keto esters with B-(3-pinanyl)-9-borabicyclo[3.3.1]nonaner. Synthesis of α-hydroxy esters of 100% optical purity, *J. Am. Chem. Soc.*, 106, 1531, 1984.
130. **Gombos, J., Haslinger, F., and Schmidt, U.,** Notiz über eine einfache Herstellung von (S)-Propylenoxid, *Chem. Ber.*, 109, 1645, 1976.

131. **Golding, B. T., Hall, D. R., and Sakrikar, S.,** Reaction between vicinal diols and hydrogen bromide in acetic acid; synthesis of chiral propylene oxide, *J. Chem. Soc. Perkin Trans.*, 1, 1214, 1973.
132. **Seebach, D. and Pohmakotr, M.,** Synthesis of (+)-(S,S)-(cis-6-methyltetrahydropyran-2-yl)acetic acid and of (−)-(R,R)-dideoxy-pyrenophorine using a new d^5-reagent, *Helv. Chim. Acta*, 62, 843, 1979.
133. **Mali, R. S., Pohmakotr, M., Weidmann, B., and Seebach, D.,** A short synthesis of (R,R)-(−)-pyrenophorine from (S)-propylene oxide and a 3-pentenoic acid d^5-reagent, *Justus Liebigs Ann. Chem.*, 2272, 1981.
134. **Schmidt, U., Gombos, J., Haslinger, E., and Zak, H.,** Hochstereoselektive Totalsynthese des nonactins, *Chem. Ber.*, 109, 2628, 1976.
135. **Hintzer, K., Weber, R., and Schurig, V.,** Synthesis of optically active 2S- and 7S-methyl 1,6-dioxaspiro [4.5] decane, the pheromone components of *Paravespula vulgaris* (L.) from S-ethyl lactate, *Tetrahed. Lett.*, 22, 55, 1981.
136. **Utimoto, K., Uchida, K., Yamaya, M., and Nozaki, H.,** A simple stereoselective synthesis of (R)-recifeiolide by adoption of (R)-methyl oxirane as a chiral source, *Tetrahed. Lett.*, 3641, 1977.
137. **Corey, E. J., Trybulski, E. J., Melvin, L. S., Jr., Nicolaou, K. C., Secrist, J. A., Lett, R., Sheldrake, P. W., Falck, J. R., Brunelle, D. J., Haslanger, M. F., Kim, S., and Yoo, S.,** Total synthesis of erythromycins, 3. Stereoselective routes to intermediates corresponding to $C_{(1)}$ to $C_{(9)}$ and $C_{(10)}$ to $C_{(13)}$ fragments of erythronolide B, *J. Am. Chem. Soc.*, 100, 4618, 1978.
138. **Corey, E. J., Kim, S., Yoo, S., Nicolaou, K. C., Melvin, L. S., Jr., Brunelle, D. J., Falck, J. R., Trybulski, E. J., Lett, R., and Sheldrake, P. W.,** Total synthesis of erythromycins. 4. Total synthesis of erythronolide B, *J. Am. Chem. Soc.*, 100, 4620, 1978.
139. **Coke, J. L. and Shue, R. S.,** Nucleophilic ring opening of optically pure (R)-(+)-1,2-epoxybutane. Synthesis of new (R)-2-butanol derivative, *J. Org. Chem.*, 38, 2210, 1973.
140. **Hungerbühler, E., Seebach, D., and Wasmuth, D.,** Chirale Reagentien aus Weinsäure. 1-Benzoyloxy-3,4-epoxy-2-butanol, ein vielseitiges Zwischenprodukt für die Enantiomerensynthese, *Angew. Chem.*, 91, 1025, 1979.
141. **Seuring, B. and Seebach, D.,** Synthese und Bestimmung der absoluten Konfiguration von Norpyrenophorin, Pyrenophorin und Vermiculin, *Justus Liebigs Ann. Chem.*, 2044, 1978.
142. **Seebach, D.,** Synthese und Bestimmung der absoluten Konfiguration von Pyrenophorin und Vermiculin, *Angew. Chem.*, 89, 270, 1977.
143. **Eliel, E. L.,** *Stereochemistry of Carbon Compounds*, McGraw-Hill, New York, 1962.
144. **Holleman, A. F.,** *dl*-Tartaric acid, in *Organic Synthesis, Collective*, Vol. I, John Wiley & Sons, New York, 497, 1946.
145. **Mori, K.,** Synthesis of exobrevicomin, the pheromone of Western Pine beetle, to obtain optically active forms of known absolute configuration, *Tetrahedron*, 30, 4223, 1974.
146. **Meyer, H. and Seebach, D.,** Synthese einiger Pilzmetabolite mit 4-methoxy-5,6-dihydro-2-pyron Structur, *Justus Liebigs Ann. Chem.*, 2261, 1975.
147. **Mori, K. and Tamada, S.,** Stereocontrolled synthesis of all of the four possible stereoisomers of erythro-3,7-dimethylpentadec-2-yl acetate and propionate, the sex pheromone of the pine sawflies, *Tetrahedron*, 35, 1279, 1979.
148. **Musich, J. A. and Rapoport, H.,** Synthesis of anthopleurine, the alarm pheromone from *Anthopleura elegantissima*, *J. Am. Chem. Soc.*, 100, 4865, 1978.
149. **Fronza, G., Fuganti, C., and Grasselli, P.,** Synthesis of N-benzoyl-L-ristosamine, *Tetrahed. Lett.*, 21, 2999, 1980.
150. **Ogura, K., Yamashita, M., and Tschuchilashi, G.,** Synthesis of (R)- and (S)-4-hydroxy-2-cyclopentenones, *Tetrahed. Lett.*, 759, 1976.
151. **Seebach, D., Kalinowski, H.-O., Bastani, B., Crass, G., Daum, H., Dörr, H., DuPrez, N. P., Ehrig, V., Langer, W., Nüssler, C., Oei, H.-A., and Schmidt, M.,** Herstellung von Hilfstoffen für die asymmetrische synthese aus Weinsäure. Addition von Butyllithium an Aldehyde in Chiralem Medium, *Helv. Chim. Acta*, 60, 301, 1977.
152. **Carmack, M. and Kelley, C.,** The synthesis of the optically active Cleland reagent [(−)-1,4-dithio-L_g-threitol], *J. Org. Chem.*, 33, 2171, 1968.
153. **Tsuziki, Y.,** Synthese einiger Brückenderivate der Weinsäure durch Acetylisierung ihrer Ester mit Aceton, Acetaldehyd und Formaldehyd, *Bull. Chem. Soc. Jpn.*, 11, 362, 1936.
154. **Lorette, N. B. and Howard, W. L.,** Acetone dibutylacetate, *Org. Syn.*, 42, 1, 1962.
155. **Wegner, R. M.,** Synthesis of cyclosporine. I. Synthesis of enantiomerically pure (2S,3R,4R,6E)-3-hydroxy-4-methyl-2-methylamino-6-octonoic acid starting from tartaric acid, *Helv. Chim. Acta*, 66, 2308, 1983.
156. **Branca, Q. and Fischli, A.,** Eine chiral ökonomische Totalsynthese von (R)- und (S)-Muskon via Epoxysulfoncyklofragmentierung, *Helv. Chim. Acta*, 60, 925, 1977.
157. **Evans, D. A., Sack, C. E., Kleschick, W. A., and Taber, T. R.,** Polyether antibiotic synthesis. Total synthesis and absolute configuration of the ionophore A-23187, *J. Am. Chem. Soc.*, 101, 6789, 1979.

158. **Meyers, A. I., Babiak, K. A., Campbell, A. L., Comins, D. L., Fleming, M. P., Henning, R., Heuschmann, M., Hudspeth, J. P., Kane, J. M., Reider, P. J., Roland, D. M., Shimizu, K., Tomioka, K., and Walkup, R. D.,** Total synthesis of (−)-maysine, *J. Am. Chem. Soc.*, 105, 5015, 1983.

159a. **Collum, D. B., McDonald, J. H., and Still, W. C.,** Synthesis of the polyether antibiotic monensin. 1. Strategy and degradations, *J. Am. Chem. Soc.*, 102, 2117, 1980.

159b. **Collum, D. B., McDonald, J. H., and Still, W. C.,** Synthesis of the polyether antibiotic monensin. 2. Preparation of intermediates, *J. Am. Chem. Soc.*, 102, 2118, 1980.

159c. **Collum, D. B., McDonald, J. H., and Still, W. C.,** Synthesis of the polyether antibiotic monensin. 3. Coupling of precursors and transformation to monensin, *J. Am. Chem. Soc.*, 102, 2120, 1980.

160. **Ridley, D. D. and Strawlow, M.,** The stereospecific asymmetric reduction of functionalised ketones, *J. Chem. Soc. Chem. Commun.*, 400, 1975.

161. **Levene, P. A. and Walti, A.,** 1-Propylene glycol, *Organic Synthesis, Collective,* Vol. II, John Wiley & Son, New York, 1946, 545.

162. **Aberhart, D. J.,** Microbial preparation of (S)-(+)-2,3-dihydroxy-3-methylbutanoic acid by *syn* dihydroxylation of 3-methylcrotonic acid, *J. Org. Chem.*, 45, 5218, 1980.

163. **Cohen, M., Lopresti, R., and Saucy, G.,** A novel total synthesis of (2R,4′S,8′R)-α-tocopherol (vitamin E). Construction of chiral chromanes from an optically active nonaromatic precursor, *J. Am. Chem. Soc.*, 101, 6710, 1979.

164. **Adams, R. and Hauserman, F. B.,** The total structure of monocrotalinee. XIII. Synthesis of dihydroanhydromonocrotalic acid, *J. Am. Chem. Soc.*, 74, 694, 1952.

165. **Schmid, M. and Barner, R.,** Totalsynthese von natürlichem α-Tocopherol: Aufbau der Seitenkette aus (−)-(S)-3-Methyl-γ-butyrolacton, *Helv. Chim. Acta*, 62, 464, 1979.

166. **Mori, K.,** Synthesis of Optically active forms of frontalin, the pheromone of *Dendroctonus* bark beetles, *Tetrahedron,* 31, 1381, 1975.

167. **Kinze, G. W.,** Bark Beetle attractants. Identification, synthesis and field bioassay of a new compound isolated from *Dendroctonus, Nature,* 221, 477, 1969.

168. **Fujimoto, Y., Yadav, J. S., and Sih, C. J.,** (S)-Citramalic acid, a useful synthon of the synthesis of 15-deoxy-16(S)-hydroxy-16-methylprostaglandins, *Tetrahed. Lett.*, 21, 2481, 1980.

169. **Černý, M., Elbert, T., and Pacák, J.,** Preparation of 1,6-anhydro-2,3-dideoxy-2,3-epimino-β-D-mannopyranose and its conversion to 2-amino-1,6-anhydro-2-deoxy-β-D-mannopyranose, *Collect. Czech. Chem. Commun.*, 39, 1752, 1974.

170. **Černý, M. and Staněk, J., Jr.,** 1,6-Anhydro derivatives of aldohexoses, *Adv. Carbohyd. Chem.*, 34, 23, 1977.

171. **Guthrie, R. D. and Murphy, D.,** Nitrogen-containing carbohydrate derivatives. IV. Some azido and epimino sugars, *J. Chem. Soc.*, 5228, 1963.

172. **Baer, E.,** L-α-Glycerophosphoric acid (barium salt), *Biochem. Prep.*, 2, 31, 1952.

173. **Baldwin, J. J., Raab, A. W., Mensler, K., Arison, B. H., and McClure, D. E.,** Synthesis of (R)- and (S)-epichlorhydrin, *J. Org. Chem.*, 43, 4876, 1978.

174. **Nelson, W. L., Wennerstroom, J. E., and Sankar, S. R.,** Absolute configuration of glycerol derivatives. 3. Synthesis and circular dichroism spectra of some chiral 3-aryloxy-1,2-propanediols and 3-aryloxy-1-amino-2-propanols, *J. Org. Chem.*, 42, 1006, 1977.

175. **Morpian, C. and Tisserand, M.,** A possible model for a new chiral glyceride synthesis. 1. Synthesis of 1-O-aroyl-2-O-tosyl-*sn*-glycerols, *J. Chem. Soc. Perkin Trans.*, 1, 1379, 1979.

176. **Ollis, W. D., Smith, C., and Wright, D. E.,** The orthosomycin family of antibiotics. I. The constitution of flambamycin, *Tetrahedron,* 35, 105, 1979.

177. **Barner, R. and Scoyd, M.,** Totalsynthese von natürlichen α-Tocopherol: Aufbau des Chromanringsystems aus Trimethylhydrochinon und einem optisch aktivem C_4 bezw. C_5 Synthon, *Helv. Chim. Acta,* 62, 2384, 1979.

178. **Schmid, M. and Barner, R.,** Totalsynthese von natürlichen α-Tocopherol: Aufbau der Seitenkette aus (−)-(S)-3-Methyl-γ-butyrolacton, *Helv. Chim. Acta,* 62, 464, 1979.

179. **Klyne, W. and Buckingham, J.,** *Atlas of Stereochemistry,* Chapman and Hall, London, 1974.

180. **Baer, E. and Fischer, H.,** Synthese optisch-aktiever Glyceride, *Naturwissenschaften,* 25, 588, 1937.

181. **Baer, E. and Fischer, H.,** Conversion of D(+)-acetone glycerol into its enantiomorph, *J. Am. Chem. Soc.*, 67, 944, 1945.

182. **Kanda, P. and Wells, M. A.,** A simplified procedure for the preparation of 2,3-O-isopropylidene-*sn*-glycerol from L-arabinose, *Lipid Res.*, 21, 257, 1980; *Chem. Abstr.*, 93, 72149y, 1980.

183. **Jung, M. E. and Shaw, T. J.,** Total synthesis of (R)-glycerol acetonide and the antiepileptic and hypotensive drug (−)-γ-amino-β-hydroxybutyric acid (GABDB). Use of vitamin C as a chiral starting material, *J. Am. Chem. Soc.*, 102, 6304, 1980.

184. **Baer, E. and Fischer, H.,** Studies on acetone-glyceraldehyde. VII. Preparation of *l*-glyceraldehyde and *l*(−)-acetone glycerol, *J. Am. Chem. Soc.*, 61, 761, 1939.

185. **Angyal, S. J. and Hoskinson, R. M.,** L-Mannitol from L-inositol, *Meth. Carbohydr. Chem.*, 2, 87, 1964.

186. **Lok, C. M., Ward, J. P., and van Dorp, D. A.,** The synthesis of chiral glycerides starting from D- and L-serine, *Chem. Phys. Lipids,* 16, 115, 1976.
187. **Mori, K.,** Absolute configuration of (+)-ipsdienol, the pheromone of *Ips paraconfusus* L., as determined by the synthesis of its (−)-isomer, *Tetrahed. Lett.,* 1609, 1976.
188. **Mori, K.,** Synthesis and absolute configuration of (−)-ipsenol (2-methyl-6-methylene-7-octen-4-ol), the pheromone of Ips paraconfusus L., *Tetrahed. Lett.,* 2187, 1975.
189. **Mori, K.,** Synthesis of optically active forms of ipsenol, the pheromone of Ips bark beetles, *Tetrahedron,* 32, 1101, 1976.
190. **Mori, K., Takigawa, T., and Matsuno, T.,** Synthesis of optically active forms of ipsdienol and ipsenol, the pheromone components of Ips bark beetles, *Tetrahedron,* 35, 933, 1979.
191. **Mori, K.,** Synthesis of (1S:2R:4S:5R)-(−)-α-multistriatin, the pheromone in the smaller European elm bark beetle *Scolytus multistriatus, Tetrahedron,* 32, 1979, 1976.
192. **Stork, G. and Takahashi, T.,** Chiral synthesis of prostaglandin (PGE_1) from D-glyceraldehyde, *J. Am. Chem. Soc.,* 99, 1275, 1977.
193. **Miller, J. G., Kurz, W., Untch, K. G., and Stork, G.,** Highly stereoselective total syntheses of prostaglandins via stereospecific sulfenate-sulfoxide transformations. 13-*cis*-15β-prostaglandins E_1 to prostaglandins E_2, *J. Am. Chem. Soc.,* 96, 6774, 1974.
194. **Takayama, H., Ohmori, M., and Yamada, S.,** Facile, stereoselective synthesis of (24R)-24,25-dihydroxyvitamin D_3 using D-glyceric acid as chiral synthon, *Tetrahed. Lett.,* 21, 5027, 1980.
195. **Rokach, J., Young, R. N., and Kakushima, M.,** Synthesis of leucotriene — new synthesis of natural leucotriene A_4, *Tetrahed. Lett.,* 22, 979, 1981.
196. **Corey, E. J. and Kang, J.,** Stereospecific total synthesis of 11(R)-HETE (2), lipoxygenation product of arachidonic acid via the prostaglandin pathway, *J. Am. Chem. Soc.,* 103, 4618, 1981.
197. **Jurczak, J., Bauer, T., Filipek, S., Tkacz, M., and Zygor, K.,** Asymmetric induction in the high-pressure cycloaddition of 2,3-o-isopropylidene-D-glyceraldehyde to 1-methoxybuta-1,3-diene, *J. Chem. Soc. Chem. Commun.,* 540, 1983.
198. **Verheyden, J. P. H., Richardson, A. C., Bhatt, R. S., Grant, B. D., Fitch, W. L., and Moffat, J. G.,** Chiral syntheses of antibiotics anisomycin and pentenomycin from carbohydrates, *Pure Appl. Chem.,* 50, 1363, 1978.
199. **Sueda, M., Ohrui, H., and Kuzuhara, H.,** Stereoselective synthesis of (+)-cerulenin from D-glucose, *Tetrahed. Lett.,* 2039, 1979.
200. **Ohrui, H. and Emoto, S.,** Stereoselective synthesis of (+)-(−)-tetrahydrocerulenin from D-glucose. The correct absolute configuration of natural cerulenin, *Tetrahed. Lett.,* 2095, 1978.
201. **Pougny, J. R. and Sinaÿ, P.,** Syntheses of (+)- and (−)-tetrahydrocerulenin from D-xylose. Revised stereochemistry of natural (+)-cerulenin, *Tetrahed. Lett.,* 3301, 1978.
202. **Pietraszkiewicz, M. and Sinaÿ, P.,** A total synthesis of natural cerulenin from D-glucose, *Tetrahed. Lett.,* 4741, 1979.
203. **Stork, G., Takahashi, T., Kawamoto, I., and Suzuki, T.,** Total synthesis of prostaglandin $F_{2\alpha}$ by chirality transfer from D-glucose, *J. Am. Chem. Soc.,* 100, 8272, 1978.
204. **Hanessian, S. and Lavalee, P.,** A stereospecific, total synthesis of thromboxane B_2, *Can. J. Chem.,* 55, 562, 1977.
205. **Kelly, A. G. and Roberts, J. S.,** A simple, stereocontrolled synthesis of a thromoxane B_2 synthon, *J. Chem. Soc. Chem. Commun.,* 228, 1980.
206. **Corey, E. J., Shibasaki, M., and Knolle, J.,** Simple, stereocontrolled synthesis of thromboxane B_2 from D-glucose, *Tetrahed. Lett.,* 1625, 1977.
207. **Ferrier, R. J. and Prasit, P.,** Routes to prostaglandins from sugars, *Pure Appl. Chem.,* 55, 565, 1983.
208. **Hanessian, S., Dextraze, P., Fougerousse, A., and Guindon, Y.,** Synthèse stereocontrollée des precurseurs des 11-oxaprostaglandins, *Tetrahed. Lett.,* 3983, 1974.
209. **Fitzsimmons, B. J. and Fraser-Reid, B.,** Annulated pyranosides as chiral synthons for carboxylic systems. Enantiospecific routes to both (+)- and (−)-chrysanthemumdicarboxylic acids from a single progenitor, *J. Am. Chem. Soc.,* 101, 6123, 1979.
210. **Ogawa, T., Takasaka, N., and Matsui, M.,** Synthesis of (−)-*cis*-rose oxide from D-glucose, *Carbohydr. Res.,* 60, C4, 1978.
211. **Hick, D. R. and Fraser-Reid, B.,** Synthesis of one enantiomer, the other enantiomer, and a mixture of both enantiomers of frontalin from a derivative of methyl-α-D-glucopyranoside, *J. Chem. Soc. Chem. Commun.,* 869, 1976.
212. **Fitzsimmons, B. J., Plaumann, D. E., and Fraser-Reid, B.,** Chiral synthons for multistriatin, *Tetrahed. Lett.,* 3925, 1979.
213. **Redlich, H. and Francke, W.,** Optisch aktives chalcogran (2-Ethyl-1,6-dioxaspiro[4.4]nonan), *Angew. Chem.,* 92, 640, 1980.
214. **Ogawa, T., Kawano, T., and Matsui, M.,** A biomimetic synthesis of (+)-biotin from D-glucose, *Carbohydr. Res.,* 57, C31, 1977.

215. **Hanessian, S. and Frenette, R.,** Total stereospecific synthesis of (+)-azimic and (+)-carpamic acids from D-glucose, *Tetrahed. Lett.*, 3391, 1979.
216. **Tatsuta, K., Nakagawa, A., Maniwa, S., and Kinoshita, M.,** Stereospecific total synthesis and absolute configuration of macrocyclic lactone antibiotic A 26771B, *Tetrahed. Lett.*, 21, 1479, 1980.
217. **Joullié, M. M., Wang, P., and Semples, J. E.,** Total synthesis and revised structural assignment of (+)-furanomycin, *J. Am. Chem. Soc.*, 102, 887, 1980.
218. **Semple, J. E., Wang, P. C., Lysenko, Z., and Joullié, M.,** Total synthesis of (+)-furanomycin and stereoisomers, *J. Am. Chem. Soc.*, 102, 7505, 1980.
219. **Ziegler, F. E., Gillian, P. J., and Chakrabury, V. R.,** Synthesis of a functionalized D-glucose. A synthon for the carbomycins (mangamycins), *Tetrahed. Lett.*, 3371, 1979.
220. **Nicolaou, K. C., Pavia, M. R., and Seitz, S. P.,** Synthesis of 16-membered ring macrolide antibiotics. I. Stereoselective construction of the "right wing" of the carbomycins and leucomycins from D-glucose, *Tetrahed. Lett.*, 2327, 1979.
221. **Ohrui, H. and Emoto, S.,** Stereoselective synthesis of (+)- and (−)-avenaciolide from D-glucose. The correct absolute configuration of natural avenaciolide, *Tetrahed. Lett.*, 3657, 1975.
222. **Anderson, R. C. and Fraser-Reid, B.,** A synthesis of optically active avenaciolide from D-glucose. The correct stereochemistry of the natural product, *J. Am. Chem. Soc.*, 97, 3870, 1979.
223. **Anderson, R. C. and Fraser-Reid, B.,** A synthesis of naturally occurring (−)-isovenaciolide, *Tetrahed. Lett.*, 2865, 1977.
224. **Curtis, D. V., Laidler, D. A., Stoddart, J. F., and Jones, G. R.,** Synthesis of configurationally chiral cryptands and cryptates from carbohydrate precursors, *J. Chem. Soc. Chem. Commun.*, 833, 1975.
225. **Lafont, D., Sinou, D., and Descotes, G.,** Hydrogenation asymetrique de precurseurs d'aminoacides a l'amide de mono- et diphosphines derivees de sucres, *J. Organometal. Chem.*, 169, 87, 1979.
226. **Cullen, W. R. and Sugi, Y.,** Asymmetric hydrogenation catalyzed by diphosphinite rhodium complex derived from a sugar, *Tetrahed. Lett.*, 1635, 1978.
227. **Cullen, W. R. and Sugi, Y.,** Enantioselective catalytic hydrogenation of α-acetamidoacrylic acids. Diphosphinite-rhodium catalysts derived from a sugar, *Chem. Lett.*, 39, 1979.
228. **Redlich, H., Schneider, B., and Francke, W.,** Offenkettige Zuckerdithioacetale als Bausteine in der Naturstoffsynthese. II. Synthese von (1S)-(3R)-(5R)-1,3-Dimethyl-2,9-dioxabicyclo[3.3.1]nonan aus D-Glukose, *Tetrahed. Lett.*, 21, 3009, 1980.

228a. **Hecht, S. M., Ruprecht, K. M., and Jacobs, P. M.,** Synthesis of L-erythro-β-hydroxyhistidine from D-glucosamine, *J. Am. Chem. Soc.*, 101, 3982, 1979.

229. **Wang, P. C. and Joullié, M. M.,** Synthesis of (2R,4S,5S)-epiallomuscarine, (2S,3R,5S)-isoepiallomuscarine, and (2S,3S,4S,5S)-3-hydroxyepiallomuscarine from D-glucose, *J. Org. Chem.*, 45, 5359, 1980.
230. **Tulshian, D. B. and Fraser-Reid, B.,** A synthetic route to the C4 octadienic esters of trichothecenes from D-glucose, *J. Am. Chem. Soc.*, 103, 474, 1981.
231. **Kakinuma, K., Otake, N., and Yonehara, H.,** Synthesis of detoxininolactone derivatives and the revised absolute stereochemistry of detoxinine, *Tetrahed. Lett.*, 21, 167, 1980.
232. **Ohrui, H. and Emoto, S.,** Stereospecific synthesis of (+)-biotin, *Tetrahed. Lett.*, 2765, 1975.
233. **Ireland, R. E. and Vevert, J.-P.,** A chiral total synthesis of (−)- and (+)-nonactic acids from carbohydrate precursors and the definition of the transition for the enolate Claisen rearrangement in heterocyclic systems, *J. Org. Chem.*, 45, 4259, 1980.
234. **Ireland, R. E. and Vevert, J.-P.,** Synthese totale des acides (+)- et (−)-nonatique a partir de carbohydrates, *Can. J. Chem.*, 59, 572, 1981.
235. **Tam, T. F. and Fraser-Reid, B.,** Chiral models of the furenone moiety of germacranolide sesquiterpene, *J. Org. Chem.*, 45, 1344, 1980.
236. **Corey, E. J., Clark, D. A., Goto, G., Marfat, A., Mioskowski, C., Samuelson, B., and Hammarström, S.,** Stereospecific total synthesis of a "slow reacting substance" of anaphylaxis, leukotriene C_1, *J. Am. Chem. Soc.*, 102, 1436, 1980.
237. **Stork, G. and Raucher, S.,** Chiral synthesis of Prostaglandins from carbohydrates. Synthesis of (+)-15-(S)-prostaglandin A_2, *J. Am. Chem. Soc.*, 98, 1583, 1976.
238. **Cohen, N., Banner, B., and Lopresti, R. J.,** Synthesis of optically active leukotriene (SRS-A) intermediates, *Tetrahed. Lett.*, 21, 4163, 1980.
239. **Hardegger, E. and Lohse, F.,** Über Muscarine. Synthese und absolute Configuration des Muscarins, *Helv. Chim. Acta*, 40, 2383, 1957.
240. **Cox, H. C., Hardegger, E., Kögl, F., Liecht, P., Lohse, F., and Salemnik, C. A.,** Über die Synthese von racemischen Muscarin, seine Spaltung in die Antipoden und Herstellung von (−)-Muscarin aus D-Glucosamin, *Helv. Chim. Acta*, 41, 229, 1958.
241. **Wang, P. C., Lysenko, Z., and Joullié, M. M.,** A facile synthesis of D-epiallomuscarine, *Tetrahed. Lett.*, 1657, 1978.
242. **Mubarak, A. M. and Brown, D. M.,** A simple, stereospecific synthesis of (+)-muscarine, *Tetrahed. Lett.*, 21, 2453, 1980.

243. **Ireland, R. E. and Daub, J. P.,** Synthesis of chiral subunits for macrolide synthesis. The Prelog-Djerassi lactone and derivatives, *J. Org. Chem.*, 46, 479, 1981.
244. **Costa, S. S., Lagrange, A., Olesker, A., Lukacs, G., and Thang, T. T.,** Chiral synthon for the total synthesis of macrolide antibiotics, *J. Chem. Soc. Chem. Commun.*, 721, 1980.
245. **Just, G. and Payette, D. R.,** The synthesis of the (−)-enantiomer of (+)-anhydromyriocin, the γ-lactone derived from myriocin (thermozymocidin), *Tetrahed. Lett.*, 21, 3219, 1980.
246. **Fraser-Reid, B., Magdzinski, L., and Molino, B.,** New strategy for carbohydrate-based syntheses of multichiral arrays: Pyranosidic homologation. 3, *J. Am. Chem. Soc.*, 106, 731, 1984.
247. **Knierzinger, A. and Vasella, A.,** Synthesis of 6-epithienamycin, *J. Chem. Soc. Chem. Commun.*, 9, 1984.
248. **Corey, E. J. and Snider, B. B.,** Preparation of an optically active intermediate for the synthesis of prostaglandin, *J. Org. Chem.*, 39, 256, 1974.
249. **ApSimon, J.,** *The Total Synthesis of Natural Products*, Vol. 1, Wiley-Interscience, Toronto, 1973.
250. **Meyers, A. I., Williams, D. R., Erickson, G. W., White, S., and Dreulinger, M.,** Enantioselective alkylation of ketones via chiral nonracemic lithioenamines. An asymmetric synthesis of α-alkyl and α,α′-dialkyl cyclic ketones, *J. Am. Chem. Soc.*, 103, 3081, 1981.
251. **Meyers, A. I., Williams, D. R., White, S., and Erickson, G. W.,** An asymmetric synthesis of acyclic and macrocyclic α-alkyl ketones. The role of (E)- and (Z)-lithioenamines, *J. Am. Chem. Soc.*, 103, 3088, 1981.
252. **Donaldson, R. E. and Fuchs, P. L.,** A triply convergent total synthesis of *l*(−)-prostaglandin E_2, *J. Am. Chem. Soc.*, 103, 2108, 1981.
253. **Saddler, J. C., Donaldson, R. E., and Fuchs, P. L.,** Enantioconvergent syntheses of two classes of chiral cyclopentyl sulfone synthons, *J. Am. Chem. Soc.*, 103, 2110, 1981.
254. **Saddler, J. C. and Fuchs, P. L.,** Enantiospecific synthesis of γ-substituted enones. Organometallic S_N2 conjugate-addition reactions of epoxy vinyl sulfones, *J. Am. Chem. Soc.*, 103, 2112, 1981.
255. **Jakovac, I. J., Lok, G. N. P., and Jones, J. B.,** Preparation of useful chiral lactone synthons via stereospecific enzyme-catalyzed oxidation of meso-diols, *J. Chem. Soc. Chem. Commun.*, 515, 1980.
256. **Boland, W., Mertes, K., Jaenicke, L., Müller, D. G., and Fölster, E.,** Absolute configuration of multifidene and viridiene, the sperm releasing and attracting pheromones of Brown Algae, *Helv. Chim. Acta*, 66, 1905, 1983.
257. **Bridges, A. J., Raman, P. S., Ng, G. S. Y., and Jones, J. B.,** Enzymes in organic synthesis. 31. Preparation of enantiomerically pure bicyclic [3.2.1] and [3.3.1] chiral lactones *via* stereospecific horse liver alcohol dehydrogenase catalyzed oxidations of *meso* diols, *J. Am. Chem. Soc.*, 106, 1461, 1984.
258. **Schmidt, U. and Schölm, R.,** Optisch aktive Pyrrolidin Derivate aus 2-Glutaminsaure. Bemerkungen zur Diboran-Reduktion von Ammonium und Alkali-carboxylaten, *Synthesis*, 752, 1978.
259. **Enders, D. and Eichenauer, H.,** Asymmetric synthesis of α-substituted ketone by metallation and alkylation of chiral hydrazones, *Angew. Chem. Int. Ed. Engl.*, 15, 549, 1976.
260. **Enders, D. and Eichenauer, H.,** Asymmetrische Synthesen *via* Metallierte chirale Hydrazone. Enantioselektive Alkylierung von cyclischen Ketone und Aldehyden, *Chem. Ber.*, 112, 2938, 1979.
261. **Eichenauer, H., Friedrich, E., Lutz, W., and Enders, D.,** Regiospezifische und enantioselektive Aldol-Reaktionen, *Angew. Chem.*, 90, 219, 1978.
262. **Enders, D. and Eichenauer, H.,** Enantioselective Alkylation of aldehydes via metallated chiral hydrazones, *Tetrahed. Lett.*, 191, 1977.
263. **Mukaiyama, T., Asami, M., Hanna, J., and Kobayashi, S.,** Asymmetric reduction of acetophenone with chiral hydride reagent prepared from lithium aluminum hydride and (S)-2-(anilinomethyl)-pyrrolidine, *Chem. Lett.*, 783, 1977.
264. **Evans, D. E., Bartoli, J., and Shih, T. L.,** Enantioselective aldol condensation. 2. Erythro-selective chiral aldol condensations via boron enolates, *J. Am. Chem. Soc.*, 103, 2127, 1981.
265. **Evans, D. A., Ennis, M. D., and Mathre, D.,** Asymmetric alkylation reactions of chiral imide enolates. A practical approach to the enantioselective synthesis of α-substituted carboxylic acid derivatives, *J. Am. Chem. Soc.*, 104, 1737, 1982.
266. **Evans, D. E., Ennis, M. D., Le, T., Mandel, N., and Mandel, G.,** Asymmetric acylation reactions of chiral imide enolates. The first direct approach to the construction of chiral β-dicarbonyl synthons, *J. Am. Chem. Soc.*, 106, 1154, 1984.
267. **Krishnamurthy, S., Vogel, F., and Brown, H. C.,** Lithium 3-isopinocamphenyl-9-borabicyclo[3.3.1]nonyl hydride. A new reagent for the asymmetric reduction of ketones with remarkable consistency, *J. Org. Chem.*, 42, 2534, 1977.
268. **Midland, M. M., McDowell, D. C., Hatch, R. L., and Tramontano, A.,** Reduction of α,β-acetylenic ketones with β-3-pinanyl-9-borabicyclo[3.3.1]nonane. High asymmetric induction in aliphatic systems, *J. Am. Chem. Soc.*, 102, 867, 1980.

269. **Midland, M. M., Greer, S., Tramontano, A., and Zderic, S. A.,** Chiral trialkylborane reducing agents. Preparation of 1-deutero primary alcohols of high enantiomeric purity, *J. Am. Chem. Soc.*, 101, 2352, 1979.
270. **Valentine, D., Jr. and Scott, J. W.,** Asymmetric synthesis, *Synthesis*, 329, 1978.
271. **Červinka, O. and Bělovský, O.,** Some factors influencing the course of asymmetric reduction of optically active alkoxy lithium aliminium hydrides, *Collect. Czech. Chem. Commun.*, 30, 3897, 1967.
272. **Morisson, J. D. and Mosher, H. S.,** *Asymmetric Organic Reactions*, Prentice Hall, Englewood Cliffs, N.J., 1971.
273. **Jacques, A., Fouquey, C., and Viterbo, R.,** Enantiomeric cyclic binaphtyl phosphoric acids as resolving agents, *Tetrahed. Lett.*, 4617, 1971.
274. **Noyori, R., Tomino, I., and Tanimoto, Y.,** Virtually complete enantioface differentiation in carbonyl group reduction by a complex aluminium hydride reagent, *J. Am. Chem. Soc.*, 101, 3129, 1979.
275. **Noyori, R., Tomino, I., and Nishizawa, M.,** A highly efficient synthesis of prostaglandin intermediates possessing the 15S configuration, *J. Am. Chem. Soc.*, 101, 5843, 1979.
276. **Morrison, J. E.,** *Asymmetric Synthesis*, Vol. 1—3, Academic Press, New York, 1983.
277. **Brinkmeyer, R. S. and Kapoor, V. M.,** Asymmetric reductions. Reduction of acetylenic ketones with chiral hydride reagent, *J. Am. Chem. Soc.*, 99, 8339, 1977.
278. **Johnson, W. S., Brinkmeyer, R. S., Kapoor, V. M., and Yarnell, T. M.,** Asymmetric total synthesis of 11-α-hydroxyprogesterone via a biomimetic polyene cyclization, *J. Am. Chem. Soc.*, 99, 8341, 1977.
279. **Solladié, G., Greck, C., Demailly, G., and Solladié-Cavallo, A.,** Reduction of β-ketosulfoxides. A highly efficient asymmetric synthesis of both enantiomers of methyl carbinols from the corresponding esters, *Tetrahed. Lett.*, 23, 5047, 1982.
280. **Meyers, A. I. and Mihelich, E. D.,** The synthetic utility of 2-oxazoline, *Angew. Chem. Int. Ed. Engl.*, 15, 270, 1976.
281. **Meyers, A. I.,** Asymmetric carbon-carbon bond formation from chiral oxazolines, *Acc. Chem. Res.*, 11, 375, 1978.
282. **Hoobler, M. A., Bargbreiter, D. E., and Newcomb, M.,** Origins of stereoselectivity in asymmetric syntheses using chiral oxazolines, *J. Am. Chem. Soc.*, 100, 8182, 1978.
283. **Meyers, A. I., Snyder, E. S., and Ackermann, J. H.,** Stereochemistry of metallation and alkylation of chiral oxazolines. A ^{13}C nuclear magnetic resonance study of lithio oxazolines, *J. Am. Chem. Soc.*, 100, 8186, 1978.
284. **Meyers, A. I. and Smith, R. K.,** A total asymmetric synthesis of (+)-ar-turmerone, *Tetrahed. Lett.*, 2749, 1979.
285. **Johnson, C. R.,** The utilization of sulfoximines and derivatives as reagents for organic synthesis, *Acc. Chem. Res.*, 6, 341, 1973.
286. **Kennewell, P. D. and Taylor, J. B.,** The sulfoximides, *Chem. Soc. Rev.*, 4, 189, 1975.
287. **Field, L.,** Some developments in synthetic organic sulfur chemistry since 1970, *Synthesis*, 713, 1978.
288. **Mioskowski, Ch. and Solladié, G.,** New stereospecific synthesis of chiral α-sulfinylesters of known absolute configuration, *Tetrahed. Lett.*, 3341, 1975.
289. **Mioskowski, Ch. and Solladié, G.,** Asymmetric synthesis of 3-hydroxy acids using chiral α-sulfinylester enolate ions, *J. Chem. Soc. Chem. Commun.*, 163, 1977.
290. **Furukawa, N., Takahashi, F., Yoskimura, T., and Oar, S.,** Reaction of sulfoximides with diazomalonate in the presence of Cu-salt. A new synthesis and stereochemistry of optically active oxosulfonium ylides, *Tetrahedron*, 35, 317, 1979.
291. **Seebach, D. and Aebi, J. A.,** α-Alkylation of threonine, *Tetrahed. Lett.*, 24, 3311, 1983.
292. **Seebach, D. and Weber, T.,** α-Alkylation of a cysteine derivatives without racemization and without the use of a chiral auxiliary, *Tetrahed. Lett.*, 14, 3315, 1983.
293. **Seebach, D., Boes, M., Naeff, R., and Schweizer, W. B.,** Alkylation of amino acids without loss of the optical activity: Preparation of α-substituted proline derivatives. A case of self-reproduction of chirality, *J. Am. Chem. Soc.*, 105, 5390, 1983.
294. **Enders, D. and Papadopoulos, K.,** Asymmetric synthesis of β-substituted δ-ketoesters via Michael-additions of SAMP/RAMP-hydrazones to α,β-unsaturated esters. Virtually complete 1,6-asymmetric induction, *Tetrahed. Lett.*, 24, 4967, 1983.
295. **Meyers, A. I., Harre, M., and Garland, R.,** Asymmetric synthesis of quaternary carbon center, *J. Am. Chem. Soc.*, 106, 1147, 1984.
296. **Oppolzer, W., Moretti, R., Godel, T., Meunier, A., and Löher, H.,** Asymmetric 1,4-additions of coordinated $MeCu{\cdot}BF_3$ to chiral enolates. Enantioselective syntheses of (S)-(−)-citronellic acid, *Tetrahed. Lett.*, 24, 4971, 1983.
297. **Uskoković, M. R., Lewis, R. L., Partridge, J. J., Despreaux, C. W., and Pruess, D. L.,** Asymmetric synthesis of alloheteroyohimbine alkaloids, *J. Am. Chem. Soc.*, 101, 6742, 1979.
298. **Nelson, N. A. and Scahill, T. A.,** Synthesis of a key chiral intermediate for 12-hydroxy-prostaglandins, *J. Org. Chem.*, 44, 2790, 1979.

299. **Rosi, R., Salvatori, P. A., Carpita, A., Niccoli, A.,** Synthesis of the pheromone components of several species of *Trogoderma (Coleoptera: Dermestidae)*, *Tetrahedron,* 35, 2039, 1979.
300. **Rossi, R. and Carpita, A.,** Insect pheromones. Synthesis of chiral sex pheromone components of several species of *Trogoderma (Coleoptera: Dermestidae)*, *Tetrahedron,* 33, 2447, 1977.
301. **Pirkle, W. H. and Boeder, C. W.,** Synthesis and absolute configuration of (−)-methyl(E)-2,4,5-tetradecatrienoate, the sex attractant of the male Dried bean weevil, *J. Org. Chem.,* 43, 2091, 1978.
302. **Farnum, D. G., Veysoglu, T., Cardé, A. M., Duhl-Emswiller, B., Pancoast, T. A., Reitz, T. J., and Cardé, R. T.,** A stereospecific synthesis of (+)-dispalure, sex attractant of the gypsy moth, *Tetrahed. Lett.,* 4009, 1977.
303. **Johnson, F. and Witehead, A.,** The stereochemistry of 2-substituted cyclohexanone amines and the corresponding Schiff's bases, *Tetrahed. Lett.,* 3825, 1964.
304. **Schaffer, H. J. and Jain, V. K.,** The synthesis of dehydrocycloheximide and the conversion of *cis*-2,4-dimethylcyclohexanone to its *trans*-isomer, *J. Org. Chem.,* 29, 2595, 1964.
305. **Barner, R. and Schmid, M.,** Total synthese von natürlichen α-Tocopherol. 4. Mitteilung. Aufbau des Chromanringsystem aus Trimethylhydrochinon und einem optisch aktiven C_4 bezw. C_5 Synthon, *Helv. Chim. Acta,* 62, 2384, 1979.
306. **Maycock, C. D. and Stoodley, R. J.,** Studies related to thiiranes. Part 1. Synthesis of Chiral thiiranecarboxylates, *J. Chem. Soc. Perkin Trans.,* 1, 1852, 1979.
307. **Stork, G. and Schoofs, A. R.,** Concerted intramolecular displacement in allylic systems. Displacement of an allylic ester with a carbanion, *J. Am. Chem. Soc.,* 101, 5081, 1979.
308. **Anderson, R. J., Adams, K. G., Chinn, H. R., and Henrick, C. A.,** Synthesis of the optical isomers of 3-methyl-6-isopropenyl-9-decen-1-yl acetate, a component of the California Red scale pheromone, *J. Org. Chem.,* 45, 2229, 1980.
309. **Mori, K. and Ueda, H.,** Synthesis of optically active forms of faranal, the trail pheromone of Pharao's ant, *Tetrahed. Lett.,* 22, 461, 1981.
310. **Tomioka, K., Ishiguro, T., and Koga, K.,** First, asymmetric total synthesis of (+)-stenagacin and determination of absolute stereochemistry, *Tetrahed. Lett.,* 21, 2973, 1980.
311. **Meyers, A. I. and Smith, R. K.,** A total asymmetric synthesis of tumerone, *Tetrahed. Lett.,* 2749, 1979.
312. **Meyers, A. I., Smith, R. K., and Whitten, C. E.,** Highly stereoselective addition of organolithium reagents to chiral oxazoline. Asymmetric synthesis of 3-substituted alkanoic acids and 3-substituted lactones, *J. Org. Chem.,* 44, 2250, 1979.
313. **Mori, K., Ebata, T., and Sakakibara, M.,** Synthesis of (2S,3R,7RS)-stegobinone [2,3-dihydro-2,3,5-trimethyl-6-(1-methyl-2-oxobutyl)-4H-pyran-4-one] and its (2R,3S,7RS)-isomer, *Tetrahedron,* 37, 709, 1981.
314. **Rossiter, B. E., Katsuki, T., and Sharpless, K. B.,** Asymmetric epoxidation provided shortest route to four chiral epoxy alcohols which are key intermediates in syntheses of methymycin, erythromycin, leukotriene C_1 and dispalure, *J. Am. Chem. Soc.,* 103, 464, 1981.
315. **Mori, K. and Iwasawa, H.,** Stereoselective synthesis of optically active forms of δ-multistriatin, the attractant for European population of the smaller European elm bark beetle, *Tetrahedron,* 36, 87, 1980.
316. **Mori, K. and Akao, H.,** Synthesis of optically active alkynyl alcohols and α-hydroxy esters by microbial asymmetric hydrolysis of the corresponding acetates, *Tetrahedron,* 36, 91, 1980.
317. **Johnson, M. R., Nakata, T., and Kishi, Y.,** Stereo- and regioselective methods for the synthesis of three consecutive asymmetric units in many natural products, *Tetrahed. Lett.,* 4343, 1979.
318. **Johnson, M. R. and Kishi, Y.,** Cooperative effect by a hydroxy and ether oxygen in epoxidation with a peracid, *Tetrahed. Lett.,* 4347, 1979.
319. **Hasan, I. and Kishi, Y.,** Further studies on stereo-specific epoxidation of allylic alcohols, *Tetrahed. Lett.,* 21, 4229, 1980.
320. **Schmid, G., Fukuyama T., Akasaka, K., and Kishi, Y.,** Total synthesis of monensin. 1. Stereocontrolled synthesis of the left half of monensin, *J. Am. Chem. Soc.,* 101, 259, 1979.
321. **Fukuyama, T., Wang, C. L. J., and Kishi, Y.,** Total synthesis of monensin. 2. Stereocontrolled synthesis of the right half of monensin, *J. Am. Chem. Soc.,* 101, 260, 1979.
322. **Fukuyama, T., Akasaka, K., Karanewsky, D. S., Wang, C. L. J., Schmid, G., and Kishi, Y.,** Total synthesis of monensin. 3. Stereocontrolled total synthesis of monensin, *J. Am. Chem. Soc.,* 101, 262, 1979.
323. **Guerrier, L., Royer, J., Grierson, D. S., and Husson, H. P.,** Chiral 1,4-dihydropyridine equivalents. A new approach to the asymmetric synthesis of alkaloids. The enantiospecific synthesis of (+)- and (−)-coniine and dihydropinidine, *J. Am. Chem. Soc.,* 105, 7754, 1983.
324. **Demuth, M., Chandrasekhar, S., and Schaffner, K.,** Total synthesis with tricyclooctanone building blocks. Loganin aglucon 6-acetate, *J. Am. Chem. Soc.,* 106, 1092, 1984.
325. **Lu, L. D.-L., Johnson, R. A., Finn, M. G., and Sharpless, K. B.,** Two new asymmetric epoxidation catalysts. Unusual stoichiometry and inverse enantiofacial selection, *J. Org. Chem.,* 49, 728, 1984.

326. **Takano, S., Otaki, S., and Ogasawara, K.,** Enantioselective synthesis of an ant venom alkaloid (−)-[3S-(3β,5β,8α)]-3-heptyl-5-methylpyrrolizidine, *J. Chem. Soc. Chem. Commun.*, 1172, 1983.
327. **Wrobel, J. E. and Ganem, B.,** Total synthesis of (−)-vertinolide. A general approach to chiral tetronic acids and butenolides from allylic alcohols, *J. Org. Chem.*, 48, 3761, 1983.
328. **Mori, K. and Otsuka, T.,** Synthesis of both the enantiomers of erythro-6-acetoxy-5-hexadecanolide, the major component of a mosquito oviposition attractant pheromone, *Tetrahedron,* 39, 3267, 1983.
329. **Roush, W. R., Brown, R. J., and Mare, M. D.,** The synthesis of carbohydrates. 2. Regiochemical control of nucleophilic ring opening of acylated 2,3-epoxy alcohols, *J. Org. Chem.*, 48, 5083, 1983.
330. **Fuganti, C., Grasselli, P., Pedrocchi-Fanton, G., Servi, S., and Zirotti, C.,** Carbohydrate like chiral synthons, Preparation of (R)-γ-hexanolide, (5R,6S,7S)-6,7-isopropylidenoxy-δ-octanolide and (+)-exo-brevicomin from (2S,3S,4R) 2,3-isopropylidendioxy-4-benzyloxyhept-6-ene, *Tetrahed. Lett.*, 24, 3753, 1983.
331. **Szabo, W. A. and Lee, H. T.,** Chiral materials and reagents, *Aldrichim. Acta,* 13, 13, 1980.
332. **Mori, K. and Tamada, S.,** Stereocontrolled synthesis of all of the four possible stereoisomers of erythro-3,7-dimethylpentadec-2-yl acetate and propionate, the sex pheromone of the Pine sawflies, *Tetrahedron,* 35, 1279, 1979.
333. **Kitahara, T., Mori, K., and Matsui, M.,** A total synthesis of (+)-brefeldin A, *Tetrahed. Lett.*, 3021, 1979.
334. **Meyers, A. I. and Amos, R. A.,** Studies directed toward the total synthesis of sterptogramin antibiotics. Enantiospecific approach to the nine-membered macrocycle of griseoviridin, *J. Am. Chem. Soc.*, 102, 870, 1980.
335. **Stork, G., Nakahara, Y. Y., and Greenlee, W. J.,** Total synthesis of cytochalasin B, *J. Am. Chem. Soc.*, 100, 7775, 1978.
336. **Ireland, R. E., Thaisrivongs, S., and Wilcox, C. S.,** Total synthesis of lasalocid A, (X537A), *J. Am. Chem. Soc.*, 102, 1155, 1980.
337. **Ireland, R. E.,** unpublished results.
338. **Whistler, R. L. and BeMiller, J. N.,** ''α''-D-Glucosaccharino-1,4-lactone. 2-*C*-Methyl-D-*ribo*-pentono-1,4-lactone from D-fructose, *Meth. Carbohydr. Chem.*, 2, 484, 1963.
339. **Ireland, R. E., McGarvey, G. J., Anderson, R. C., Badoud, R., Fitzsimmons, B., and Thaisrivongs, S.,** A chiral synthesis of the left-side aldehyde for lasalocid A, *J. Am. Chem. Soc.*, 102, 6179, 1980.
340. **Westley, J. W., Evans, R. H., Jr., Williams, T., and Stempel, A.,** Pyrolytic cleavage of antibiotic X-357A and related reactions, *J. Org. Chem.*, 38, 3431, 1973.
341. **Nakata, T., Schmid, G., Vranesic, B., Okigawa, M., Smith-Palmer, T., and Kishi, Y.,** A total synthesis of lasalocid A, *J. Am. Chem. Soc.*, 100, 2933, 1978.
342. **Nakata, T. and Kishi, Y.,** Synthetic studies on polyether antiobiotics. III. A stereocontrolled synthesis of isolasalocid ketone from acyclic precursors, *Tetrahed. Lett.*, 2745, 1978.
343. **Woodward, R. B.,** Recent advances in the chemistry of natural products, *Pure Appl. Chem.*, 17, 519, 1968.
344. **Woodward, R. B.,** Recent advances in the chemistry of natural products, *Pure Appl. Chem.*, 25, 283, 1971.
345. **Woodward, R. B.,** The total synthesis of vitamin B_{12}, *Pure Appl. Chem.*, 33, 145, 1973.
346. **Fleming, I. and Woodward, R. B.,** Exo-2-hydroxyepicamphor, *J. Chem. Soc. (C),* 1289, 1968.
347. **Fleming, I. and Woodward, R. B.,** A synthesis of (−)-(R)-trans-β-(1,2,3-trimethylcyclopent-2-enyl) acrylic acid, *J. Chem. Soc. Perkin Trans.*, 1, 1653, 1973.
348. **Eschenmoser, A.,** Ein neuer syntetischer Zugang zum Corrinsystem, *Angew. Chem.*, 81, 301, 1969.
349. **Shastri, M. H., Patil, D. G., Patil, U. D., and Sukh Dev,** Monoterpenoids-V. (+)-Carvone from (+)-car-3-ene, *Tetrahedron,* 41, 3083, 1985.
350. **Hirama, M., Noda, T., and Ito, S.,** Convenient synthesis of (S)-citronellol of high optical purity, *J. Org. Chem.*, 50, 127, 1985.
351. **Hanessian, S., Murray, P. J., and Sahoo, S. P.,** A tactically novel alternative to acyclic stereoselection based on the concept of a replicating chiron-1,5-C-methyl substitution, *Tetrahed. Lett.*, 26, 5627, 1985.
352. **Hanessian, S., Murray, P. J., and Sahoo, S. P.,** A tactically novel alternative to acyclic stereoselection based on the concept of a replicating chiron—1,3 and 1,4-C-methyl substitution, *Tetrahed. Lett.*, 26, 5627, 1985.
353. **Hanessian, S., Sahoo, S. P., and Murray, P. J.,** A tactically novel alternative to acyclic stereoselection based on the concept of replicating chiron—access to 1,3-polyols, *Tetrahed. Lett.*, 26, 5631, 1985.
354. **Olsen, R. K., Bhat, K. L., and Wardle, R. B.,** Synthesis of (S)-(-)-3-piperidinol from L-glutamic acid and (S)-malic acid, *J. Org. Chem.*, 50, 896, 1985.
355. **Chamberlain, A. R. and Chung, J. Y. L.,** Synthesis of optically active pyrrolizidinediol: (+)-Heliotridene, *J. Am. Chem. Soc.*, 105, 3653, 1983.
356. **Mori, K. and Watanabe, H.,** Synthesis of the enantiomers of 3-hydroxy-1,7-dioxaspiro [5.5] undecane, a minor component of the olive fly pheromone, *Tetrahed. Lett.*, 25, 6025, 1984.

357. **Hills, L. R. and Ronald, R. C.,** Total synthesis of (−)-grahamimycin, *J. Org. Chem.*, 50, 470, 1985.
358. **Valverde, S., Herrandon, B., and Martin-Lomas, M.,** The use of L-tartaric acid in the synthesis of enantiomerically pure compound: synthesis of 4-O-benzyl-2,3-dideoxy-L-threo-hex-2-eno-1,5-lactone, *Tetrahed. Lett.*, 26, 3731, 1985.
359. **Nicolaou, K. C., Papahatjis, D. P., Claremon, D. A., Magolda, R. L., and Dolle, R. E.,** Total synthesis of ionophore antibiotic X-14547A, *J. Org. Chem.*, 50, 1440, 1985.
360. **Kuehne, M. F. and Podhorez, D. E.,** Studies in biomimetic alkaloid syntheses. 12. Enantioselective total synthesis of (−)- and (+)-vincadiformine and of (−)-tabersonine, *J. Org. Chem.*, 50, 924, 1985.
361. **Takano, S., Tanaka, M., Seo, K., Hirama, M., and Osagawara, K.,** General chiral route to irregular monoterpenes via a common intermediate: synthesis of (S)-lavandulol, cis-(1S,3R)-chrysanthemol, (1S,2R)-rothrockene and (R)-santolinatriene, *J. Org. Chem.*, 50, 931, 1985.
362. **Freskos, J. N. and Swenton, J. S.,** Annelation reactions of levoglucosenone. Chiral intermediates for the synthesis of naphtho [2.3-c] pyran-5,10-quinone antibiotics, *J. Chem. Soc. Chem. Commun.*, 658, 1985.
363. **Wilcox, C. S. and Thomasco, L. M.,** New syntheses of carbocycles from carbohydrates. Cyclization of radicals derived from unsaturated halo sugars, *J. Org. Chem.*, 50, 546, 1985.
364. **Harada, T., Izumi, Y.,** Improved modified Raney nickel catalyst for enantioface-differentiating (asymmetric) hydrogenation of methyl acetoacetate, *Chem. Lett.*, 1195, 1978.
365. **Tai, A., Nahakata, M., Harada, T., Izumi, Y., Kusumoto, S., Inage, M., and Shiba, T.,** A facile method for preparation of the pure3-hydroxytetradecanoic acid by an application of asymmetrically modified nickel catalyst, *Chem. Lett.*, 1125, 1980.
366. **Ireland, R. E., Häbich, D., and Norbeck, D. W.,** The convergent synthesis of polyether ionophore antibiotics: The synthesis of the monesin spirketal, *J. Am. Chem. Soc.*, 107, 3271, 1985.
367. **Ireland, R. E. and Norbeck, D. W.,** The convergent synthesis of polyether ionophore antibiotics: The synthesis of the monesin bis(tetrahydropyran) *via* the Calisen rearrangement of an ester enolate with a β-leaving group, *J. Am. Chem. Soc.*, 107, 3279, 1985.
368. **Ireland, R. E., Norbeck, D. W., Mandel, G. S., and Mandel, N. S.,** The convergent synthesis of polyether ionophore antibiotics: an approach to the synthesis of the monesin tetrahydropyran-bis(tetrahydrofuran) via ester enolate Claisen rearrangement and reductive decarboxylation, *J. Am. Chem. Soc.*, 107, 3285, 1985.

INDEX

A

Acidic hydrolysis, 159—160
Acorone, 94
Acyclic compounds, 20
Acyloin, 15
  condensation, 52, 53
  reaction, 135, 136
Adrenosterone, 33
Agastocha formosanum oil, 186
Alcohol, stereoselective synthesis of, 6—9
Aldol, 15
Aldosterone, 183
Aliphatic chemistry, Cope rearrangement in, 61
Alkylation methods
  α-alkylation, 89, 91—92
  ipso alkylations, 92—93
  α,α′-dialkylation, 89
  oxidative coupling of phenols, 92—93
Alkylations
  inverse, 94—95
  modification of, 30
  S_N, 23
α-Alkynones, thermal cyclization of, 38
Allylic acetates, nucleophilic displacement in, 94
Allylpalladium complex, 92
Allyltrimethylsilanes, 67
Aluminum hydrides, 204
Amino acids
  cysteine, 193—194
  glutamic acid, 190—193
  leucine, 193
  norvaline, 189—190
  phenylalanine, 190
α-Aminoadipic acid, 194
α-Aminobutyric acid, 190
Amitermes evuncifer, 186
Annulation, see also *ortho*-Condensed systems; Monotopic annulations
  bicyclic systems and, 107
  Danheiser, 26, 69
Anthopleurine, 196
Antibiotic A26711B, 200
Antibiotics, see also specific antibiotics
  macrolide, 202
  natural polyether, 209
L-Arabinose, 202
Aromadendrene, 59
Auxiliary rings, as precursors of substituents, 157—163
  via cleavage of bridged systems, 161—163
  via cleavage of *ortho*-condensed systems, 158—161
(−)-Avenaciolide, 201
(+)-Azimic acid, 200

B

Baldwin rules, 18, 20
Barton reaction, 68
Batrachotoxine, 183
Bayer-Villiger reaction, 162
N-Benzoyl-L-ristosamine, 196
β-*trans*-Bergamotene, 130
Biomimetic polyene cyclization, 35, 53
Biotin, 200
D-(+)-Biotin, 193
Birch reduction, 161
Bredt's rule, 108
Brefeldin A, 206
Bridged systems
  auxiliary rings in, 161—163
  bicyclic
    cyclization outside annulation sites, 114, 115
    cyclization at bridgehead positions, 107—114
  polycyclic, 128—130
  tricyclic, 114—117
    rearrangements in synthesis of, 127—128
    syntheses by bridging a central ring, 125—127
    syntheses including formation of central ring, 117—125
Bromochloromethane, 170—171
Bromoepoxide moiety, 10
Bromohydrin, 101, 162
Buchi synthesis, 59
Bullnesol, 60
(+)-*trans*-Burserane, 193

C

Camphor, 188—189
Cantharidin, 130
Capnellene, 69
Carbohydrates
  L-arabinose, 202
  as chiral synthons, 197—198
  glucose, 199—201
  glyceraldehyde, 199
  mannose, 201, 202
  D-ribose, 202
Carbomycin, 200
Carbon-carbon bond, 15
Carbon-heteroatom bond, 15
Carboxychrysanthemic aicd, 200
β-Carene, 188
(+)-Carpamic acid, 200
Carvone
  as chiral synthon, 185—186
  preparation of, 209
Central ring
  cyclobutane, 123
  of tricyclic systems, 116
Cephalosporin, 193

(+)-Cerulenin, 199
Chalcogran, 192, 194, 200
Chiral cryptates, 201
Chirality
  axial, 4
  central, 4
  formation of new elements of, 3—4
    axis → center chirality induction, 5
    center → axis chirality induction, 5, 6
    center → center chirality induction, 4—5
    plane → center chirality induction, 6
  planar, 4
Chiral shift reagent, 1
Chiral synthons, 1, 183—184
  amino acids
    cysteine, 193—194
    glutamic acid, 190—193
    leucine, 193
    norvaline, 189—190
    phenylalanine, 190
  carbohydrates, 197—198
    L-arabinose, 202
    glucose, 199—201
    glyceraldehyde, 199
    mannose, 201, 202
    D-ribose, 202
  hydroxy acids
    citramalic acid, 197
    dimethyl acrylic acid, 197
    ethyl acetoacetate, 197
    lactic acid, 194—196
    malic acid, 194
    (S)-2-methyl-3-hydroxypropionic acid, 197
    tartaric acid, 196—197
  miscellaneous, 202—205
  syntheses involving several
    brefeldin A, 206
    cytochalasin B, 207—208
    lasalocid A, 208—209
    molecular fragment of griseoviridine, 206, 207
    monensin, 209—211
    sexual pherome of pine sawflies, 205, 206
    vitamin B_{12}, 209, 212, 213
  terpenes
    camphor, 188—189
    β-carene, 188
    carvone, 185—186
    β-citronellol, 187
    limonene, 184
    β-pinene, 187—188
    pulegone, 186—187
Chloramines, 68
Chorismic acid, 152
Citramalic acid, 197
β-Citronellol, 187
Claisen rearrangement, 5, 62, 68, 140, 166, 200, 205
  in stereoselective formation of carbon-carbon bonds, 153, 155
ortho-Condensed systems, 52
  auxiliary rings in, 158—161
  bicyclic, 35
  cis, trans-annulation of
    cis- and *trans*-annulated decalins, 63—65
    cis- and *trans*-annulated hydrindanes, 63
    cis- and *trans*-annulated hydroazulenes, 65—66
    introduction of angular methyl group, 66—69
  cyclizations at annulation site
    ditopic annulations, 38—52
    monotopic annulations, 23—38
  rearrangements to bridged systems, 113
  ring fission in, 157—158
  types of, 107, 108
Conia cyclization, 99
Copaene, 119
Cope-Claisen rearrangement, 62, 155
Cope rearrangement, 23, 58, 61—62, 139, 140, 142
  oxy-Cope rearrangement, 41, 61, 155
  in stereoselective formation of carbon-carbon bonds, 153, 155
  in synthesis of spirocyclic ketones, 99
Corey synthesis, 140
  of (S)-12-hydroxy-5,8,14-*cis*-10-*trans*-eikosatetraenic acid, 194
  of leukotriene C-1, 202
  of prostaglandins, 162
Coriolin, 27
Cortisole, 183
Cyanide ion, 67
Cyclic systems
  modification of, 15—16
    combined ring contraction and expansion, 59—60
    Cope rearrangement, 61—62
    fragmentation of cross-piece bond in ring, 60—61
    ring contractions, 54—55
    ring expansions, 55—58
  stereoselective substitution in, 19—20
    de novo construction of stereospecifically substituted ring, 145—148
    stereoselective attachment of substituents, see also Substituents, 148—170
  synthesis of
    polycyclic systems, 18
    ring-closure reactions in, 17—18
    ring formation, 15—16
    systems with one ring, 16—17
Cyclization
  outside annulation sites, 52—54
  at bridgehead positions
    ditopic cyclization, 110—112
    fragmentation methods, 113—114
    monotopic cyclizations, 107—110
    rearrangements in synthesis of bridged systems, 112—113
  Diels-Alder, 111—112
  ditopic (methods of), 54, 122—125
  monotopic, 117—122
  outside spiroatom, 99, 100
Cycloadditions
  conia, ene, and related cyclizations, 99

[2 + 1], 39, 96
[2 + 2], 39—43, 96—97, 122, 124
[3 + 2], 44
asymmetric Diels-Alder reaction, 48
heterosubstituted dienes, 46—48
intramolecular Diels-Alder reaction, 48—52
[4 + 2], 44—46, 97—99, 122, 124
Cycloartenol, 33
Cyclobutanes
stereospecifically substituted, 145—147
tricyclic structures derived from, 116
Cyclobutanone ring, retro-Dieckmann fission of, 159
Cyclodecane, 168, 169
Cycloheptatriene, 129
Cyclohexadiene, 171
Cyclohexane-cis-1,2-dicarboxylic acids, 172
Cyclohexane rings, 30
Cyclohexanes, stereospecifically substituted, 146—148
Cyclohexene, π system in, 169
Cyclohexenones, 94
Cyclooctane, 168
Cyclooctene, 169
Cyclopalladation, 172
Cyclopentane annulation, 24
Cyclopentanes
stereospecifically substituted, 146, 147
tricyclic structures derived from, 116
Cyclopropane ring
stereoselective cleavage of, 158
vicinal substituents on, 160
Cyclopropanes, 145, 146
Cyclosativene, 128
Cysteine, 193—194
Cytochalasin B, 207—208

D

Danheiser annulation, 26, 69
Danheiser cyclization, 171
Daunosamine, 152—153
Debromination, photochemical, 93, 94
Decalins, *cis*- and *trans*-annulated, 63—65
Decalin skeletons, 23, 28
Decalin systems, 44—45
stereospecific fragmentation of hydroxy tosylates with, 138, 139
synthesis of, 2, 47
Demyanov rearrangement, 55, 100
Deoxynorpatchoulenol, 121
Diamantane, 129
Diazoketones, 110
Diazotation, 56
Dichloroketene, 43
Dieckmann condensation, 15, 52, 53
Diels-Alder addition, 102, 130
Diels-Alder cyclization, 112
Diels-Alder reaction, 2, 17, 18, 23, 43, 44, 208
asymmetric, 48
outside annulation sites, 63
chirality in, 199
for construction of bicyclic bridged systems, 110—111
intramolecular, 48—52, 65, 66, 69, 124, 125
photo-induced, 69
of quinone, 45
regioselectivity of, 97
retro, 46, 150
stereochemistry of, 162
for stereoselective substitution in a cyclohexane derivative, 160
of ynamine, 163
Dienones, photorearrangements of cross-conjugated, 101
Dimethyl acrylic acid, 197
Diol monotosylates, 60
DIOP, 204
cis-Disubstituted derivative 32
trans-Disubstituted derivatives, 162
Ditopic annulations, 38
[2 + 1] cycloadditions, 39
[2 + 2] cycloadditions, 39—43
[3 + 2] cycloadditions, 43, 44
asymmetric Diels-Alder reaction, 48
heterosubstituted dienes, 46—48
intramolecular Diels-Alder reaction, 48—52
[4 + 2] cycloadditions, 44—46
Ditopic processes, classification of, 18
Ditopic ring closure, 15
(−)-DOPA, 194
Double bond, 15
addition of ynamines to, 159
cyclopentenone, 151
stereoselective hydroboration of, 205
Dreiding models, 109

E

Ecdysone, 183
Enamide cyclization, photoinduced, 101—102
Enantioselective synthesis, 1
Ene reaction
cyclizations accomplished by, 15
intramolecular, 36—38
Enolate, 121
Enolates, 48
α,β-Enone system, 9, 27, 30, 31
Enzymatic reactions, stereoselective, 2
Ephedrine, 1, 2
(+)-Epoxy acid, 203
α,β-Epoxy ketones, 55
Erythronolide B, 196
Estrone, 183
Estrone methyl ether, 62
Ethyl acetoacetate, 197

F

Favorskii rearrangement, 54—55

Fortamine, 153
Frontalin, 200
Fructose, 211
Furane ring, 69
Furanomycin, 200
Fürst-Plattner rule, 11, 149

G

Glucose, 198—201
Glutamic acid
  as chiral synthon, 190—193
  synthetic use of, 191, 192
L-Glutamic acid, 209
Glyceraldehyde, 199
Grandisol, 160
Grignard reaction, 23
Grignard reagents
  catalyzed cross-coupling with allyl bromide of, 6
  reactions of cyclic anhydrides with, 94
  stereoselective reaction of carbonyl group with, 8
Griseoviridine, 206, 207
Grob fragmentation, 142
  γ-diolmonomesylates, 101, 102
  of 1,3-diolmonotosylates, 138

H

β-Haloketone, 27
Halolactonization, 156
Helenanolide, 53
(+)-Heliotridende, 209
Henry reaction, 70
Heteroatoms
  cyclohexane ring substituted by, 47
  stereoselective introduction of, 148
Heterodienones, 39
Himachalene, 50
Hirsutene, 60, 69
Homoallylic processes, 166
Huang-Minlon reduction, 121
Humulene, 137
Hydrindanes, 63
Hydrindane skeletons, 23, 28
Hydroazulenes, 65—66
Hydroxy acids
  citramalic acid, 197
  dimethyl acrylic acid, 197
  ethyl acetoacetate, 197
  lactic acid, 194—196
  malic acid, 194
  (S)-2-methyl-3-hydroxypropionic acid, 197
  tartaric acid, 196—197
Hydroxylactone, 191
11α-Hydroxyprogesterone, 36
Hypochlorites, 68

I

Illudol, 41
Iminium salts, 94
Indole alkaloids, 204
(−)-Isoavenaciolide, 201
Isoclovene, 122
Isocomene, 35, 36
9-Isocyanopupukenane, 124
threo-Isomer, 137
Isoprenoids
  biogenesis of, 33
  synthesis of, 36
(−)-Isosteganone, 193

J

(+)-Juvabione, 184

K

Kametani syntheses, of A-aromatic steroids, 51
Ketalization, 41, 42
Ketone, 203
Khusimonine, 189
Kinetic steric control, 2—3, 8—10

L

Lactic acid, 194—196
(−)-Lactone, 203
Lactone, 203
Lasalocid A, 208—209
Lasalocide A, 187
Leucine, 190, 193
Leukomycin A, 200
Leukotriene, 202
Lewis acids, 43, 49, 61, 70, 135, 171
Lewis acids catalysis, 94
Limonene, 184—185
Longicyclene, 129
Longifolene, 120
Lycopodium lucidulum, 186

M

Macrocyclic compounds, 19
Macrolide antibiotics, 202
Malic acid, 194
Malonate esters, 89
Mannose, 201, 202, 211
Markovnikov rule, 11, 149
de Mayo reaction, 40
de Mayo retro-aldolization, 138, 139
McMurry coupling, 15
McMurry dicarbonyl coupling, 135, 142

McMurry synthesis, of humulene, 137
Meerwein-Ponndorf reduction, 5
Menispermum coculus, 186
Mentha pulegium, 186
Methionine, 190
Methylcyclopentenones, 70
Methyl group, angular, 66—69
(S)-2-Methyl-3-hydroxypropionic acid, 197
Michael addition, 23—24, 29, 89, 91, 112
  intramolecular, 108, 122, 146, 147
  modification of, 30, 31
  palladium-catalyzed stereospecific equivalent of, 171
Modhephene, 36, 38
Monensin, 209—211
Monosaccharide, 198
Monotopic annulations, 23
  carbocation addition to double bond, 33—36
  ene reaction, 36—38
  five-carbon polar annulation, 33
  four carbon polar annulation, 27—33
  thermal cyclization of α-alkynones, 38
  three-carbon polar annulation, 24—27
Monotopic ring closure, 15
Multistriatin
  stereoselective synthesis of, 200
  synthesis of, 199

N

Nazarov reaction, 29
Nonactine, 195
Norvaline, 189—190

O

Olefin, 11
Optically active compounds, 183
Oxazolines, 204
Oxidative coupling, 93
β-Oxirane ring, 48
Oxygen function, stereoselective introduction of, 156
Oxymercuration, 158
Ozonolysis, 102

P

α- and β-Panasinensin, 40, 41
Patchoulenol, 126
Patchouli alcohol, 125, 126
Pentalenene, 35
Pentalenolactone, 35
Perhydroazulene skeletons, 23
Phenylalanine, 190
Pheromones
  composition of, 192
  synthesis of, 184, 187
  gypsy moth, 196
  of Kupferstecher, 194
  multistriatin, 199
  of optically active insects, 190
  pine sawflies, 205, 206
  of wasp workers, 195—196
Picrotoxin, 186
Pinacol rearrangement, 100
β-Pinene, 187—188
Pine sawflies, sexual pheromone of, 205, 206
(S)-(−)-3-Piperidinol, 209
Podocarpic acid, 69
Polycyclic systems, 18
Posner chiral sulfoxide methodology, 171
Potassium hexachlorotungstenate, 135
Prins reaction, 35, 69
Progesterone, 183
Prostaglandins
  chemistry of, 7
  Corey synthesis of, 162
  synthesis of, 156, 197, 200
  Woodward synthesis of, 2, 3
11-oxa-Prostaglandins, 200
Pseudodilution principle, 135—136
(+)-Pseudoephedrine, 1, 2
Ptilocaulin, 26
Pulegone, 186
Pumiliotoxin C, 189
Pyrovallerolactone, 60
Pyrrolidine derivatives, 203

Q

(+)-Quebrachamine, 193

R

Rearrangements, see also specific rearrangements, 100—102
  in bridged systems, 113
  in synthesis of tricyclic bridged systems, 127—128
(R)-Recifeiolide, 196
Reserpine, 3, 10
D-Ribose, 202
Ring-closure reactions, 17—18
Ring construction, enantioselective *de novo*, see also Substituents, 170—172
Ring contraction
  ring expansion and, 59—60
  strategies for, 54—55
Ring expansions, 55—56
  reactions of ketones with diazocompounds, 56
  rearrangements of vinylcyclopropane derivatives, 58
  ring contraction and, 59—60
  tandem oxidative cleavage of double bond aldolization, 57, 58
  Tiffeneu-Demyanov, 56, 57

Rings, medium and large
cyclization reactions, 135—138
fragmentation reactions, 138—141
"zip" reactions, 140—142
Ring system, formation of, 15—16
Robinson annulation, 23, 27—33, 44, 69
(−)-*cis*-Roseoxide, 200
syn-Rule, 49

S

(−)-β-Santalene, 111
Sericenine, 142
Sesquiterpenes, 53, 142
Seychellene, 126
Simmons-Smith reaction, 39, 171
Sinularene, 126
Sodium amalgam, reduction with, 1
Spiroatom, cyclizations outside, 99, 100
Spirocyclic compounds
alkylation methods, 89—93
cyclization outside spiroatom, 99, 100
cycloadditions, 96—99
inverse alkylations and related methods, 94—95
rearrangements, 100—102
Spirocyclic ketones, 99
Spirocyclic systems, 89, 90
Spirovetivane sesquiterpenes, 92
(−)-Steganacin, 193
(−)-Steganone, 193
Stereoselectivity
problem of
control element in, 9—12
stereosynthesis of alcohol, 6—9
synthesis, 1
Steric control
conformational kinetic and thermodynamic, 2—3
kinetic, 8, 10
kinetic and thermodynamic, 2
Steroids
acid-catalyzed rearrangement of, 100
chemistry of, 59
synthesis of, 203
Steroid skeleton, synthesis of, 51
Strophantidin, 183
Substituents
stereoselective attachment of
by addition to double bonds, 148—150
auxiliary rings as precursors of substituents, 157—163
inversion of configuration, 166—167
in medium and large rings, 167—170
neighboring groups as control elements, 156—157
[3 + 3] sigmatropic rearrangements, 153—156
transpositions, 164—166
in vicinal positions, 150—153
transposition of, 164—166
Substitution
nucleophilic, 15, 17
reactions, 148
Synthons, ring construction involving two, see also Chiral synthons, 16—17

T

(−)-Tabersonine, 210
Tandem oxidative cleavage, of double bond aldolization, 55—58
Tartaric acid, as chiral synthon, 196—197
Terpenes
camphor, 188—189
β-carene, 188
carvone, 185—186
β-citronellol, 187
limonene, 184—185
β-pinene, 187—188
pulegone, 186—187
synthesis of, 59
Tertiary center, changing configuration at, 167
Testosterone, 183
Thermodynamic control, 2—3, 10
Thorpe condensation, 15, 52, 53
Thorpe reaction, 135
Tiffeneu-Demyanov ring expansion, 56, 57
Transposition of substituents
1,2-transpositions, 164
1,3-transpositions, 165—166
Trialkylaluminum-alkylidene iodide, 171
Trialkylsiloxydienes, 47
Tricyclic ketone, 126
Trimethylsilylenone, 29
Trost's protocol, 25
Trost's spiroannulation procedure, 101
Tyrosine, 194

U

Unsaturated alcohols, 94

V

(−)-Velbamine, 193
Vernolepine, 47
Verruculotoxin, 190
β-Vetivone, 101
Vicinal aminoalcohols
Demyanov rearrangement of, 55
diazotation of, 56
Vicinal hydroxyl groups, 152
(+)-Vincadiformine, 210
Vinylcyclopropanes, 58, 70
Vinylogous Dieckmann condensation, 20
Vinylsilane, 29, 30
Vitamin B_{12}, 209, 212, 213
Vulgarenone, 127

W

Wagner-Meerwein rearrangement, 59, 67—68, 70, 100, 112
Walden inversion, 17
Westphalen rearrangement, 68
Wichterle reaction, 29
Wiesner rule, 40—41, 69
Wittig coupling, 191
Wittig methylenation, 39
Wittig reaction, 7, 24—26, 99, 135, 201
  of aldehyde enolate, 92
  in synthesis of strained bridged systems, 108
Wolf rearrangement, 55
Woodward-Hoffman scheme, 44, 140

X

Xylose, as source of chirality, 211

Y

Ylangene, 119
Ynamines
  addition to double bond of, 159
  Diels-Alder reaction of, 163

Z

"Zip" reaction, 135, 140—142